The Molecular Biology of *Physarum polycephalum*

NATO ASI Series

Advanced Science Institutes Series

A series presenting the results of activities sponsored by the NATO Science Committee, which aims at the dissemination of advanced scientific and technological knowledge, with a view to strengthening links between scientific communities.

The series is published by an international board of publishers in conjunction with the NATO Scientific Affairs Division

A Life Sciences **B Physics**	Plenum Publishing Corporation New York and London
C Mathematical and Physical Sciences	D. Reidel Publishing Company Dordrecht, Boston, and Lancaster
D Behavioral and Social Sciences **E Engineering and Materials Sciences**	Martinus Nijhoff Publishers The Hague, Boston, and Lancaster
F Computer and Systems Sciences **G Ecological Sciences**	Springer-Verlag Berlin, Heidelberg, New York, and Tokyo

Recent Volumes in this Series

Volume 99—Cellular and Molecular Control of Direct Cell Interactions
edited by H.-J. Marthy

Volume 100—Recent Advances in Nervous System Toxicology
edited by Corrado L. Galli, Luigi Manzo, and Peter S. Spencer

Volume 101—Chromosomal Proteins and Gene Expression
edited by Gerald R. Reeck, Graham A. Goodwin, and Pedro Puigdomènech

Volume 102—Advances in Penicillium and Aspergillus Systematics
edited by Robert A. Samson and John I. Pitt

Volume 103—Evolutionary Biology of Primitive Fishes
edited by R. E. Foreman, A. Gorbman, J. M. Dodd, and R. Olsson

Volume 104—Molecular and Cellular Aspects of Calcium in Plant Development
edited by A. J. Trewavas

Volume 105—The Physiology of Thirst and Sodium Appetite
edited by G. de Caro, A. N. Epstein, and M. Massi

Volume 106—The Molecular Biology of *Physarum polycephalum*
edited by William F. Dove, Jennifer Dee, Sadashi Hatano, Finn B. Haugli, and Karl-Ernst Wohlfarth-Bottermann

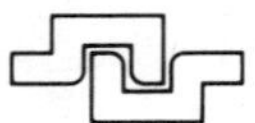

Series A: Life Sciences

The Molecular Biology of *Physarum polycephalum*

Edited by

William F. Dove

University of Wisconsin
Madison, Wisconsin

Jennifer Dee

University of Leicester
Leicester, England

Sadashi Hatano

Nagoya University
Nagoya, Japan

Finn B. Haugli

University of Tromsø
Tromsø, Norway

and

Karl-Ernst Wohlfarth-Bottermann

University of Bonn
Bonn, Federal Republic of Germany

Plenum Press
New York and London
Published in cooperation with NATO Scientific Affairs Division

Proceedings of a NATO Advanced Research Workshop on
The Molecular Biology of *Physarum polycephalum,*
held July 8–13, 1985,
in Madison, Wisconsin

Library of Congress Cataloging in Publication Data

NATO Advanced Research Workshop on the Molecular Biology of *Physarum polycephalum* (1985: Madison, Wis.)
The molecular biology of *Physarum polycephalum.*

(NATO ASI series. Series A, Life sciences; v. 106)
"Proceedings of a NATO Advanced Research Workshop on the Molecular Biology of *Physarum polycephalum* held July 8–13, 1985, in Madison, Wisconsin"—T.p. verso.
Includes bibliographical references and index.
1. *Physarum polycephalum*—Congresses. 2. Fungal molecular biology—Congresses. I. Dove, W. F. (William F.), 1936– . II. North Atlantic Treaty Organization. Scientific Affairs Division. III. Title. IV. Series.
QK635.P5N38 1985 589.2′9 86-4963
ISBN 0-306-42267-0

A Division of Plenum Publishing Corporation
233 Spring Street, New York, N.Y. 10013

Printed in the United States of America

PREFACE

One landmark in the long history of biological studies on the "slime mold" *Physarum polycephalum* was the introduction of chemically defined growth conditions for the plasmodial phase of this organism in the laboratory of Harold P. Rusch in Wisconsin in the 1950s. A number of investigators began working with *Physarum* in that era, then dispersed over the world. In the 1950s to 1960s, the regular meetings of *Physarum* workers in North America were commonly held in Wisconsin.

Strong new scientific initiatives in *Physarum* have grown up independently, from the disciplines of genetics, cytology, photobiology, and biophysics, in countries scattered over the world from Japan to Poland, Germany, France, the Netherlands, Norway, Spain, Turkey, and Great Britain. Infusion of the technical power of contemporary molecular biology--in particular, gene cloning and monoclonal antibodies--has brought these dispersed investigators into mutual communication. It was therefore timely and appropriate to assemble the *Physarum* community again in Wisconsin after a hiatus of 20 years, at a conference in the Friedrick Conference Center at the University of Wisconsin, Madison, from July 8 to 13, 1985.

The Scientific Affairs Division of the North Atlantic Treaty Organization (NATO) has for a number of years sponsored two sorts of international research meetings - Advanced Study Institutes and Advanced Research Workshops. Institutes are distinguished by being forums for the didactic dissemination of a consolidated body of knowledge. By contrast, Workshops are designed for the active investigation of emergent fields of inquiry, to formulate plans for future advancement. Dr. M. diLullo has provided guidance for this program. This publication is the result of such a Workshop. In effect, all participants have authored this volume, and are listed in the back. In practice, active investigators for each sub-domain have taken responsibility to formulate position papers, citing where possible the informal contributions made. We regret any omission of credit, but hope that the major impact of this publication is to convey the workshop spirit of our conference, which was rather like an extended laboratory research meeting. This extension has continued during formulation of this publication, with some outside contributors indicated by asterisks in the back.

Support from NATO has been complemented by generous assistance given by the University of Wisconsin--the Graduate School, the McArdle Laboratory, and the Laboratory of Genetics. Further special assistance was given by P-L Biochemicals, Promega-Biotec, and ICN Radiochemicals. Each of these contributions was essential to make this Workshop as complete a forum as possible within the space constraints of our conference center; we were able to include investigators from most countries and disciplines and from all levels of seniority. Without a critical mass of interacting investigators from around the world, research in *Physarum* will not match the intrinsic challenge offered by the biology of this organism.

Our week in Wisconsin served to focus extensive outside energy. In the Spring of 1985 the organizing committee, session chairmen, and outside rapporteurs formulated a set of working documents to identify issues. In the Autumn of 1985, these and other authors formulated the set of research and position papers that constitute this volume. Beyond those who are identified as corresponding authors, the following investigators have made focal contributions to this process: Phil Anderson, Steve Barclay, Gary Borisy, Ariel Loewy, Murdoch Mitchison, Hans Ris, Friedrich Wanka, and Michel Wright.

Realization of these scientific efforts has depended in the end upon the effective editorial assistance of Bette Sheehan, Ilse Riegel, and Carol McLeester in Madison and Pat Vann and Mary Stevenson of Plenum in New York. Linda Clipson has provided her customary care in dealing with illustrations. Mary Jo Markham and Kristen Luick have been both skillful and cheerful in collating drafts from many dialects onto the word-processor.

W. F. Dove, Madison
J. Dee, Leicester
S. Hatano, Nagoya
F. B. Haugli, Tromsø
K.-E. Wohlfarth-Bottermann, Bonn

December, 1985

CONTENTS

MOTILITY: EXPERIMENTAL INVESTIGATIONS

TECHNICAL FEASIBILITIES

Chapter 1: INTRODUCTION TO PHYSARUM

Helmut W. Sauer

Department of Biology
Texas A&M University
College Station, TX, USA

THE LIVES OF PHYSARUM

In the following, we attempt to demonstrate that *Physarum* is an adequate model system for studies of the eukaryotic mitotic cycle, cell differentiation, and motility. A schematic presentation of the life cycle of this organism is shown in Fig. 1-1, at the end of this chapter. (See Sauer and Pierron, 1983, for a recent review).

Physarum polycephalum belongs to the true slime molds, or Myxomycetes. In the field, these organisms exist as at least three distinct forms of life. In one, which is unique to the Myxomycetes, *Physarum* lives as a slimy mass of protoplasm, usually in the dark. This "Urschleim" is able to sense a food source, migrates toward microorganisms, and feeds on them by phagocytosis. It can grow to considerable size, a thin sheet several feet in diameter. Microscopic inspection reveals that this organism contains millions of nuclei within a single cell. This cellular organization is known as a plasmodium. The many nuclei of a plasmodium divide in natural synchrony every 8-10 hr, this synchrony offers unique opportunities for an experimental analysis of the cell cycle. Another form of *Physarum* occurs as a colony of black, 1- to 2-mm-tall, fruiting bodies, each with a stalk and a head. These sessile sporangia are exposed to the light and serve for propagation of the species. They are also responsible for the species designation *Physarum polycephalum*, the multiheaded. The third form of *Physarum* consists of microscopic myxamoebae about 12-15 μm in diameter. These contain a single nucleus and grow and multiply like other soil amoebae.

These three extremely different life forms and styles are

programmed by the same genome in a unique temporal sequence: from amoeba to plasmodium and on to fruiting bodies, then back to more amoebae. Amoebae hatch from spores, which are specialized cells formed in the head of the fruiting body.

The complete life cycle of Physarum can be studied in culture in the laboratory (see Fig. 1-1). The development of a plasmodium into many sporangia and of amoebae into a plasmodium are irreversible and essential steps in the life cycle of Physarum. Sporulation, the formation of fruiting bodies, is a synchronous differentiation process that is triggered by illumination. Plasmodium formation is usually the product of sexual fusion between two haploid amoebae, followed by zygote formation and uncoupling of nuclear divisions from cell division. However, a selfing (or apogamic) amoeba can change directly into a haploid plasmodium. In addition, there are two facultative steps that represent reversible differentiation processes. An amoeba can quickly change into a flagellate in a moist environment, and both an amoeba and a plasmodium can form dormant cysts under generally unfavorable conditions. An amoeba changes into a uninucleate microcyst, and a plasmodium changes into many multinucleated macrocysts, which are held together in a crust of dried slime, also called sclerotium. When macrocysts form in submerged liquid culture, they are referred to as spherules.

The exact phylogenetic position of the Myxomycetes is not clear, but the multinucleated plasmodium and the multicellular sporangium indicate a complexity above unicellular organisms. On comparing the life cycle of different Myxomycetes, one can speculate on their possible evolution. It seems that all Myxomycetes collected in the field can occur as either sexual (or heterothallic) or nonsexual (or apogamic) forms, also known as selfers. In the latter case, a plasmodium can arise in a clone derived from a single amoeba. Starting with an amoebo-flagellate, its mode of proliferation and encystment equals the life cycle of other unicellular organisms. Next, uncoupling of cell division from nuclear division resulted in the plasmodial organization, or a macroscopic undifferentiated creature. This is quite a different kind of organism that can behave as a predator of Physarum amoebae. Cell division without nuclear division, or plasmotomy, allows for one form of vegetative propagation of plasmodia. Cellularization of a plasmodium can result in two different kinds of cells, the multinucleated macrocyst or the mostly uninucleated spore. In primitive Myxomycetes, like Ceratiomyxa, individual prespores are formed at the periphery of the plasmodium. In more advanced forms, like Echinostelium, the plasmodium undergoes cellular morphogenesis and produces a single fruiting body. Large plasmodia, like those of Physarum, form many fruiting bodies; this implies a further process of scaling before segregation into small beads of protoplasm and their development into fruiting bodies. An additional feature, rarely seen in multicellular organisms, is fusion of vegetative cells (plasmodia) with one another. Therefore, the macroscopic form of *Physarum* represents a unique cellular organization, a composite of plasmodium and syncytium formation.

In summary, the life cycle of *Physarum* can be characterized as a multipotent developmental system. This is typical for higher organisms, which encounter several developmental choices, rather than the two alternate decisions observed in most simple eukaryotic systems, which result in growth or just one form of change. Sauer (1982) has developed a more detailed discussion of the universal biological phenomena of eukaryotic systems represented by the lives of *Physarum*.

HISTORICAL NOTES ON *PHYSARUM* RESEARCH

Biologists have had an interest in *Physarum* for many different reasons ranging from biosystematics to model studies in molecular, cellular, and developmental biology of a universal eukaryotic system (see von Stosch, 1965; Aldrich and Daniel, 1982; and Sauer, 1982, for reviews).

Physarum was described a long time ago (Schweinitz, 1833), later classified as a Myxomycete (Wallroth, 1833) and defined as a true slime mold, displaying the plasmodial organization and alternative vegetative and generative life cycle phases (Jahn, 1928). The taxonomy of the Myxomycetes has been comprehensively treated in a book titled "The Mycetozoans" (Olive, 1975), which indicated the intermediate position of these organisms among fungi, plants, and animals. The general biology of the plasmodial slime molds has been reviewed (Martin 1940; Alexopoulos, 1960) and updated (Martin and Alexopoulos, 1969; Gray and Alexopoulos, 1968). The precise interrelationships between the various life forms of *Physarum* were elucidated, and meiosis was detected within the maturing spores. In certain strains of *Physarum*, like Colonia, an asexual life cycle (apogamy) was described (see Fig. 1-1). These results, and research concerning the developmental physiology of the Myxomycetes before introduction of axenic culture conditions, are the subject of a noteworthy review (von Stosch, 1965). The life cycle strategies of the plasmodial slime molds have also been considered from the viewpoint of a biosystematist (Collins, 1979) as an evolutionary process of speciation, resulting in different reproductive systems.

For a while the timing of meiosis was subject to some controversy. It now seems clear that meiosis usually follows a presporangial mitosis, but occasionally it can occur earlier, before sporangia are fully formed, i.e., independent of the morphological events. Typically, meiosis is recognized by the formation of synaptonemal complexes inside the spore nucleus, followed by the two meiotic divisions, both of which are intranuclear like mitosis in plasmodia. In most cases three of the four nuclei produced by meiosis are destroyed, thus leaving one nucleus in the mature spore. Following meiosis, the chromosome number, about 90 in the diploid nucleus, has been reduced to about 43, and the DNA content of 1-1.3 pg per diploid nucleus measures 0.5-0.6 pg per mature spore and haploid amoeba. Since the DNA content of the nucleus in the plasmodium represents the G_2 phase, it has been argued that an S phase must have occurred during spore maturation and that amoebae spend most of their life in the G_2 phase. Furthermore, two centrioles appear in mature spores, and mitosis in amoebae is of the open kind with a bipolar cytoplasmic spindle. This contrasts with the intranuclear spindle and the absence of centrioles during the closed mitosis in the plasmodium (see Mohberg, 1982 for details).

The unique opportunity to study cellular motility in the form of a vigorous protoplasmic shuttle streaming in *Physarum* had also been noted long ago (Vouk, 1910). Ingenious experiments were performed, measuring the motile force generated by plasmodial strands suspended in a pressurized double chamber (Kamiya, 1959). In the first biochemical study of the plasmodium, an actomyosin-like material had been isolated from *Physarum* (Loewy, 1952). The first isolation of actin was reported by Hatano and Oosawa (1966). Actin was the first protein of *Physarum* whose amino acid sequence was determined and found to be very similar to the cytoplasmic actin of mammalian cells (Vandekerckhove and Weber, 1980). Since that time, an ongoing effort has been under way, unraveling the rhythmic contraction-relaxation cycles of the actomyosin-ATPase system in the ectoplasm, by a combination of microscopic and ultrastructural methods together with physical, physiological, and biochemical methods. This work, as well as the role of calcium ions and models of intracellular oscillators, has been reviewed (Wohlfarth-Bottermann, 1979) and compared with other motile systems (Hatano et al., 1979) as covered in Chapters 8-17.

The migration of the plasmodium was characterized as amoeboid in nature (Lewis, 1942), but it is faster than in normal amoebae because of the shuttle streaming. It has also been observed that migration was briefly interrupted at about the time of mitosis (Anderson, 1964), which coincides with an almost complete cessation of the shuttle streaming during mitosis (Guttes and Guttes, 1963). Studies of plasmodial migration toward exogenous chemical stimuli were facilitated by a simple biological assay of chemotaxis (Carlile, 1970). As a model system for cellular motility, *Physarum* has a long history.

Synchronous mitosis is perhaps the most spectacular event in the plasmodium of *Physarum*. Although it had been known for a long time (Howard, 1932), it was not explored until *Physarum* was rediscovered in the 1950s by Harold P. Rusch in his search for a model system to study cancer. He realized that its unique life strategy provided a good opportunity to investigate, independently, events associated with either growth or differentiation (for review, see Rusch, 1980). Sufficient homogeneous material for biochemical investigations would be available because the organisms could be maintained in axenic culture in the laboratory. Rusch pursued two goals in focusing his research on Physarum. First, he considered a cancer cell not so much a cell that had lost control over its proliferation, but rather as a cell that had failed to differentiate properly. Consequently, he set out to isolate from a sporulating plasmodium fractions that would inhibit growth in growing plasmodia and induce differentiation into fruiting bodies, and then to test these substances on mammalian cells in culture. The other goal was to understand the mechanism by which starvation initiates developmental pathways of differentiation. This general phenomenon in lower and higher organisms could be explored in Physarum in two ways, resulting in either cysts or spores. The sporulation pathway includes the phenomenon of competence, which is brought about by nutrient stress, and that of irreversible commitment induced by illumination. The first task toward achieving these goals was the development of an axenic medium that supported growth and differentiation. This was accomplished after it was discovered that hematin was essential for growth and niacin for sporulation (Daniel and Rusch, 1961, 1962). For determining the growth parameters in the laboratory, the macroplasmodium was fragmented into small multinucleated microplasmodia and kept as a shaken suspension in the culture medium. From that time on, many new experiments became possible, and much biochemical ground work was done. The first decade of this line of research has been summarized on two occasions (Rusch 1969, 1970). Here are a few findings of general significance.

Each microplasmodium is synchronous with respect to nuclear divisions, but there is no synchrony among microplasmodia; this indicates a lack of extracellular synchronizing signals. By use of the tendency of genetically similar plasmodia to fuse naturally with one another, a procedure was developed to create the synchronous macroplasmodium, by fusion of a large number of microplasmodia on the surface of a filter paper into a single giant "cell". It maintains completely balanced growth during several mitotic cycles, with a generation time of 8-10 hr at 26°C (Guttes and Guttes, 1964). Phase contrast microscopy of a small piece of the macroplasmodium allows determination of mitotic stages and of the degree of synchrony of the division of up to 10^{10} nuclei (Guttes et al., 1961). Many aliquots of a single plasmodium can be cut, like pieces of a pie, and handled with ease.

A series of fusion experiments was performed with pairs of macroplasmodia, each in a different phase of its synchronous mitotic cycle (Rusch et al., 1966). The main result was the re-synchronization of the mitotic cycle between the two populations of nuclei.

Those advanced in their cycle waited for the other set of nuclei to catch up. Consequently, all nuclei in the mixed plasmodium, homogenized by shuttle streaming, divided at a time precisely predictable from the difference between the cell cycle phases and the relative masses of the two partners that produced the time-heterokaryon. This result established an assay for mitogens and opened the ongoing search for cytoplasmic factors that shorten the generation time and induce mitosis.

The first determination of DNA synthesis with radioisotopes (Nygaard et al., 1960) established the S phase of the mitotic cycle of Physarum. Interestingly, this period began immediately after mitosis and indicated that the G_1 phase of the typical cell cycle may not be an essential component in a growing cell.

In a classical paper, it was shown for the first time that DNA replication in Physarum follows a sequential pattern during consecutive mitotic cycles (Braun et al., 1965). In these experiments, a portion of the DNA was labeled for a short period with ^{3}H-thymidine during one S phase, and was continuously labeled with ^{14}C-BUDR (bromodeoxyuridine) during the next S phase, which also increased the density of the replicated DNA molecules. Density-gradient analysis of DNA isolated at different times of the second S phase revealed coincidence of the distribution of the two radioisotopes only if the DNA was prepared after the pulse-labeling in the first S-phase. All samples prepared in the second S phase, before the ^{3}H-thymidine pulse of the previous S phase, contained two distinct populations of light ^{3}H- and heavy ^{14}C-labeled DNA. Therefore, the fraction of the genome replicating early in one S phase continued to replicate early in subsequent mitotic cycles.

The first enzyme with periodic activity changes in phase with the S phase, and a useful general marker of the DNA-division cycle of eukaryotic cells as conceptualized by Mitchison (Mitchison, 1971), was thymidine kinase (Sachsenmaier and Ives, 1965).

An efficient method to purify nuclei was developed, and similarities of the Physarum histones with those of other eukaryotes were established (Mohberg and Rusch, 1971). Isolated nuclei provided the starting material to initiate studies of chromatin organization and in vitro synthesis of DNA and RNA. The average DNA content per haploid genome amounts to about 0.25 pg. However, the nuclear DNA content changes during long-term culture as a macroplasmodium. The organism may have to be "purified" by a sexual cycle after experiencing senescence. As a practical consequence, the best way to keep a strain of Physarum genetically stable is to generate it periodically from frozen gametes (see Chapter 2).

An essential role for light in the process of sporulation had been demonstrated earlier (Baranetzki, 1876; Gray 1938). A point of no return after illumination indicated irreversibility of the sporulation process. Later, a regimen of 4 days in sporulation medium followed by a 4-hr illumination yielded a high incidence of sporulation in parallel cultures over a period of 16 hr, during which hundreds of sporangia developed per plasmodium, with a good synchrony (Daniel and Baldwin, 1964). A detailed account of the morphological and cytological sequence of sporulation was given (Guttes et al., 1961). Massive nuclear degradation and elimination was also detected. Consequently, only a fraction of the nuclei survive which, despite extreme nutrient stress and catabolic metabolism in the plasmodium, undergo a presporangial mitosis. This mitosis is triggered by light and occurs about 13 hr post-illumination, just before cellularization in the head of the developing fruiting body. Metabolic changes and a general decrease in macromolecular content during starvation, illumination, and sporulation have been carefully studied. A thoughtful discussion explored how an altered flux of matter and energy might generate the developmental information necessary to program sporulation (Daniel, 1966).

The genetic approach yielded new insights into the life cycle of *Physarum* (see Dee, 1982, for review). The very first paper (Dee, 1960) outlined the feasibility of Mendelian crosses among haploid amoebal clones and described a mating-type locus (now called *matA*) involved in the formation of plasmodia from the mating of amoebae (see Fig. 1-1). It was then found that a multiallelic (not the accustomed female/male) mating-type system of *Physarum* controls the fusion of appropriate amoebae. It is now clear that a plasmodium can usually develop if the gametes carry different alleles of two mating-type loci (*matA* and *matB*). It seems that both loci act on the principle of hetero-reactivity; i.e., the gametes should express different alleles of *matB* in order to fuse efficiently with each other, and different alleles of *matA* to allow the resulting zygote to develop into a plasmodium.

Direct biochemical confirmation for the ploidy change in heterothallic strains in *Physarum* and the haploid life cycle of apogamic strains (see Fig. 1-1) came from comparative determinations of the DNA content and chromosome counts of isolated nuclei from amoebae and plasmodia (see Mohberg, 1982).

Other genetically defined loci control fusion of plasmodia (the *fus*-genes). In contrast to the mating-type system, successful fusion usually requires a complete set of identical alleles of the *fus*-genes to make plasmodia fusion compatible, although some *fus* alleles are recessive (see Poulter and Dee, 1968). Still other loci (the *let*-genes) can cause lethal interactions between incompatible plasmodia (Carlile and Gooday, 1978).

While formal genetic studies can be done in a mixed culture of amoebae with food bacteria, an essential step toward biochemical genetic studies with Physarum was the development of an axenic culture medium (McCullough and Dee, 1976) and the detection of genes (*axe*) that confer growth in axenic medium. A combination of mutagenizing large numbers of haploid amoebae and the selection procedures of microbial genetics with traditional genetical techniques was initiated, and several auxotrophic, drug-resistant, and conditionally lethal mutants were detected. The Colonia strain, CL, was useful for the genetic analysis because of its facultative selfing genotype (apogamy), thus enabling the immediate detection of recessive mutations. The CL strain also allowed crossing with other heterothallic strains and complementation analysis (see Fig. 1-1 and Dee, 1982, for details).

Physarum research up to the 1970s laid the groundwork for exploring the life cycle of an interesting myxomycete, described the morphology and physiology of distinct developmental stages, worked out the procedures to culture *Physarum* from spore to spore in the laboratory, and accumulated valuable data concerning processes of general interests, like the mitotic cycle and DNA replication, cellular motility, and differentiation. In addition, the amoebal-plasmodial transition has been explored genetically.

PROGRESS IN *PHYSARUM* RESEARCH

In the decade between the seventies and the early eighties, investigations became concentrated on several critical questions of general biological importance. This research has been summarized in three recent books. "Growth and Differentiation of *Physarum polycephalum*" (Dove and Rusch, 1980) contains six concise essays, cross referenced by most of the *Physarum* research community, in order to seek a consensus of what is factually known and to point out solvable critical questions. The essays cover the search for a suitable cellular experimental system and the life cycle and review the nuclear replication cycle, transcription, nuclear proteins, the genetic approach, and developmental phases. "Cell Biology of *Physarum* and *Didymium*" (Aldrich and Daniel, 1982) is a two volume comprehensive multiauthored monograph with over 25 chapters. It also contains detailed methodological sections concerning the culture of *Physarum* and the various adaptations to biochemical techniques necessary to cope with the inherent problems that arise during the study of the molecular biology of *Physarum*. This book allows interested researchers to focus on a critical issue and to initiate new investigations without having to go back to the original literature. "Developmental Biology of *Physarum*" (Sauer, 1982) attempts to inform the interested biologist about the opportunities of this developmental system and to integrate knowledge gained from *Physarum*, other simple eukaryotes, tissue culture cells, and classical embryological systems.

From our current view, the life cycle of Physarum is a multipotent system with five separate phenomena of cell differentiation. Of those, three are reversible: macrocyst formation (spherulation), microcyst formation, and the amoeba-flagellate transformation. The two essential and irreversible phases are the amoebal-plasmodial transition and sporulation. This repertoire of distinct developmental themes is common to all higher eukaryotic organisms, like Drosophila and Xenopus. These events can be physically dissociated from one another with Physarum, which should provide unambiguous results concerning growth and differentiation. Most other unicellular model systems, like yeast and the cellular slime mold Dictyostelium, display only two alternate cellular states of growth and dormancy.

Unlike other simple eukaryotes, the Physarum amoeba has several developmental options from which to select according to environmental stimuli that can be manipulated in the laboratory; it can divide and produce a clone, change into a flagellate, become a dormant cyst, or develop into a gamete and change into a plasmodium (see Chapter 6).

Differentiation of the plasmodium results in either spherules or sporangia. Spherulation was first described as a differentiation process in the absence of growth in response to starvation and other physiological stresses. For the sporulation pathway, experiments with conditioned medium have indicated secretion of a "sporulation factor," which could be responsible for making starved plasmodia "competent" to respond to light and become irreversibly committed to differentiate (see Chapter 7).

The cell nucleus has revealed a genome organization that is typical for eukaryotic organisms. Considerable information is available on a histone H4 gene, the actin and tubulin loci, a family of highly repeated presumptive transposable elements constituting 20% of the genome, and the extrachromosomal nuclear rDNA (see Chapters 2 and 3).

Investigations of the mitotic cycle have described several discontinuous markers at the molecular level (see Chapter 5) and have established the mechanism, synchrony, and chronology of DNA replication (see Chapter 4). Three general points can be raised in this context.

As long as sufficient nutrients are supplied, the developmental program of unicellular organisms equals their proliferative cell cycle, and it seems that G_1 phase is not required for any cell cycle controlled by endogenous (endocrine) factors alone.

From this perspective, the lack of a G_1 phase from the proliferative mitotic cycles of Physarum plasmodia, in the absence of differentiation and under control of endogenous factors, represents an intermediate state between the early embryo and the adult multicellular organization. If the growing plasmodium is a case of a pure and extended growth phase, one might predict the presence and expression of certain genes, which function as proto-oncogenes in embryos or, in an "activated" form, as oncogenes in higher organisms. It has recently been claimed that oncogenes can be grouped broadly into two categories. By singly transfecting a primary fibroblast culture with either _ras_-genes or _myc_-genes, which represent the "transformation" and "immortalization" class of oncogenes respectively, cancerous growth could not be achieved. However, if the _ras_- and _myc_-genes, attached to strong viral promoters, were used in co-transfection experiments, malignant transformation of primary fibroblasts could be obtained in some cases. From these observations, a cooperation between two different types of oncogenes has been concluded, in which the _myc_-gene is perhaps causally involved in the establishment of the immortal state of some cancer cells (Land et al., 1983). Although this view is open to controversy, and because oncogenes have also been considered neither necessary nor sufficient to explain oncogenesis (Duesberg, 1985), it will be interesting to investigate the role, if any, of the putative _myc_-related gene in the cell cycle of _Physarum_ (see Chapter 5).

A final point concerns the possible implication of the cell cycle in the process of cellular commitment and differentiation. In light of the presumed invariant chronology of genome replication in _Physarum_ (see Chapter 4), a sequential expression of the genetic cell cycle program could be envisioned if the transcription of certain genes were coupled to their replication. Replication-transcription coupling has been clearly demonstrated in the early part of the S phase in _Physarum_, where the majority of newly activated transcription units are found within replicons, only a few minutes after metaphase, when chromatin was condensed, free of RNA polymerase B and transcriptionally inactive. Typically, both copies of the newly replicated transcription units are active, as seen in electron microscopic-chromatin spreads, and the origin of replication must be within the transcribed region (see Chapter 4). Detection of these novel genomic units, comprising a replicon and transcription units(s), offers a new look at the "quantal" cell cycle, previously suggested to move a cell in a multicellular system from one state of differentiation to the next (see Chapter 4). The possibility of switching a set of cell type-specific genes from a late to an early compartment of S phase has recently been pointed out by Taylor (1984) as a mechanism of cellular commitment (or determination). The coupling of replication of these genes, or their control gene(s), with their activation may lead to an understanding of this elusive mechanism in the development of multicellular organisms. As a multipotent developmental system with naturally exclusive alternate states, _Physarum_ allows for a test of this hypothesis, particularly at those two points where exocrine factors have been implicated in irreversible commitment -- during sporulation and the amoebal-plasmodial transition.

Much progress has been made in the structural-functional analysis of Physarum motility. One can argue that Physarum plasmodia have become for studies of non-muscle motility function the equivalent of the squid axon model utilized by neurobiologists (see Chapters 8,9, and 17).

CURRENT OPPORTUNITIES WITH PHYSARUM

Although modern Physarum research was initiated as model studies of biological questions of general importance, such as motility or growth and differentiation, the scientific community has been slow to accept this organism as a universal model of a eukaryotic cell. That was expected because of the idiosyncrasies of the life cycle of Physarum, initial difficulties in handling the biological material, and considerable problems regarding genetic and molecular techniques. At that time other simple eukaryotic systems, such as Saccharomyces, Tetrahymena, Chlamydomonas, and the cellular slime mold Dictyostelium, were already well-established in the laboratory, were being explored by state-of-the-art technologies, and were standard topics in biology textbooks.

The unique life history of the Myxomycete Physarum allows for physical, genetic, and conceptual dissection of a multipotent eukaryotic system. One advantage lies in the alternate phases of growth and differentiation in axenic culture, the separate investigation of microscopic and macroscopic life forms, and their interconversion under controlled conditions in the laboratory.

Genetic analysis has been performed by mutagenesis and selection with large numbers of cloned haploid amoebae, combined with traditional crossing and complementation procedures, and finally extended to include standard recombinant DNA techniques. The amoeba-flagellate transition is an example of cell polarization due to rearrangement of the intracellular architecture. One can comprehensively investigate cytomorphogenesis of an unorganized protoplasmic drop into a viable organism with two endogenously controlled rhythms, which cause the shuttle streaming and synchronous nuclear divisions. The differential control of cell fusion, exerted by the matB locus and by the fus genes, as well as the effect of exocrine factors on the amoebal-plasmodial transition and sporulation and the chemo-, photo-, and thermotactic responses, provides experimental situations for investigation of cell recognition and the reception and transmission of external stimuli.

Differences and similarities between the reversible differentiation process of encystment and irreversible sporulation may give new insights into the establishment of a developmental program. Of particular interest is the response to oxygen stress, perhaps a general mechanism that results in "competence" as a necessary prerequisite of differentiation. Other critical questions concern the light-induced state of "commitment" and the search for regulatory DNA sequences responsive to blue light.

One central issue concerns the control of proliferative cell cycles. The precise natural synchronies of nuclear division and DNA replication are uniquely suited to devise new experiments combining cell biological, molecular, and genetical techniques in order to define components conserved in all growing eukaryotic cell cycles, yet modulated during early embryogenesis and by differentiation in complex multicellular systems. Furthermore, a mechanism of programming gene expression may be discovered that employs alterations of the temporal compartmentation of the S phase to create a new state of differentiation.

ACKNOWLEDGMENTS

Financial support of research in the Author's laboratories since 1967 by the German Research Foundation (DFG, Sa 139-1-13) the Volkswagen Stiftung and by grants from the NSF (PCM 8411124) and NIH-BRSG (2-S07-RR07097-18) are each gratefully acknowledged.

REFERENCES

Aldrich, H. C., and Daniel, J. W., eds., 1982, "Cell Biology of Physarum and Didymium", 2 vols., Academic Press, New York.

Alexopoulos, C. J., 1960, Gross morphology of the plasmodium and its possible significance in the relationships among myxomycetes, Mycologia, 52:1.

Anderson, J. D., 1964, Regional differences in ion concentration in migrating plasmodia, in: "Primitive Motile Systems in Cell Biology," R. D. Allan and N. Kamiya, eds., p. 125, Academic Press, New York.

Baranetzki, J., 1876, Influence de la lumiere sur les plasmodia des myxomycetes, Mem. Soc. Sci. Nat. Cherbourg, 19:321.

Braun, R., Mittermayer, C., and Rusch, H. P., 1965, Sequential temporal replication of DNA in Physarum polycephalum, Proc. Natl. Acad. Sci., U.S.A., 53:924.

Burland, T. G., 1978, Temperature-sensitive mutants of Physarum polycephalum -- expression of mutations in amoebae and plasmodia. Ph.D. Thesis, University of Leicester, UK.

Carlile, M. J., 1970, Nutrition and chemotaxis in the myxomycete Physarum polycephalum: the effect of carbohydrates on the plasmodium, J. Gen. Microbiol., 63:221.

Carlile, M. J., and Gooday, G. W., 1978, Cell fusion in myxomycetes and fungi; in: "Membrane Fusion," G. Poste and G. L. Nicolson, eds., p. 219, Elsevier/North-Holland Biomedical Press, Amsterdam, New York.

Collins, O. R., 1979, Myxomycete biosystematics: some recent developments and future research opportunities, Bot. Rev., 45:145.

Daniel, J. W., 1966, Light-induced synchronous sporulation of a myxomycete, in: "Cell Synchrony, Studies in Biosynthetic Regulation," I. L. Cameron and G. M. Padilla, eds., p. 117, Academic Press, New York.

Daniel, J. W., and Baldwin, H. H., 1964, Methods of culture for plasmodial mycomycetes, in: "Methods in Cell Physiology," D. M. Prescott, ed., Vol. 1, p. 9, Academic Press, New York.

Daniel, J. W., and Rusch, H. P., 1961, The pure culture of Physarum polycephalum on a partially defined soluble medium, J. Gen. Microbiol., 25:47.

Daniel, J. W., and Rusch, H. P., 1962, Method for inducing sporulation of pure cultures of the myxomycete Physarum polycephalum, J. Bacteriol., 83: 234.

Dee, J., 1960, A mating-type system in an acellular slime-mould, Nature (London), 185:780.

Dee, J., 1982, Genetics of Physarum polycephalum in: "Cell Biology of Physarum and Didymium", H. C. Aldrich and J. W. Daniel, eds., Vol. 1, p. 211, Academic Press, New York.

Dove, W. F., and Rusch, H. P., eds., 1980, "Growth and Differentiation in Physarum polycephalum," Princeton University Press, Princeton.

Duesberg, P. H., 1985, Activated proto-oncogenes: sufficient or necessary for cancer?, Science, 228:669.

Gray, W. D., 1938, The effect of light on the fruiting of myxomycetes, Am. J. Bot., 25:511.

Gray, W. D., and Alexopoulos, C. J., 1968, "Biology of the Myxomycetes", Ronald Press, New York.

Guttes, E., and Guttes, S., 1963, Arrest of plasmodial motility during mitosis of Physarum polycephalum, Exp. Cell. Res., 30:242.

Guttes, E., and Guttes, S., 1964, Mitotic synchrony in the plasmodia of Physarum polycephalum and mitotic synchronization by coalescence of microplasmodia, in: "Methods in Cell Physiology", D. M. Prescott, ed., Vol. 1, p. 43, Academic Press, New York.

Guttes, E., and Guttes, S., and Rusch, H. P., 1961, Morphological observations on growth and differentiation of Physarum polycephalum grown in pure culture, Develop. Biol., 3:588.

Hatano, S., Ishikawa, H., and Sato, H., eds., 1979, "Cell Motility: Molecules and Organization", University Park Press, Baltimore.

Hatano, S., and Oosawa, F., 1966, Isolation and characterization of plasmodium actin, Biochim. Biophys. Acta, 127:488.

Howard, F. L., 1932, Nuclear division in plasmodia of Physarum, Ann. Bot. (London), 46:461.

Jahn, E., 1928, Myxomycetes, in: "Die natürlichen Pflanzenfamilien," A. Engler and K. Prantl, eds., Vol. 2, p. 304, Engelmann, Leipzig.

Kamiya, N., 1959, Protoplasmic streaming, Protoplasmatologia, 8(3a):1.

Land, H., Parada, L. F., and Weinberg, R. A., 1983, Cellular oncogenes and multistep carcinogenesis, Science, 222:771.

Lewis, W. H., 1942, The relation of the viscosity changes of protoplasm to ameboid locomotion and cell division, in: "The Structure of Protoplasm", W. Seifritz, ed., p. 163, The Iowa State College Press, Ames.

Loewy, A. G., 1952, An actomyosin-like substance from the plasmodium of a myxomycete, J. Cell. Comp. Physiol., 40: 127.

Martin, G. W., 1940, The myxomycetes, Bot. Rev., 6:356.

Martin, G. W., and Alexopoulos, C. J., 1969, "The Myxomycetes", University of Iowa Press, Ames.

McCullough, C.H.R., and Dee, J., 1976, Defined and semi-defined media of the growth of amoebae Physarum polycephalum, J. Gen.Microbiol., 95: 151.

Mitchison, J. M., 1971, "The Biology of the Cell Cycle", Cambridge University Press, Cambridge.

Mohberg, J., 1982, Ploidy throughout the life cycle in Physarum polycephalum, in: "Cell Biology of Physarum and Didymium," H. C. Aldrich, and J. W., Daniel, eds., Vol. 1, p. 253, Academic Press, New York.

Mohberg, J., and Rusch, H. P., 1971, Isolation and DNA content of nuclei of Physarum polycephalum, Exp. Cell Res., 66:305.

Nader, W. F., Edlind, T. D., Huettermann, A., and Sauer, H. W., 1985, Cloning of Physarum actin sequences in an exonuclease deficient bacterial host, Proc. Natl. Acad. Sci., U.S.A., 82:2698.

Nygaard, O. F., Guttes, S., and Rusch, H. P., 1960, Nucleic acid metabolism in a slime mold with synchronous mitosis, Biochim. Biophys. Acta, 38:298.

Olive, L. S., 1975, "The Mycetozoans", Academic Press, New York.

Poulter, R. T. M., and Dee, J., 1968, Segregation of factors controlling fusion between plasmodia of the true slime mould Physarum polycephalum, Genet. Res. Camb., 12:71.

Rusch, H. P., 1969, Some biological events in the growth processes, in: "Biological Organization -- Cellular and Subcellular", C. H. Waddington, ed., p. 263, Pergamon Press, New York.

Rusch, H. P., 1970, Some biochemical events in the life cycle of Physarum polycephalum, in: "Advances in Cell Biology", D. M. Prescott, L. Goldstein, and E. McConkey, eds., Vol. 1, p. 297, Appleton-Century-Crofts, New York.

Rusch, H. P., 1980, The search, in: "Growth and Differentiation in Physarum polycephalum", W. F. Dove and H. P. Rusch, eds., p. 1, Princeton University Press, Princeton.

Rusch, H. P., Sachsenmaier, W., Behrens, K., and Gruter, V., 1966, Synchronization of mitosis by the fusion of the plasmodia of Physarum polycephalum, J. Cell Biol., 31:204.

Sachsenmaier, W., and Ives, D. H., 1965, Periodische Aenderungen der Thymidinkinase-Aktivität im synchronen Mitosecyclus von Physarum polycephalum, Biochem. Z., 343:399.

Sauer, H. W., 1982, "Developmental Biology of Physarum", Cambridge University Press, Cambridge, New York.

Sauer, H. W., and Pierron, G., 1983, Morphogenesis and differentiation in Physarum, in: "Fungal Differentiation", J. F. Smith, ed., p. 73, Marcel Dekker, New York.

Schweinitz, L. D., 1822, Synopsis Fungorum, Naturforschende Ges., Leipzig.

Taylor, J. H., 1984, Origins of replication and gene regulation, Mol. Cell Biol., 61:99.

Vandekerckhove, J., and Weber, K., 1978, The amino acid sequence of Physarum polycephalum actin, Nature (London), 276:720.

von Stosch, H. A., 1965, Wachstums- und Entwicklungsphysiologie der Myxomyceten, in: "Handbuch der Pflanzenphysiologie", U. Ruhland, ed., Vol. 15, p. 641, Springer, Berlin, Heidelberg and New York.

Vouk, V., 1910, Untersuchungen über die Bewegung der Plasmodien. Die Rhythmik der Protoplasmaströmung, Sitzungsb. Kais. Akad. Wiss. Wien, 119:853.

Wallroth, C. F. W., 1833, Flora cryptogamica Germaniae II. Nurnberg.

Wohlfarth-Bottermann, K.-E., 1979, Oscillatory contraction activity in Physarum, J. Exp. Biol., 81:15.

Fig. 1-1. (facing). The life cycle of *Physarum polycephalum* (Redrawn from Burland, T. G., 1978, with permission). The essential stages in the life cycle of *Physarum* are the amoeba, which differentiates into a plasmodium, which in turn differentiates into sporangia with spores from which more amoebae germinate.

The amoebae derived from a plasmodium can either proliferate or differentiate into flagellates or dormant cysts. Alternatively, if two amoebae carry different alleles of the mating-type genes, they can fuse and produce a diploid zygote, which develops into the plasmodium, a giant single cell with many nuclei. The plasmodium exhibits naturally synchronous mitotic cycles during its growth phase. The plasmodium can either differentiate into a sclerotium, which contains dormant macrocysts, or else into many sporangia. Sclerotium formation is a reversible process and results from starvation or other unfavorable environmental conditions. Sporulation is an irreversible process that is triggered by illumination following an extended period of starvation.

Cellularization in the head of the sporangium produces prespores, which differentiate into mature spores with ornate thick walls and usually a single haploid nucleus, a result of meiosis.

The outer circuit shows the life cycle of a typical heterothallic strain (*matA1*, *matA2*, etc.), alternating between haploid amoebal and diploid plasmodial phases. (The open and closed circles denote nuclei carrying different *matA* alleles). The inner circuit shows the life cycle of *matAh* strain, where both amoebal and plasmodial phases are haploid. Note that *matAh* amoebae can undergo sexual fusion with heterothallic amoebae to form diploid plasmodia.

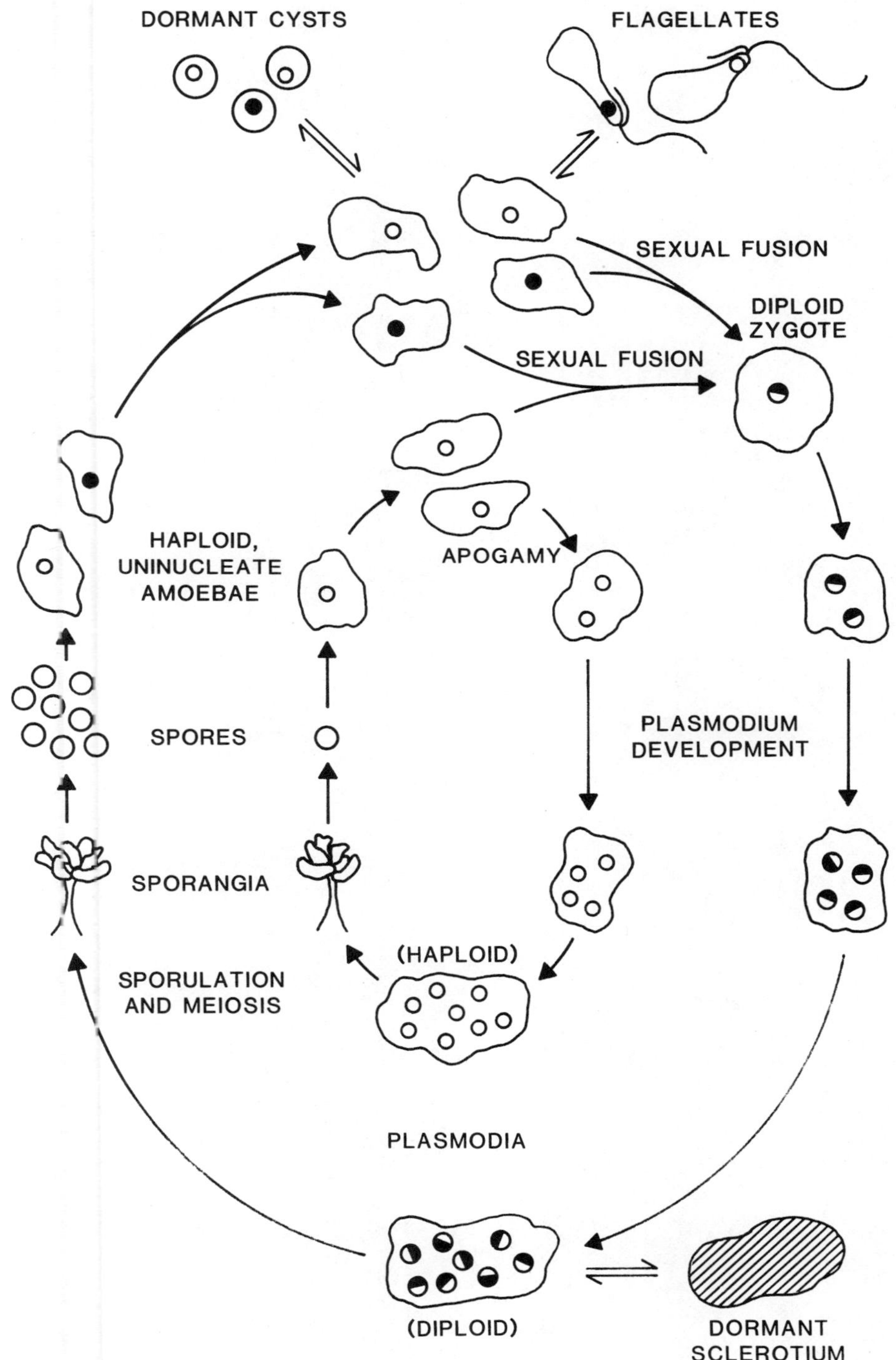
DORMANT CYSTS
FLAGELLATES
SEXUAL FUSION
DIPLOID
ZYGOTE
SEXUAL FUSION
HAPLOID,
UNINUCLEATE
AMOEBAE
APOGAMY
SPORES
PLASMODIUM
DEVELOPMENT
SPORANGIA
(HAPLOID)
SPORULATION
AND MEIOSIS
PLASMODIA
(DIPLOID)
DORMANT
SCLEROTIUM

Chapter 2: GENETIC ANALYSIS IN *PHYSARUM POLYCEPHALUM*

Timothy G. Burland

McArdle Laboratory for Cancer Research
University of Wisconsin
Madison WI, USA

with contributions from: P. Anderson, R. W. Anderson, R. Braun, J. Dee, K. Foster, K. Gull, F. B. Haugli, T. G. Laffler, D. Pallotta, E. C. A. Paul, H. Sauer, T. Schedl, and V. Vogt

THE NATURAL LIFE CYCLE

Detailed knowledge of the life cycle is a prerequisite for meaningful genetic analysis of any organism. The life cycle of *Physarum polycephalum* as we know it in the laboratory is summarized in Fig. 1-1.

The dormant sclerotium (Fig. 1-1) is one form of the organism commonly used for storage of natural isolates in the laboratory. Under moist, nutrient conditions, the sclerotium hatches to yield a multinucleate, diploid syncytium, the plasmodium. Although nutrients are available, the plasmodium continues to grow, with synchronous nuclear divisions occurring in the absence of cytokinesis. When starved in the dark, the plasmodium develops again into a dormant sclerotium, but if exposed to light, the starving plasmodium sporulates, producing dormant sporophores that are multi-headed (hence *polycephalum*). During sporulation, meiosis occurs, although the precise timing of meiosis in relation to sporophore development is unclear (Laane and Haugli, 1976). Cytological evidence suggests that three of the four nuclear products of meiosis are usually degraded and that most mature spores are uninucleate. Under moist conditions, haploid, uninucleate myxamoebae hatch from the spores. Like plasmodia, the amoebae are capable of indefinite growth if nutrients are available, but amoebal nuclear division is followed by cytokinesis, so that the amoebal population remains haploid and uninucleate. Upon suspension in water, amoebae transform to nonproliferating flagellates, which in turn transform back to nonflagellated amoebae when returned to drier conditions. When starved, amoebae develop reversibly into dormant cysts.

The life cycle is completed when amoebae develop into plasmodia (Fig. 1-1). All natural isolates of *Physarum polycephalum* so far studied are heterothallic (Collins, 1981), wherein the amoebae obtained by sporulating a plasmodium carry either of two different alleles of the mating type locus, *matA* (for example *matA1* and *matA2*). Amoebae of either mating type are self-sterile, and plasmodium development occurs only when amoebae of different mating types are mixed together. Cellular and nuclear fusion occurs between two amoebae of different mating type to form a uninucleate, heterozygous diploid zygote. This zygote then develops into a multinucleate plasmodium, with successive nuclear divisions occurring in the absence of cytokinesis. So far, when a new plasmodium has been isolated from nature, upon sporulation it has produced amoebae carrying two new alleles of *matA*, each allele being self-sterile but compatible with all other known alleles; thus, for example, *matA1* amoebae will mate with any of the amoebae carrying the known alleles *matA2* through *matA16*.

THE EXPERIMENTALLY MODIFIED LIFE CYCLE

As noted by Dee, the heterothallic life cycle facilitates simple mendelian genetic analysis of mutant or naturally variant amoebae, and the diploid plasmodial phase allows genetic complementation and dominance/recessiveness to be tested for traits that are expressed in plasmodia. The ability to perform routine mendelian genetics has also permitted the construction of isogenic strains of amoebae carrying various combinations of useful genetic markers, a vital asset for genetic analysis (Dee, 1982). Further, the haploid, uninucleate amoebae are readily manipulated in classical experiments on microbial mutant selection. However, these features are insufficient for all of the genetic techniques we need to apply in our experiments.

One particular interest, for example, is the question of whether mutations detected in amoebae are expressed in plasmodia. This can be tested in mutant heterothallic strains by first crossing the mutant with a nonmutant amoeba of different mating type, then choosing a mutant recombinant carrying the second mating type to cross back to the original mutant amoeba, producing a diploid plasmodium homozygous for the mutation. For one mutant, this is a reasonable project, but to examine fifty mutants in this way is impracticable. Fortunately, a strain of amoebae was developed, called CL ("Colonia Leicester"), that produces plasmodia in clones (Cooke and Dee, 1974). Mutant amoebae of CL can thus be allowed to produce plasmodia in clones, and these plasmodia can then be tested directly for mutant phenotype; no crosses are necessary to test for expression of mutations in plasmodia. The CL strain of amoebae was developed not only for its ability to produce plasmodia in clones, but also for its ability to grow and sporulate well; CL amoebae can also be crossed with heterothallic amoebae. Plasmodia develop from CL amoebae by apogamy--haploid amoebae develop clonally into haploid plasmodia without fusion of amoebae (Cooke and Dee, 1974); their life cycle is illustrated in the inner circle of Fig. 1-1. It is thought that the few viable spores obtained from the haploid CL plasmodia arise from meiosis in the small percentage of diploid nuclei that can be detected in these syncytia (Laffler and Dove, 1977). Most strains used in genetic analysis have been back-crossed to make them nearly isogenic with CL, and are referred to as "Colonia-background" strains. The mating type allele carried by CL amoebae is called <u>matAh</u> to distinguish it from the heterothallic mating types.

While solving one problem, the characteristics of the <u>matAh</u> allele introduce another. One of the advantages of the natural life cycle of <u>Physarum polycephalum</u> is the ability to culture entirely separately two different vegetative cell types, the amoeba and the plasmodium. The selfing of <u>matAh</u> amoebae is so efficient that it is difficult to obtain growing populations of amoebae that are free of plasmodia. The efficient selfing complicates crosses of <u>matAh</u> amoebae with heterothallic strains; when, for example, <u>matAh</u> and <u>matA1</u> amoebae are mixed together, predominantly <u>matAh</u> selfed plasmodia develop, with only a few crossed <u>matAh</u>/<u>matA1</u> plasmodia. The two sorts of plasmodia can be distinguished by genetic markers, but a more efficient crossing technique is desirable. Selfing of <u>matAh</u> amoebae can be curtailed by incubation at 30°C, but Cooke and Dee (1975) devised a superior method. A mutant strain of CL, called CLd, was isolated, which was drastically delayed in its selfing behavior; plasmodium development occurs only after extended incubation and at high cell density, effectively allowing pure culture of amoebae. Yet, when selfed plasmodia are desired from CLd amoebae, they can be readily obtained by extended incubation. It was later shown that CLd carries a mutation in the <u>npfC</u> gene at the complex mating type locus (see Chapter 6), and that selfing of CLd amoebae occurs only after the low-frequency reversion of this mutation.

One other potential problem for crossing amoebae together in a mixture of heterothallic and <u>matAh</u> strains could be the somatic fusion of selfed <u>matAh</u> plasmodia with crossed plasmodia. In practice, this problem is circumvented by ensuring that the two strains to be crossed carry different alleles of <u>fusA</u>, one of several loci controlling plasmodial fusions (note that <u>fusA</u> has no effect on amoebal fusion, and <u>matA</u> has no effect on plasmodial fusion). In most Colonia-background strains, all known <u>fus</u> loci except <u>fusA</u>

carry the same allele (fusB1, fusC1, and probably other unidentified loci), while different Colonia-background strains are available that carry either fusA1 or fusA2. A fusA1 plasmodium will fuse only with a fusA1 or fusA1/fusA1 plasmodium, a fusA2 plasmodium will fuse only with a fusA2 or fusA2/fusA2 plasmodium, and a fusA1/fusA2 plasmodium will fuse only with a fusA1/fusA2 plasmodium, provided that the alleles of other fus loci are identical (Poulter and Dee, 1968). Thus, from a mixture of, for example, matAh fusA2 and matA3 fusA1 amoebae, any selfed matAh fusA2 plasmodia will not fuse with crossed matAh/matA3 fusA2/fusA1 plasmodia, and the two sorts of plasmodia are readily distinguished by tests with plasmodia of known fus genotype. It is essential to use Colonia-background strains if this technique is to be readily applied, insomuch as each of the multiple fus loci in Physarum is naturally polymorphic, and because the tests would be further complicated by the presence of different kil loci (loci controlling plasmodial killing reactions) in other natural isolates (Carlile, 1973). Even two amoebae from the same natural plasmodial isolate may carry different alleles of both fus and kil loci.

The efficiency of crossing amoebae in the laboratory has been further improved following the discovery of the rac (rapid crossing) locus (Dee, 1979), later called the matB locus (Youngman et al., 1981). This is another locus for which multiple alleles are found in nature (Kirouac-Brunet et al., 1981), and it controls the rate of crossing, apparently at the stage of amoebal cell fusions. Amoebae mate much more rapidly if the two strains to be mated carry different alleles of the matB locus. While this phenomenon facilitates more efficient crossing, Dee emphasized a most important use of this discovery, which is in the construction of diploid amoebae heterozygous for selected markers. The strategy for producing such diploids is simple both in essence and in practice (Anderson, 1979; Shipley and Holt, 1982; Anderson and Youngman, 1985). The two haploid amoebae from which the diploid is to be constructed should carry identical alleles of matA and different alleles of matB. Since amoebae carrying different matB alleles will fuse readily, binucleate heterokaryons will frequently be formed. Nuclear fusion in such a cell, either during interphase or, more likely, during the open mitosis of the amoeba (Burland et al., 1981), may produce a diploid cell line. These diploids will not develop into plasmodia, because they will be homozygous at matA--plasmodium development would occur only if the amoebae were heterozygous at matA. The diploid status of such cells can then be tested with appropriate genetic markers. In this way, genetic complementation can be tested between mutations that express their phenotypes in amoebae. R. W. Anderson stated at the meeting that the method for constructing diploid amoebae was routine, but Burland and Haugli had found some difficulties for particular mutant combinations.

THE ROLE AND PRESENT STATUS OF GENETIC ANALYSIS IN PHYSARUM

In a talk on "The role and present status of genetic analysis in Physarum", Dee stressed that genetics is principally another technique in Physarum, and not an end in itself. She emphasized that genetic analysis in this organism is now routine, but believed that many workers consider the techniques to be complicated. These fears are probably due not to any complication in the theoretical aspects or in the methods used, but rather in the inherent versatility of Physarum for genetical research. While simple mapping and recombination analysis take advantage of the heterothallic nature of the natural life cycle, apogamic strains are available that circumvent this process, allowing the effects of recessive mutations to be tested in plasmodia without the need for backcrossing. Yet apogamic strains are able also to participate in the heterothallic life cycle (see Fig. 1-1); this feature is a valuable asset for the experimenter, but perhaps confusing to the newcomer to Physarum genetics. Further, genetic complementation and dominance/recessivity relationships can be tested in three different ways, according to the preference of the investigator: first, by crossing together two amoebae of different matA type and testing the resultant diploid, heterozygous plasmodium; second, by fusing together two compatible haploid (or two diploid) plasmodia and testing the resulting heterokaryon; and third, by fusing together two amoebae of the same matA type but of different matB type and testing the resulting diploid amoebae. It should be pointed out that the ability to compare plasmodial complementation tests in both heterokaryons and heterozygous diploids is a luxury few organisms offer us in the laboratory at the level of technical simplicity feasible with Physarum.

Two major genetic precautions need to be observed, however. First, the versatility of Physarum genetics can be exploited fully only if the strains used are of Colonia genetic background. Second, strains should be stored in an inactive state to preclude genetic change, preferably as frozen or desiccated stocks of amoebae.

One consequence of these requirements would be that investigators who wish to begin to use genetic analysis in addition to biochemical studies should consider changing the laboratory strain on which they work. Most investigators currently use a plasmodial strain such as M3B or M3CVIII etc., which ultimately are derived from the diploid natural isolate Wisconsin 1 (Wis 1) (Mohberg, 1982). The problem here is twofold. Wis 1 and its derivatives are heterozygous for most of the markers used in genetic analysis, and almost certainly heterozygous for many unknown genes; this heterozygosity would complicate many aspects of detailed genetic analysis. Further, Wis 1 and its derivatives can be perpetuated in the laboratory as syncytial plasmodia only by serial subculture. Thus, many mutational changes have probably occurred (even considering that the process has been retarded by laboratories keeping their stocks as dormant microsclerotia), and the syncytial nature of plasmodia precludes recloning to purify genetically homogeneous lines. It is worth noting at this point one clearly established problem found in at least one of the plasmodial sublines of Wis 1. Kubbies and Pierron (1983) showed by flow cytometry that the nuclei in M3CV plasmodia consist of two distinct populations, one apparently normal and the other with abnormal DNA content and aberrant replication pattern. A different strain of plasmodia (TU291 from Haugli) had a single population of nuclei with respect to both these parameters. Such aberrations may account for some of the variation in chromosome number observed in plasmodia (Mohberg et al., 1973). Results like this vindicate the proposal that, wherever possible, propagation of Physarum in the laboratory by serial subculture of plasmodia, even when slowed down by use of sclerotia, should be avoided.

A solution to this problem would be for investigators who wish to work on plasmodia to use a diploid strain made freshly by mating together two amoebae of different mating type and different matB type, but otherwise of Colonia genetic background. Two appropriate amoebal strains for this purpose, both made in Dee's laboratory, would be LU648 (matA1 matB1 fusA1 leu^+) and LU215 (matA3 matB3 fusA1 leu-1); these two strains could be stored separately as frozen amoebal stocks, with samples mated together every few months to produce genetically characterized and stable plasmodia. LU215 and LU648 amoebae, and detailed instructions for culturing them, are available either from Dee or from Burland.

FUNCTION OF TUBULINS IN PHYSARUM: INSIGHTS FROM GENETIC ANALYSIS

With regard to the use of genetics in elucidating tubulin function in Physarum, Burland pointed out that the subject of his talk was not genetics, nor was it Physarum; rather, the subject was tubulin function, and the data had been obtained using, in part, genetic analysis in Physarum. This emphasized Dee's point that genetic analysis in Physarum was not an end in itself, but rather one of several techniques available to study biological problems.

Alpha and beta tubulins are related polypeptides that form alpha/beta tubulin dimers; these dimers, or "protomers", polymerize to form microtubules, 25 nm-diameter fibers that function in mitosis, meiosis, motility, and, as part of the cytoskeleton, in the determination of cell shape. Alpha and beta tubulins are ubiquitous among eukaryotes and are encoded by separate genes. Burland pointed out that one of the most interesting aspects of tubulin biology is that, in many eukaryotes, there are multiple genes for alpha and for beta tubulin and multiple forms of alpha and beta tubulin polypeptides. Putting these observations together with the known multiple functions of microtubules, one important question is whether different tubulin genes or proteins have specific function--e.g., in mitosis vs. cell shape determination.

Details on the characterization of tubulin genes, proteins and microtubules in Physarum have already been published (Burland et al., 1983, 1984; Schedl et al., 1984; Roobol et al., 1984). Whereas microtubules are utilized in the mitotic spindle, cytoskeleton, and centrioles in the amoebae, and in flagellar axonemes in flagellates, microtubules have been found in the plasmodium only in the mitotic and meiotic spindles. Remarkably, more tubulin isotypes can be detected on two-dimensional gels of plasmodial proteins than on gels of amoebal and flagellate proteins (Fig. 2-1). At first sight, this does not encourage the notion that certain tubulin isotypes may have functional specificity; fewer microtubular functions are found in the plasmodial phase of the life cycle, where tubulin isotypes seem more diverse. Yet it is difficult to accept that Physarum has multiple tubulin polypeptides and multiple tubulin sequences in the genome (at least four unlinked loci for alpha tubulin and three unlinked loci for beta tubulin) if the multiple gene products are functionally redundant.

One way to address the question of function of individual tubulin genes would be to obtain a series of mutants, each with a mutation in a different tubulin gene and each mutation having an identifiable phenotype. Sheir-Neiss et al. (1978) showed that tubulin mutants could be obtained directly in Aspergillus by selecting for resistance to benomyl, an antitubulin benzimidazole drug. Physarum "BEN" mutants have been obtained by selection for amoebal resistance to the antitubulin benzimidazole methyl benzimidazole carbanate (MBC) (Burland et al., 1984). A single mutation at any one of four unlinked loci (benA, benB, benC or benD) is sufficient to confer MBC resistance on amoebae. One mutation at the benD locus, benD210, confers drug resistance on both amoebae and plasmodia, and in addition a novel beta tubulin isotype, beta 1-210, is detected on two-dimensional gels of the mutant proteins (Fig. 2-1). The beta 1-210 polypeptide is detected in amoebae in addition to a polypeptide migrating in the position of the wild-type polypeptide beta 1. In mutant plasmodia, beta 1-210 is also expressed, but there is no polypeptide migrating in the position where plasmodial beta 1 tubulin is found in wild-type (Fig. 2-1). The benD210 mutation is thus instrumental in elucidating that, although only a single beta tubulin isotype is detectable by electrophoretic analysis of amoebal proteins, this isotype consists of the products of two (or more) genes; one of these genes is expressed in plasmodia, the other is amoeba-specific. In addition, the plasmodium-specific beta 2 tubulin is unaffected by the benD210 mutation, and so is almost certainly the product of a third beta tubulin gene. Further, on the issue of tubulin function, the fact that benD210 strains are resistant to MBC suggests that the *benD* gene product is used in mitosis in both amoebal and plasmodial stages; otherwise, mitosis would be blocked by MBC, and the mutants could not grow.

While the assignment of a *ben* mutation to a tubulin structural gene is greatly assisted by the presence of an electrophoretically altered tubulin polypeptide, an independent method is available for associating mutations with known structural genes. This method involves the use of restriction fragment length polymorphisms as markers of genes for which DNA probes are available (Schedl and Dove, 1982). Polymorphisms for the lengths of genomic restriction fragments for actins were readily detected within the Wis 1 natural isolate of *Physarum*. One of the most difficult steps in making this method feasible for *Physarum* tubulin sequences was to find the right combination of amoebal strains and restriction enzymes that would give distinct sizes for different alleles of the multiple tubulin restriction fragments detectable. The strain combination that finally yielded information was CLd and MA275; as discussed above, CLd is a Colonia-background strain, but MA275 is a haploid amoebal segregant of the Wis 2 plasmodial isolate, genetically distantly related to CLd. Even using such diverged strains, however, it was possible to obtain distinct restriction fragment sizes with few restriction enzymes. Only Eco RV was found to give complete differences for alpha tubulin fragments, and Stu I was the only enzyme found that came close to giving different sizes for the set of beta tubulin restriction fragments (Schedl et al., 1984). Four unlinked loci were revealed for alpha tubulin, *altA*, *altB*, *altC* and *altD*, and three loci for beta tubulin, *betA*, *betB* and *betC*.

The restriction fragment polymorphisms that allowed the above mapping of tubulin DNA sequences in the *Physarum* genome could be put to use in asking whether any of the *ben* mutations that confer resistance to benzimidazoles map to any of the tubulin sequences identified. The BEN mutants had been isolated from CLd amoebae, so all that was needed for each mutant was to repeat the above analysis for the cross BEN mutant X MA275, and ask whether any of the tubulin restriction fragments from the BEN mutant parent co-segregated with benzimidazole resistance. Two *benD* mutations, *benD107* and *benD210*, co-segregated with the *betB1* beta tubulin allele, confirming the assignment of benD as a beta tubulin locus. Further, the *benA* allele tested co-segregated with the *betA1* beta tubulin allele, suggesting that the *benA* locus may also be allelic to a beta tubulin structural gene. However, this mapping procedure falls short of proof of allelism between *benA* and *betA*; although no recombinants were found between markers for these loci, the upper limit of distance between *benA* and *betA* at the 95% confidence level is 22 centimorgans (Schedl et al., 1984).

If *benA* and *betA* really are allelic, then the mystery of the multigene family for tubulins deepens. On the one hand, mutation at either of two expressed beta tubulin genes is sufficient to confer resistance to MBC, suggesting that the two beta tubulin genes may be functionally redundant. On the other hand, construction of a diploid amoebal strain heterozygous for the benD210 mutation shows that this mutation is recessive--i.e., the heterozygous diploid is as sensitive to MBC as wild-type. Thus, in the case of the haploid benA or benD mutants, where 50% (i.e., one or other) of expressed beta tubulin genes carry a drug-resistance allele, drug resistance is conferred; in the case of the diploid heterozygous for benD210, only 25% of expressed beta tubulin genes carry a drug-resistance allele. One possible explanation for this result, consistent with redundancy of benA and benD, is that one is simply detecting a "titration" effect, wherein the resistant beta tubulin subunits in the heterozygous diploid are diluted out by their wild-type allelic product. However, if this were true, one might expect the benD210/+ heterozygote to have a level of resistance intermediate between the levels of resistance of the wild-type and benD210 haploids, rather than the complete recessivity observed for benD210. Such an intermediate level of resistance ought to be clear, since the benD210 haploid is resistant to high levels of MBC (50 μM) compared with wild-type, which is completely inhibited by 5 μM MBC. An alternative explanation could be that in fact the benA and benD genes are not entirely redundant, but have some level of specificity, perhaps at the level of polymerization into individual microtubules such that benA and benD beta tubulins are never polymerized together in the same microtubule.

LOCATION OF A SINGLE BETA TUBULIN GENE PRODUCT IN BOTH CYTOSKELETAL AND MITOTIC-SPINDLE MICROTUBULES

An alternative way to use tubulin mutants to address the question of whether a particular organelle utilizes a particular tubulin polypeptide was described by Paul. The experiments exploited the fact that the mutant beta 1-210 tubulin produced by benD210 strains is electrophoretically distinguishable from all other tubulins detected. Even though several tubulin isotypes are distinguishable on two-dimensional gels of Physarum proteins from wild-type strains, it does not follow that a single isotype corresponds to a single gene product; it could correspond to several. Further, it is not clear whether different isotypes detected in wild-type strains are encoded by the same or by different genes. The electrophoretic shift of the mutant polypeptide, however, acts as an unequivocal marker for a single gene product.

Paul's experiments rest on another convenient aspect of the biology of Physarum, along with the development of a method for stably isolating microtubular structures from the organism. The presence of microtubules exclusively in the mitotic spindle of the plasmodium, together with its mitotic synchrony, facilitated the purification of microtubular structures from the plasmodium that were exclusively of mitotic-spindle origin. The proteins contained in these spindles were then analyzed on two-dimensional gels, and it was found that, whether the preparations were made from wild-type or benD210 strains, all of the tubulins seen in whole cell lysates were also present in spindle preparations and in similar relative proportions in both whole cells and mitotic spindles. This established that all of the plasmodial tubulin isotypes, including the beta 1-210 tubulin that identifies a single gene product, are utilized in the mitotic spindle.

The next step was to examine the tubulin polypeptide composition of another organelle in a benD210 strain, in this case the

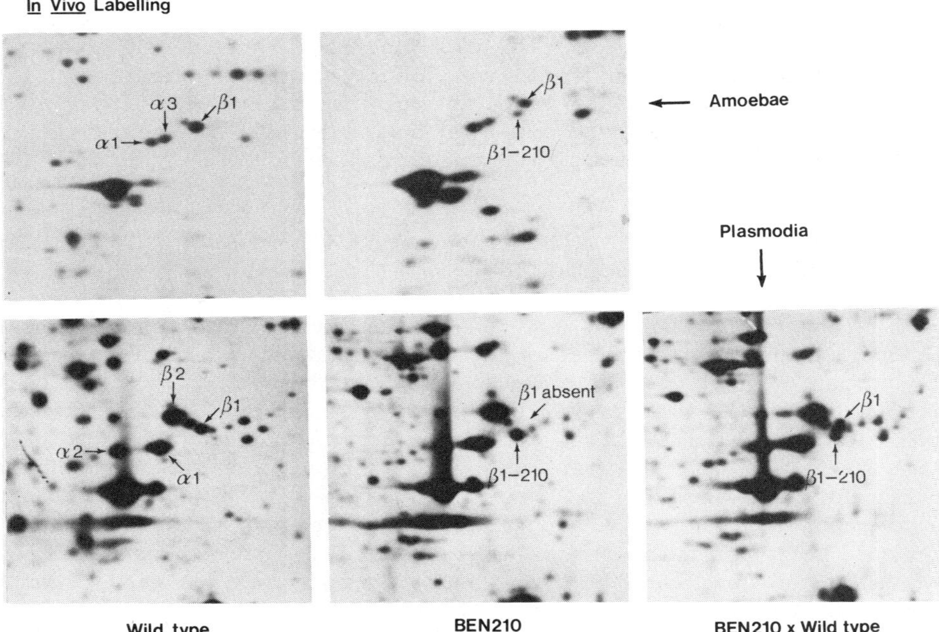

Fig. 2-1. Tubulins in wild-type and BEN210 amoebae and plasmodia. Two-dimensional electrophoresis of proteins labeled in vivo with [^{35}S], detected by fluorography. Only the actin-tubulin region of each gel is shown. Isoelectric focusing is from left (basic) to right (acidic). SDS polyacrylamide separation is from top to bottom. (Reprinted from Burland et al. (1984), with permission.)

cytoskeleton, and determine its tubulin composition. The cytoskeleton can be isolated intact from axenically grown amoebae. Although the original mutant carrying benD210 was a strain incapable of axenic growth (as is normal for amoebae of Physarum), Burland, using standard genetic procedures, constructed the strain "BEN210-AXE", which carries both the benD210 mutation and the alleles of the axe genes necessary for axenic growth. Conveniently, the mitotic index of axenically grown amoebae is very low (ca 1%), so that the vast majority of microtubules in a growing population derive from the cytoskeleton (no flagellates are present in axenic cultures). Paul presented evidence that all of the tubulins detected in whole cell lysates of benD210 amoebae, including beta 1-210, were found in cytoskeleton preparations in similar relative proportions. (The flagellate-associated alpha 3 tubulin was absent from whole cell and cytoskeletal preparations; axenic amoebal cultures do not flagellate.) Together, these experiments established that a single beta tubulin gene product, beta 1-210, was utilized in both cytoskeletal and mitotic-spindle microtubules, a conclusion consistent with the notion that the multiple tubulin gene products may be functionally redundant.

It has long been known that tubulins from mammalian brain could be purified and assembled into microtubules in vitro, but tubulins from other sources had not been purified and assembled in this way until quite recently. Physarum was the first microorganism from which tubulin was purified and assembled in vitro (Roobol et al., 1981), and this was achieved from amoebae of the axenic mutant strain CLd-AXE. Because tubulins are such a small proportion of cellular protein (ca 0.1%) in Physarum, it is necessary to grow large numbers of amoebae (ca 10^{11}) for tubulin preparation before studies on assembly of tubulin into microtubules in vitro can be contemplated. It would be extremely difficult to do this with amoebae grown on bacterial lawns on plates, the method needed to grow wild-type strains of amoebae. The availability of the axenic strain BEN210-AXE greatly simplified studies of benD210 mutant tubulins in vitro.

Paul summarized data, obtained by Foster in Gull's laboratory, from experiments that were designed to test in vitro the assembly of tubulins from the axenic benD210 strain into microtubules. The experiments were done both in the absence and presence of benzimidazole. It was already known (Quinlan et al., 1981) that assembly into microtubules of tubulins from wild-type amoebae is inhibited by several benzimidazoles. Foster's data revealed that tubulin assembly from benD210 amoebae was less sensitive than wild-type to benzimidazole, as one would predict. Of considerable interest was the finding that there was no substantial drug-induced enrichment for the mutant beta 1-210 polypeptide in the assembled microtubules; instead, the different tubulin isotypes appeared present in similar relative proportions whether assembly was done in the presence or absence of benzimidazole. The most simple explanation for this result would be that the microtubules made in vitro are composed of a mixture of the two beta tubulin isotypes, and that only a proportion of the beta tubulin molecules need to be of the mutant (beta 1-210) type to confer stability on the whole microtubule when benzimidazoles are present.

Among the questions on the talks by Burland and Paul was one from Sauer, who noted a paradox. The suggestion that the recessiveness of the benD210 mutation could indicate some degree of specificity among the tubulins, perhaps at the level of different beta tubulins assembling into distinct sets of microtubules, appears to contradict the observations that both beta tubulins are found in cytoskeletal and mitotic-spindle microtubules, and further in microtubules assembled in vitro, in the presence of MBC. There are two distinct issues here. One is the issue of whether different functions are executed by the different tubulin polypeptides. The experiments involving isolation of mitotic spindles and cytoskeletons from benD210 strains indicate no functional specificity for the beta 1-210 tubulin in these structures. The second issue concerns the possibility that the different tubulins are assembled into distinct sets of microtubules, regardless of whether the different microtubule sets have redundant or specific functions. The recessiveness of the benD210 mutation supports some level of specificity here, but the experiments on assembly of tubulins into microtubules in vitro do not. One could resolve this paradox if a mechanism conferring specificity of assembly of different tubulins into distinct sets of microtubules is lost in vitro. Burland felt that more experiments were needed to resolve the problem. These include studies of polypeptide composition of microtubular organelles isolated from diploid amoebae, heterozygous for different ben mutations, grown in the presence of benzimidazole, together with analysis of the drug resistance levels of the amoebae. The question of whether individual microtubules in vivo contain specific sets of tubulin polypeptides might be addressed by immunoelectron microscopy if antibodies can be found that react specifically with only one tubulin polypeptide.

On a related issue, Gull pointed out that the altered beta tubulin in haploid benD210 amoebae is less abundant than the co-expressed wild-type beta tubulin, so that the amount of mutant beta tubulin in the heterozygous benD210/+ diploid would be extremely small (see Fig. 2-2); therefore, he continued, Burland's conception of the heterozygous diploid having only 25% drug-resistance alleles was wrong, as the amount of "drug-resistant tubulin" was much less than 25%. Burland agreed that the level of resistant tubulin was significantly lower than the level of wild-type tubulin, but did not feel that this invalidated his argument.

DISCUSSION SESSION ON GENETICS

P. Anderson started the discussion session with a summary of strategies used in genetic analysis. Conventional genetics uses the approach of first identifying a biological process of interest, then trying to isolate mutants altered in that process and studying their phenotype. This type of attack on a problem would be completed when the product of the mutated genes could be identified and their functions assigned. Progressing from the stage of mutant phenotype to gene product is the challenging step in this approach. Alternatively, what P. Anderson described as "emerging genetics" starts with a protein product, and with either an antibody to the protein or information on the protein sequence, the gene encoding the protein can be cloned. To address the function of the gene, the cloned DNA can be altered (mutated) in vitro, reintroduced into the cell's genome, and the phenotype of the resulting mutant can be studied to deduce the function of the wild-type gene. Either method is satisfactory only if all of the steps of the approach are successfully completed. In Physarum, only the conventional approach has been used, and here only in limited cases. Successful application of the emerging genetics will require a suitable DNA transformation system to be established (see Chapter 25).

Whichever genetic approach might be used to address a particular problem, P. Anderson pointed out that a detailed linkage map would be of enormous value. For example, if a new mutation could be mapped to a known gene, progress toward elucidating the gene's product and function could be substantially simplified. Any new information of this nature would further be additive, so that the usefulness of a linkage map would increase in proportion to the total amount of information. Conventional genetic experiments in Physarum have not yet sought to establish a linkage map, not least because the haploid chromosome number is around 40 (Mohberg, 1977). However, P. Anderson was enthusiastic about the possibility of using restriction fragment length polymorphisms to establish a linkage map. He was particularly encouraged that a significant amount of work relevant to this project has already been done. The analysis of the tubulin gene families, and the mapping of ben mutations with respect to tubulin loci, has yielded over 40 amoebal strains that are progeny of the cross CLd X MA275 (or BEN mutant X MA275, which is the same for the purposes of restriction fragment polymorphisms). These same progeny strains can be used to expand the mapping done for tubulins to any genes for which DNA probes exist.

The key to this project is the large amount of restriction fragment length polymorphism that exists between CLd and MA275 amoebae. Although there had been some difficulty in finding an appropriate restriction enzyme that would reveal the polymorphisms for all of the beta tubulin genes, Schedl pointed out that five of ten randomly chosen cDNA clones, when used as probes against CLd and MA275 genomic DNA blots, gave polymorphism for at least one of only two restriction enzymes tested. Laffler reported finding polymorphism for histone H4 DNA sequences in Physarum for four of eight restriction enzymes tested. These results suggest that restriction fragment polymorphisms may be readily found for most genes. Thus, simply by use of DNA from existing segregants of the CLd X MA275 crosses, mapping of genomic sequences for any cloned gene can be undertaken. As P. Anderson stressed, it would not even be necessary to know the identity of the cloned gene to use it for mapping purposes. Provided that everyone interested in this project would use the very same set of segregants from the CLd X MA275 and BEN mutant X MA275 crosses, every time a new sequence was cloned, it could be mapped relative to all other genes identified in this way. However, Laffler pointed out that while the existing strains would facilitate mapping of restriction fragment polymorphisms, it would be necessary to perform new crosses and repeat the DNA isolations and mapping in order to determine the location of new mutations on the map.

On the subject of obtaining DNA for the mapping project, Pallotta questioned the feasibility of isolating large amounts of DNA from several strains of amoebae grown on bacterial lawns on plates, since the amoebal strains in question cannot grow axenically. Burland replied that while it would be a substantial amount of work, there was no technical limitation to this approach. In fact, half of the amoebal progeny from the CLd X MA275 and BEN mutant X MA275 crosses are selfing strains, as are the CLd and BEN parents, so half of the DNAs can be obtained from plasmodia, which are much easier to grow in large quantities.

Two possible problems with the above mapping procedure were noted. First, one would need to know the recombination frequency per base pair (bp) in Physarum. This is not yet established, but with a haploid complement of 40 chromosomes, this project would be feasible only if the recombination frequency were substantially less than 50% over the length of a whole chromosome. With a genome size of approximately 2 X 10^8 bp, the average size of a Physarum chromosome is 5 X 10^6 bp. Thus, if 1 centimorgan were equivalent to 10^6 bp, as in humans, then the mapping project would be highly feasible; if 1 centimorgan were equivalent to 10^3 bp, as in yeast, the project would be impracticable (see Pontecorvo, 1958, for estimates of centimorgans per bp). It is worth noting, however, that the large genetic distance between CLd and MA275 amoebae may fortuitously help the mapping project to succeed. If these two strains have been geographically separated for a long time, chromosomal differences such as inversions and translocations may have accumulated; such rearrangements would act to reduce the observed recombination frequency.

The second problem is one of logistics. It would be inefficient for each laboratory that cloned a DNA sequence from *Physarum* to grow 20 or 40 amoebal clones and isolate DNA from each in order to do the DNA blotting. Instead, as Vogt pointed out, persuading a central laboratory to prepare large quantities of DNA from 40 or so amoebal clones and to distribute small quantities of DNA to interested laboratories would be much more efficient. Specific funding would be required to do this, and several participants felt that a joint effort to raise funding would be valuable. There was, however, a general consensus that the McArdle Laboratory would be an appropriate central laboratory to supervise the preparation and distribution of DNA samples from the amoebal clones. This possibility is being further considered. Another suggestion for improving the overall logistics of the project was the possibility of the central laboratory sending out DNA samples that were already digested with restriction endonucleases, and perhaps even blotted to nitrocellulose. Less voiced, but of overall importance for the success of such a mapping project, is the requirement that all of the participants agree to undertake their respective roles in a disciplined way, so that the results could be reliably collected together. Despite the many possible pitfalls, however, there was overall tremendous enthusiasm for determining the feasibility of this project.

One aspect of the feasibility of the mapping project that was not resolved but that would need careful study involves the use of probability theory. One could predict the number of linkage groups likely to be found in *Physarum* from the limited data available on linkage. The study of actin and tubulin sequences has revealed four actin loci (Schedl and Dove, 1982), four alpha tubulin loci, and three beta tubulin loci. All these loci are unlinked to one another except for *ardC* and *altD*, which are about 2 centimorgans apart. All eleven loci are unlinked to *matA*. Thus, from twelve loci studied, there is one case of linkage, and use of probability theory can predict whether this result is consistent with forty linkage groups, as would be expected if each chromosome is less than 50 centimorgans, or whether more linkage groups are likely to be encountered by mapping with such a low density of markers (see Feller, 1957, for details of pertinent probability theory). There is, however, one tricky complication. One actin locus and one alpha tubulin locus are each complex - i.e., have more than one sequence (Schedl and Dove, 1982; Schedl et al., 1984). In fact, we do not know whether these multiple sequences at each locus are very tightly linked, constituting a genuine complex locus, or whether they are several centimorgans apart and really constitute additional, linked loci. In the latter case, we would have three cases of linkage among fourteen loci, rather than one case among twelve. This would alter the predictions of the number of linkage groups. Further, as P. Anderson pointed out, one might argue that close linkage of related loci like actin or tubulin is not a representative example for randomly chosen loci, and thus that statistical predictions based on linkage of related loci may prove inaccurate. More work on mapping other loci will be needed before the feasibility of the full-scale mapping project is established.

Braun, discussing technical feasibilities, suggested a possible alternative to the mapping project as outlined above. With "pulsed-field" electrophoresis, it is possible to resolve DNA molecules up to the size of whole yeast chromosomes (Schwartz and Cantor, 1984; Carle and Olsen, 1985). If this could be achieved for _Physarum_ chromosomal DNA molecules, then mapping could be done by probing blots of pulsed-field electropherograms with cloned DNA probes, and clones could be directly assigned to linkage groups. A big advantage of this method would be that DNA from only a single strain of _Physarum_ would be needed. The technique would be severely limited, however, if, as one might conclude from photographs of _Physarum_ chromosome spreads (Mohberg, 1977), the chromosomes do not differ sufficiently in size to allow resolution of all of the chromosome-sized DNA molecules. Also, the average size of _Physarum_ chromosomes is about 25 times that of yeast chromosomes. As Braun pointed out, however, it may still be possible to obtain a useful number of "linkage groups" with this method. Although successful application of this method could simplify construction of a linkage map for polymorphic DNA markers, as noted by Dee and Laffler, conventional genetic analysis would still be required to assign mutations to linkage groups.

CONCLUDING REMARKS

One can distinguish two purposes for genetic analysis. One is to elucidate aspects of genetics itself, such as the relationship between a gene and a complementation group. In this respect genetics is the purest form of biological science. To study aspects of genetics, however, requires considerable power in the practical aspects of genetic analysis of a particular organism. Thus, for example, valuable information on the fine structure of the gene could be obtained from analysis of _rII_ mutants of bacteriophage T4 (Benzer, 1961). The T4 _rII_ system is capable of measuring recombination frequencies as low as 2×10^{-4}, a frequency much lower (for T4) than the recombination frequency between adjacent nucleotide pairs. Clearly, genetic techniques used for _Physarum_ are not this powerful. A second purpose for genetic analysis is as a tool to use in elucidating the nature of a biological process, such as the regulation of a metabolic pathway or the function of particular gene products. The requirements for power in practical aspects of genetic analysis for this second purpose are much less stringent; the ability to isolate mutants and to analyze them by complementation and recombination often suffices. What is of equal importance, however, is that the biological process to be studied is interesting and tractable in the organism chosen. This is where *Physarum* will be valuable, particularly in the analysis of the cell cycle, where the natural synchrony of the plasmodium provides an experimentally tractable situation, and in the analysis of development, where the simplicity of the transitions between single cell types offers relatively simple experimental opportunities.

In the past, few investigators have used genetic analysis in *Physarum* to study biological problems. From the enthusiasm expressed at the meeting for genetic approaches, together with some significant advances in genetic techniques reported, it seems likely that genetic analysis will be used much more in future. This increased use of genetics will be of value not only for the specific biological problems addressed, but also for the whole community of biologists who use *Physarum*. The increased use of genetics will increase the resources available--new mutants, new ways of selecting mutants, refined techniques of analysis. But it is no secret that in other organisms where genetic analysis has been particularly fruitful, the published genetic resources have been shared among the investigators who use the particular organism. This sharing will need to be pursued enthusiastically among the small *Physarum* research community if significant biological problems are to be successfully resolved.

RELATED TOPICS COVERED IN OTHER SECTIONS

One of the most exciting possibilities for expanding the capabilities for genetic analysis in *Physarum* is the development of a DNA transformation system, especially if this could be used in gene disruption experiments (see Chapter 17). Progress on this subject is discussed in Chapter 25.

R. Anderson's talk on mating-type expression in *npf* mutants was of considerable interest with respect to the mechanism of mating control in *Physarum*. This subject is discussed in Chapter 6.

Chapter 3 includes basic information on the structure of the *Physarum* genome and discusses some transposon-like elements that are present. These elements may be of use in establishing an efficient DNA transformation system. Chapter 3 also covers interesting data on the inheritance of the multicopy, extrachromosomal ribosomal DNA molecules of *Physarum*, and the ability to transform yeast cells with complete *Physarum* ribosomal DNA molecules.

The feasibility of isolating conditional lethal mutants, particularly cell-cycle mutants, is addressed in Chapter 5 on the Cell Cycle.

ACKNOWLEDGMENT

I thank William F. Dove for helpful comments on the manuscript. My research is supported by program project grant No. CA-09230 and by core grant No. CA-07175 to the McArdle Laboratory from the National Cancer Institute.

REFERENCES

Anderson, R. W., 1979, Complementation of amoebal-plasmodial transition mutants in Physarum polycephalum, Genetics, 91:409.

Anderson, R. W., and Youngman, P. J., 1985, Complementation of npf mutations in diploid amoebae in Physarum polycephalum: the basis for a general method of complementation at the amoebal stage, Genet. Res. Camb., 45:21.

Benzer, S., 1961, Genetic fine structure. in: "Harvey Lectures", Vol. 56, Academic Press, New York.

Burland, T. G., Chainey, A. M., Dee, J., and Foxon, J. L., 1981, Analysis of development and growth in a mutant of Physarum polycephalum with defective cytokinesis, Dev. Biol., 85:26.

Burland, T. G., Gull, K., Boston, R. S., Schedl, T., and Dove, W. F., 1983, Cell type-dependent expression of tubulins in Physarum, J. Cell Biol., 97:1852.

Burland, T. G., Schedl, T., Gull K., and Dove, W. F., 1984, Genetic analysis of resistance to benzimidazoles in Physarum: differential expression of beta tubulin genes, Genetics, 108:123.

Carle, G. F., and Olsen, M. V., 1985, An electrophoretic karyotype for yeast, Proc. Natl. Acad. Sci. U.S.A., 82:3756.

Carlisle, M. J., 1973, Cell fusion and somatic incompatibility in myxomycetes, Ber. Dtsch. Bot. Ges., 86:123.

Collins, O. R., 1981, Myxomycete genetics, 1960-1981, J. Elisha Mitchell Soc., 97:101.

Cooke, D. J., and Dee, J., 1974, Plasmodium formation without change in nuclear DNA content in Physarum polycephalum, Genet. Res. Camb., 23:307.

Cooke, D. J., and Dee, J., 1975, Methods for the isolation and analysis of plasmodial mutants in Physarum polycephalum, Genet. Res. Camb., 24:175.

Dee, J., 1979, A gene unlinked to mating-type affecting crossing between strains of Physarum polycephalum, Genet. Res. Camb., 31:85.

Dee, J., 1982, Genetics of Physarum polycephalum, in: "Cell Biology of Physarum and Didymium", Vol 1, H. C. Aldrich and J. W. Daniel, eds, Academic Press, New York.

Kirouac-Brunet, J., Masson, S., and Pallotta, D., 1981, Multiple allelism at the matB locus in Physarum polycephalum, Can. J. Genet. Cytol., 23:9.

Kubbies, M., and Pierron, G., 1983, Mitotic cell cycle control in Physarum: Unprecedented insights via flow-cytometry, Exp. Cell Res., 149:57.

Laane, M. M., and Haugli, F. B., 1976, Nuclear behaviour during meiosis in the myxomycete Physarum polycephalum, Norw. J. Bot., 23:7.

Laffler, T. G., and Dove, W. F., 1977, Viability of Physarum polycephalum spores and ploidy of plasmodial nuclei, J. Bacteriol., 131:473.

Mohberg, J., 1977, Nuclear DNA content and chromosome numbers throughout the life cycle of the Colonia Strain of the Myxomycete Physarum polycephalum, J. Cell Sci., 24:95.

Mohberg, J., 1982, Genealogy and characteristics of some cultivated isolates of Physarum polycephalum, in: "Cell Biology of Physarum and Didymium", Vol 1, H. C. Aldrich and J. W. Daniel, eds, Academic Press, New York.

Mohberg, J., Babcock, K. L., Haugli, F. B., and Rusch, H. P., 1973, Nuclear DNA content and chromosome numbers in the Myxomycete Physarum polycephalum, Dev. Biol., 34:228.

Pontecorvo, G., 1958, "Trends in Genetic Analysis". Columbia University Press, New York.

Poulter, R. T. M., and Dee, J., 1968, Segregation of factors controlling fusion between plasmodia of the true slime mould Physarum polycephalum, Genet. Res. Camb., 12:71.

Quinlan, R. A., Roobol, A., Pogson, C. I., and Gull, K., 1981, A correlation between in vivo and in vitro effects of the microtubule inhibitors colchicine, parbendazole and nocodazole on myxamoebae of Physarum polycephalum, J. Gen. Microbiol., 122:1.

Roobol, A., Paul, E. C. A., Birkett, C. R., Foster, K. E., Gull, K., Burland, T. G., Dove, W. F., Green, L., Johnson, L., and Schedl, T., 1984, Cell types, microtubular organelles, and the tubulin gene families of Physarum, in: "Molecular Biology of the Cytoskeleton", G. G. Borisy, D. W. Cleveland, and D. B. Murphy, eds., Cold Spring Harbor Laboratory, New York.

Roobol, A., Pogson, C. I., and Gull, K., 1981, In vitro assembly of microtubule proteins from myxamoebae of Physarum polycephalum, Exp. Cell. Res., 130:203.

Schedl, T., and Dove, W. F., 1982, Mendelian analysis of the organization of actin sequences in Physarum polycephalum. J. Mol. Biol., 160:41.

Schedl, T., Owens, J., Dove, W. F., and Burland, T. G., 1984, Genetics of the tubulin gene families of Physarum, Genetics, 108:143.

Schwartz, D. C., and Cantor, C. R., 1984, Separation of yeast chromosome-sized DNAs by pulsed field gradient gel electrophoresis, Cell, 37:67.

Sheir-Neiss, G., Lai, M. H., and Morris, N. R., 1978, Identification of a gene for beta tubulin in Aspergillus nidulans, Cell, 15:639.

Shipley, G. L., and Holt, C. E., 1982, Cell fusion competence and its induction in *Physarum polycephalum* and *Didymium iridis*, *Dev. Biol.*, 90:110.

Youngman, P. J., Anderson, R. W., and Holt, C. E., 1981, Two multi-allelic mating compatibility loci separately regulate zygote formation and zygote differentiation in *Physarum polycephalum*, *Genetics*, 97:513.

Chapter 3: MOLECULAR ORGANIZATION OF THE PHYSARUM GENOME

Norman Hardman

Department of Biochemistry
University of Aberdeen
Aberdeen, SCOTLAND

with contributions from: E. M. Bradbury, R. Braun, M. E. Christensen, M. Czupryn, E. Epstein, A. M. Foote, S. Kostelny, P. Loidl, J. G. Opstelten, D. H. Pearston, V. M. Vogt, F. Wanka, J. H. Waterborg, and F. X. Wilhelm

INTRODUCTION

Physarum polycephalum, because of its unique life cycle, is a useful organism in which to address a wide variety of questions of biological interest. These include problems relating to motility, differentiation, cell cycle regulation, and ultrastructural organization of the mitotic apparatus. As reviewed in other chapters of this volume, much progress has already been made on several fronts in these and other areas. In many instances, as stressed repeatedly in this Workshop, further progress will depend on the development of procedures that provide the experimental tools necessary for rigorous molecular genetic analysis in *Physarum*. In other organisms, such procedures have depended for their success on the identification of important regulatory sequences and novel genetic elements, providing a means to develop appropriate vectors for the cloning, in vitro manipulation, reintroduction, and controlled expression of important genes. This information can be obtained only from studies of the structure and organization of genomic DNA sequences. Thus, apart from its intrinsic interest, the molecular characterization of the *Physarum* genome will have important implications for future studies of the many facets of *Physarum* biology.

This chapter reviews our current state of knowledge of the molecular organization of the *Physarum* genome, at both the DNA and the chromatin levels. The most detailed studies to date have been carried out on the nucleolar rDNA "minichromosome", since techniques were developed early for the purification of intact rDNA and for the selective isolation of nucleolar chromatin. These studies have provided a considerable amount of information on the structure and function of the rDNA molecule and provided a useful system in which to investigate the interaction of proteins that may play key roles in the function of the rDNA minichromosome. In addition to studies of rDNA, recent work on the characterization of cloned segments of genomic DNA has complemented earlier investigations aimed at defining the general pattern of organization of the *Physarum* genome and has led to the identification of a major transposon-like repetitive sequence family, with a number of interesting features, that may account for over one-half of all the repetitive DNA in *Physarum*. These and other mobile genetic elements in *Physarum* DNA may ultimately prove to be useful for developing efficient cloning vehicles for transformation of *Physarum* (see Chapter 25). Unexpected technical difficulties have been encountered in generating totally representative genomic DNA libraries from *Physarum* (see Chapter 22); this has slowed progress toward characterizing other genetic loci, though a steadily increasing number of genes have now been isolated (Chapters 22-24).

The latter part of this chapter is devoted to studies of the nuclear proteins of *Physarum*. These include the chromatin proteins, notably the histones, and the nonchromatin proteins, which constitute the fibrillar structure generally referred to as the nuclear "matrix". The chapter reviews recent structural work on the histone proteins and of the cloned histone H4 gene, and includes recent data on the cell cycle-dependent transcription of H4 genes and posttranslational modification of histones. With regard to the nuclear matrix, the attention of the Workshop focused primarily on its possible role in the organization of DNA into replicons, the functional units of replication.

STRUCTURE, FUNCTION AND INHERITANCE OF rDNA IN *PHYSARUM*

Molecular Organization of Ribosomal DNA

The genes for ribosomal RNA, together with their associated nontranscribed spacer regions, are referred to as rDNA. These genes are repeated about 200-fold in eukaryotes. In most organisms rDNA is arrayed as tandem repeats at one or more chromosomal sites, each location constituting a nucleolar organizer region. In some lower eukaryotes, rDNA is extrachromosomal and takes the form of independently replicating DNA molecules, usually with one or two rRNA transcription units on each molecule. *Physarum* and *Tetrahymena* are the best studied examples of this type of organization. *Physarum* rDNA has several features that make it a useful model that can be exploited for studying chromosome structure and gene expression:

replication origins; telomeres; specific binding proteins; and processed gene transcripts. Additionally, since a single copy of the rDNA molecule is inherited at meiosis, it may prove to be a useful model for studying gene reduction and amplification.

The Physarum rDNA molecule is about 60 kb in size. About 150 rDNA molecules are present per nucleus. It is a large palindromic molecule with an axis of symmetry at its center (Fig. 1-1). Vogt and his collaborators (Epstein and Kostelny) provided a comprehensive account of recent work on the structure and expression and replication of rDNA.

Transcription

Transcription of the pre-rRNA starts at a position 18.2 kb from both ends of the molecule (Blum et al., 1983) and proceeds outward to about 4.9 kb from the ends. The entire transcription unit [5'-external-transcribed spacer [ETS]--19S RNA--internal-transcribed spacer [ITS]--5.8S RNA--[ITS]--26S RNA (including introns)--3' end] at 13.3 kb is thus one of the largest known in any species. Separating the two transcription units is the 22 kb central nontranscribed spacer segment, which contains two replication origins per half-molecule. The spacer DNA is composed of complicated arrays of inverted-repeat sequences (Fig. 1b) punctuated by two copies per half-molecule of a 0.6 kb "unique" sequence (Ferris and Vogt, 1982; Ferris, 1985). Despite its complexity, the entire spacer is made up of only about 1.2 kb of different sequences. Specific gaps (Johnson, 1980) and probably simple repeated CA-rich sequences are found at the ends of the rDNA molecule, similar to those found at the termini of Tetrahymena rDNA and yeast chromosomes. Much of the rDNA has now been sequenced: all of the 26S RNA region (Otsuka et al., 1983); some of the ITS and much of the ETS regions (Kukita et al., 1981; Hattori et al., 1984), including the region surrounding the RNA transcription start site (Blum et al., 1983); representative sequences of all the repeated elements that make up the central spacer (Ferris, 1985); and part of the region near the rDNA termini (Bergold et al., 1983).

Only in the last several years have the molecular details of rDNA transcription been elucidated. Primarily from work in mammalian systems, Xenopus, and yeast, it has been suggested that one or more protein factors in addition to RNA polymerase I are required for correct and efficient transcription. In contrast to RNA polymerase-II transcription, Pol-I-mediated transcription is species-specific, consistent with the absence of any consensus sequences upstream of the RNA transcription start site, and probably reflecting a requirement for species-specific rDNA binding proteins. rDNA transcription in Physarum is maximal in the middle of the G2 phase of the cell cycle and is greatly reduced upon spherulation. Differences in the chromatin configuration of the transcribed and

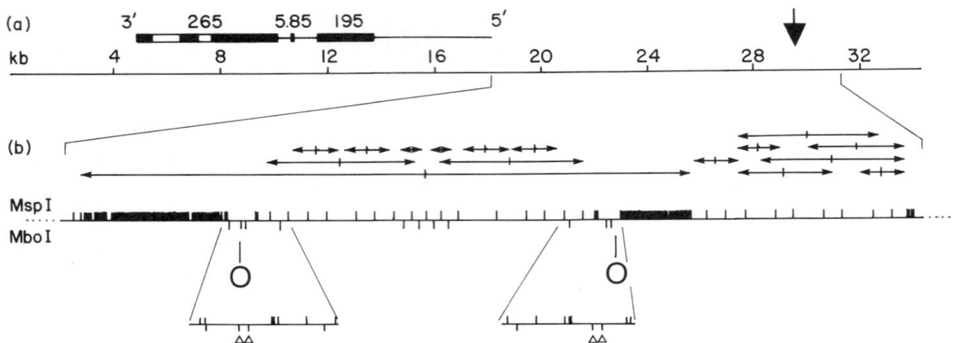

Fig. 3-1. Structural features of the extrachromosomal rDNA molecule of Physarum.
(a) Approximately one-half of the rDNA molecule is illustrated, from the left end to the central axis of symmetry at ca 29.6 kb (vertical arrow). Except for some asymmetry immediately surrounding the central axis and restriction site polymorphisms throughout the two halves of the molecule, the right-hand of the rDNA is a mirror image of the left. The rRNA primary transcript is shown, containing the 5' transcribed spacer, 19S, 5.8S, and 26S rRNA sequences. Most Physarum strains contain two intervening sequences in the 26S RNA gene, the positions of which are indicated by open boxed segments.
(b) An expanded view of the nontranscribed central spacer, extending rightward from the transcription start site. The positions of complex arrays of direct and inverted repeats are indicated by horizontal arrows, and the positions of MspI and MboI restriction sites are shown. Regions marked "O" represent the approximate locations of rDNA origins of replication. Open triangles show the positions of protected Mbol restriction sites in the rDNA molecule in vivo in nucleolar chromatin, which map close to the putative origin loci. See text for further details.
The figure is adapted from Ferris (1985) and includes data compiled from Ferris and Vogt (1982) and Kunzler et al. (1984).

nontranscribed portions of the rDNA molecule have been deduced from studies of nucleosome structure (Prior et al., 1983) and accessibility to psoralen cross-linking (Judelson and Vogt, 1982).

Intervening sequences spliced out of the primary transcripts are now recognized as ubiquitous in eukaryotic organisms. Introns can be grouped into three classes, including those that occur in (1) messenger RNAs, (2) tRNAs, and (3) genes for other RNAs and organelle-specific RNAs. In this latter class is the well-characterized intron of Tetrahymena rDNA, which was the first example of an RNA with the capacity to undergo "self-splicing" in the absence of proteins. The rDNAs of most Physarum strains have two introns in the 26S-coding region (Fig. 3-1a). This is the only other example of a nuclear DNA molecule with class (3) introns. Comparative sequence analyses have led to the conclusion that the Tetrahymena rDNA intron, the two Physarum rDNA introns, certain mitochondrial and chloroplast introns, viroids, and some plant virus-associated RNAs all share conserved sequences that are inferred to be important in the splicing mechanism. Some mitochondrial introns have also recently been directly demonstrated to self-splice, though no similar data have yet been reported for the Physarum rDNA introns. Kostelny and Vogt (personal communication) have discovered a third rDNA intron of about 0.8 kb in some strains of Physarum (the "Carolina" strain of Shipley) positioned 20 nucleotides upstream of the insertion site for intron-II in the 26S rRNA. Interestingly, a chromosomal sequence homologous to the third intron was shown by hybridization analysis to be present in Physarum genomic DNA, the study of which may provide some clue as to the origin of this rDNA intron. These observations, taken together with the existence of Tetrahymena strains lacking an rDNA intron, indicate that such introns may represent mobile genetic elements that have been capable of transposition on a relatively short evolutionary time scale.

Nucleolar Proteins and rDNA Function

Two different types of approach were reported at the Workshop in attempts to identify specific nucleolar proteins that might be important for rDNA function. Using a filter-binding assay involving radioactively labeled rDNA restriction fragments and partially purified Physarum nuclear extracts, Epstein and Vogt (personal communication) have found two proteins with specific affinities for different segments of the rDNA molecule. One of the proteins binds to a region within a few hundred bp upstream of the rRNA transcription start site and might represent a "species-specific" regulatory factor of rDNA transcription of the kind referred to above. Such an assay should provide a means to obtain highly purified protein for future in vitro studies of its possible role in rDNA transcription. With a similar approach, a second protein was identified, which binds specifically near the rDNA terminus (tentatively at a site within the terminal 500-700 bp portion of the molecule). The protein may therefore be connected in some way to the function of rDNA telomeres (see below). It will be important to determine the relation of this protein to those reported previously to be tightly bound to the rDNA terminus (Cheung et al., 1981).

An alternative approach used monoclonal antibodies to trace the location of proteins that may be potentially important for rRNA maturation (Christensen, unpublished). Christensen focused attention on one major nuclear protein, B-36, a 34,000 dalton protein primarily associated with the nucleolus. At least three different antigenic determinants were identified on B-36 with a panel of eight monoclonal antibodies. These were used as probes to demonstrate that B-36 is confined to nucleoli during active growth of *Physarum*. RNase/DNase digestion experiments suggested that the protein is probably localized in ribonucleoprotein material rather than being DNA-associated, but its biological role is not presently known. Cross-reactivity of *Physarum* antibodies with B-36 from other sources confirms earlier suggestions (Christensen et al., 1981) that the protein is subject to a high degree of evolutionary conservation, consistent with its having some important role in the nucleolus. Wider implications of the work are evidenced by the presence of the B-36 antigen in autoimmune sera of patients with *scleroderma* (Ochs et al., 1984).

Replication of rDNA

Molecules of rDNA replicate throughout most of the synchronous mitotic cycle of *Physarum*: the last two-thirds of S-phase and all of the G2-phase. Replication is "unscheduled", although the rDNA population approximately doubles in size, meaning that in one cycle some molecules replicate twice, most replicate once, and others do not replicate at all. This suggests that some homeostatic mechanism probably controls the total amount of rDNA.

The positions of rDNA replication origins have been mapped approximately by electron microscopy, and these sites correspond to the location of the two 0.6 kb "unique" sequences per half-molecule of rDNA (see above and Fig. 3-1b). In intact interphase nuclei, digestion of chromatin with restriction enzymes reveals that DNA sequences near the putative replication origins are selectively protected from digestion (Kunzler et al., 1984; see open triangles in Fig. 3-1b), this suggests that protein is tightly bound there in vivo. Braun discussed more recent data from his laboratory related to this work, indicating specific phasing of nucleosomes around the strongly protected regions that surround the putative rDNA origin of replication, which may be a further reflection of sequence-specific binding of proteins to these regions (U. Pauli, P. Kunzler, and R. Braun, unpublished). Some special structural features of the telomeric sequences on rDNA molecules must also have a role in replication, as has been inferred for linear chromosomes in other organisms such as yeast and *Trypanosoma*. One-nucleotide gaps occur near the ends of *Physarum* rDNA, at the sequences (C)CCTA (Johnson, 1980). These are very similar to the (C)CCCAA "gap" sequences in *Tetrahymena* rDNA, which also function in yeast, as evidenced by their ability to promote stable replication of linear yeast plasmids in this organism.

Apart from the interest generated by the possibility of using rDNA replication origins as elements of Physarum transformation vectors (Chapter 25), there was considerable excitement over Braun's presentation of recent work in his laboratory suggesting that Physarum rDNA can be introduced efficiently into yeast by transformation and maintained stably without selection (Kunzler, 1985). This should provide considerable new opportunities for deletion analysis and in vitro genetic manipulation of the rDNA molecule in order to determine which segments are important for replication and efficient segregation in yeast.

Inheritance of rDNA at Meiosis

Vogt reviewed evidence that Physarum rDNA is inherited after meiosis as if it were a single molecule, but the pattern of inheritance is non-Mendelian (Ferris et al., 1983). Experiments were described in which two haploid Physarum strains (A and B) carrying rDNAs, which could be distinguished by minor restriction-site polymorphisms, were mated to form a diploid plasmodium. This plasmodium was induced to sporulate, and individual progeny amoebal clones were analyzed for rRNA type. In such experiments it was found that all progeny carried rDNA of one or the other type, but not both. Furthermore, there were always more clones of one type (Type A) than the other, and the ratio of A:B clones varied with the age of the plasmodium that was induced to sporulate; the older the plasmodium, the higher the ratio of A to B clones. The ratio of rDNA-A to rDNA-B in the plasmodium itself also changed with age; relatively young plasmodia had A:B ratios as low as 4, whereas in some cases "old" plasmodia had ratios over 100.

A model was proposed to explain these data on the basis of minor differences in replication efficiencies of Type A and Type B rDNA, and the random selection, at or near meiosis, of a single "master" rDNA copy from the rDNA pool. The germinating spores thus inherit only one or the other rRNA type, with a frequency reflecting the abundance of type A or B rDNA molecules.

ANALYSIS OF NUCLEAR DNA SEQUENCES OTHER THAN rDNA

Much of the work on the general characterization of genomic sequences other than rDNA has been carried out by Hardman and collaborators. Some of the general aspects of genome organization in Physarum were summarized, and recent data were presented on the structure and sequence of a highly abundant transposon-like element in Physarum DNA.

Early studies using reassociation kinetic analysis showed

that three sequence components can be identified in Physarum nuclear DNA: a "foldback" component consisting of 6% of the DNA; a repetitive DNA component representing 31% of the DNA; a majority component (63%) comprising largely single-copy sequences (Hardman et al., 1980). Evidence from physical studies, including electron microscopic analysis, showed that all three sequence components are mutually interspersed in a majority of the genome, though about one-half of the mass of foldback duplexes are formed from "clustered" inverted repetitive DNA sequences, forming complex "bubbled hairpin" structures (Hardman et al., 1979). The physical properties and distribution of foldback sequences suggested that they probably are a representative cross-section of the repetitive DNA component and that foldback DNA structures therefore probably arise when two repetitive elements from the same sequence family, arranged in inverted orientation, happen to be close enough to be present in the same DNA fragment (Jack and Hardman, 1980). Many of the general properties of repetitive sequences in Physarum DNA closely parallel observations made in other eukaryotes that display short-period interspersion of repetitive and single-copy DNA sequences but, except for a few specific repetitive sequence families, it was not clear until recently what their origin might be. This preliminary work has now been superseded by analysis of specific families of repetitive sequences by DNA cloning, as outlined below.

Attention was called to the link between the organization and methylation status of a major sub-fraction of the repetitive DNA. It has been shown that about 20% of the nuclear DNA sequences in the Physarum plasmodium forms a methylated "compartment", consisting of long sequence tracts, with contiguously methylated HpaII restriction sites, that contain an exclusive set of repetitive sequences (Whittaker et al., 1981; see Fig. 3-2a). With specific, cloned hybridization probes containing foldback sequences, it could be demonstrated that a small number of highly abundant, specific, repetitive DNA sequences are clustered within these methylated regions of the genome, whereas other families of less-abundant repetitive sequences appear to be widely distributed among both the hypermethylated and undermethylated DNA fractions (Gerrie et al., 1983; Peoples et al., 1983). Plasmid DNA clones containing "dispersed" repetitive DNA sequences are unstable in E. coli hosts such as HB101 and generate deletions that map to the position of foldback elements (Peoples et al., 1983). Others have experienced similar problems in cloning gene sequences, possibly because of the presence of similar dispersed repetitive elements in these genetically active segments (Chapter 22). As discussed in other Workshop sessions, the answer to some of these problems may lie in prudent choice of vectors and E. coli host strains (Chapter 22).

More recently, attention has focused on the nature and organization of the repetitive sequences specifically located in the hypermethylated (M+) regions of Physarum genomic DNA. Nearly two-thirds of all the 5-methylcytosine residues in Physarum DNA are concentrated in the HpaII-resistant genomic compartment, which comprises 20% of the genome. Hence, there is a fivefold greater concentration of these modified bases in this DNA fraction (Peoples et al., 1985). With foldback DNA-containing clones (referred to above) that were shown to hybridize specifically to hypermethylated domains (Whittaker et al., 1981), it has been possible to isolate much longer M+ DNA segments by DNA cloning and to investigate the organization of the highly repetitive DNA sequences that these regions contain. Surprisingly, results showed that these DNA segments are occupied almost exclusively by just one family of long (> 6 kb) highly repetitive sequences (Peoples and Hardman, 1983; Peoples et al., 1985). Even more puzzling was the observation that elements of the repeat appeared to be scrambled in different M+ DNA segments. For example, if the sequence order in the normal repeat unit is A-B-C-D-E-F, then segments can often be seen whose structure is represented by scrambled assemblies of these segments (C-D-E-F-D-E-A-B, etc.). Such scrambled clusters of highly repeated DNA sequences are a general feature of eukaryotic genomes in a wide range of organisms, from Drosophila to chicken, but no satisfactory explanation has been found for their origin. It was therefore argued that, by studying the structure of DNA segments containing "scrambled" repeats, it might be possible to derive an explanation for scrambling and for the origin of a major portion of foldback DNA in Physarum, 50% of which originates from M+ DNA segments containing this highly repetitive sequence (Peoples et al., 1983).

Transposon-like Nature of the Major Repetitive Sequence in Physarum DNA

Pearston summarized the results of his recent studies on the structure of M+ DNA segments (Pearston et al., 1985). He concluded that the highly repetitive element that dominates M+ DNA segments, referred to as the "HpaII-repeat", is an 8.6 kb sequence with terminally redundant sequences of 277 bp (Long Terminal Repeats, LTRs). The LTRs, like those of recognized retrotransposons in other systems, are terminated by 5'TGTTGG.....CTAACA3', and contain "TATA-box"-like sequences and potential poly(A)-addition signals. Bordering the left LTR in the internal domain of the element is a 14-bp sequence identical to the corresponding region of the transposable element Copia. Adjacent to the right LTR is a 7-bp polypurine tract. In retrotransposons, these features are recognized as being important for control of (-) and (+) strand DNA synthesis during the reverse transcription phase of the transposition cycle. These structural properties imply that the HpaII-repeat family of highly repetitive elements in Physarum DNA may have evolved by retrotransposition. Such sequences have the capacity to self-replicate by a cyclical process involving transcription into RNA, reverse transcription into cDNA, and genomic integration of daughter cDNA copies. With DNA probes derived from internal portions of the HpaII-repeat,

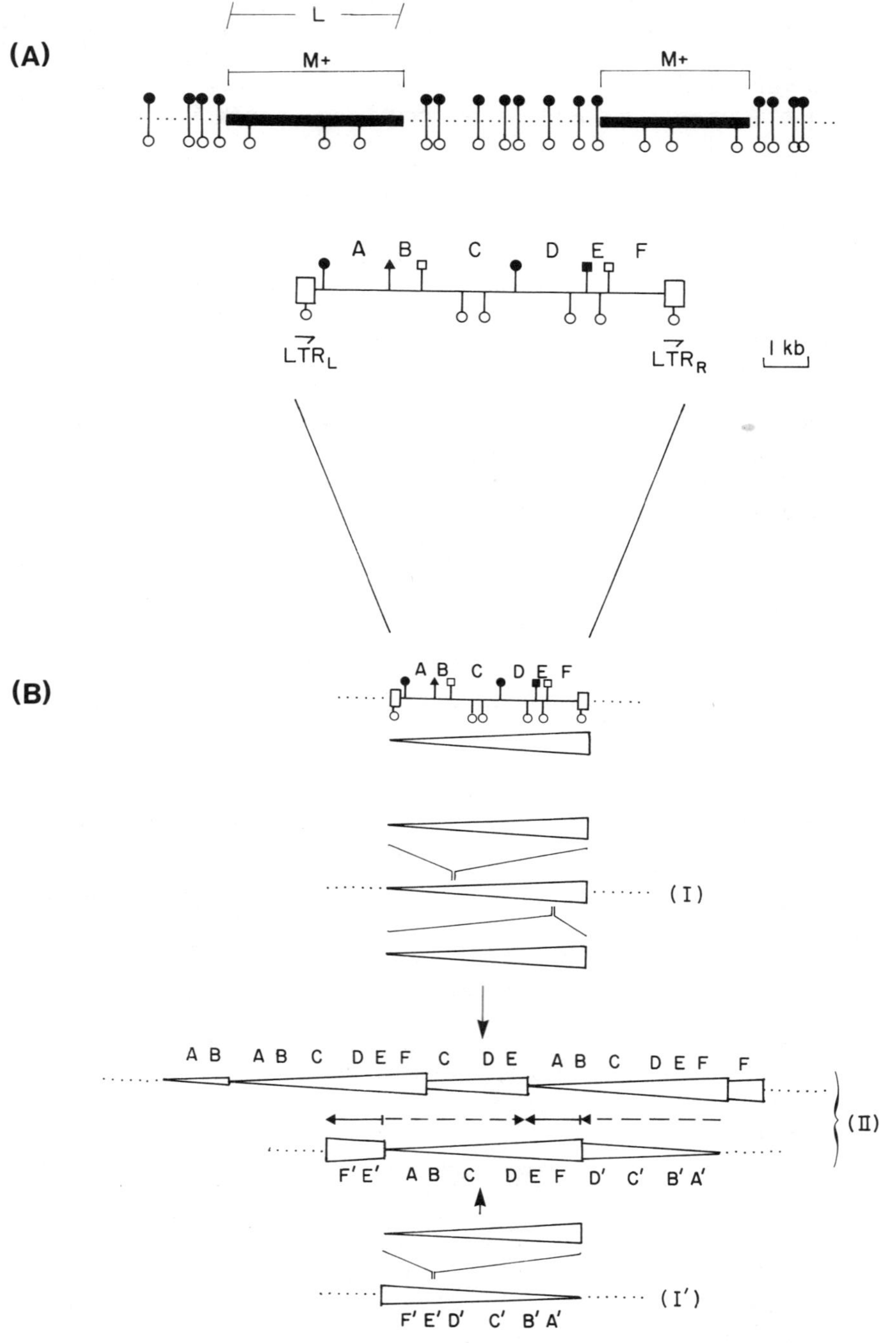
L
(A)
M+
M+
A
B
C
D
E
F
$\overrightarrow{LTR}_L$
$\overrightarrow{LTR}_R$
1 kb
(B)
A B C D E F
(I)
A B A B C D E F C D E A B C D E F F
(II)
F' E' A B C D E F D' C' B' A'
(I')
F' E' D' C' B' A'

Fig. 3-2. Organization of transposon-like elements in hypermethylated domains of *Physarum* genomic DNA.

(a) Structure of hypermethylated segments of *Physarum* nuclear DNA: Regions of hypermethylated (M+) DNA contain an approximately fivefold greater concentration of 5-methylcytosine than the remainder of the genome. These regions contain contiguous arrays of methylated CmeCGG sequences. They are thus resistant to cleavage with the restriction endonuclease HpaII (closed circles), but are cleaved with the isoschizomer MspI (open circles). The minimum length (L) of these regions is 20-50 kb, but data cannot presently exclude the possibility that they are derived from much longer methylated segments with occasional unmethylated HpaII/MspI restriction sites. These regions account for 20% of the nuclear DNA and are comprised almost entirely of one family of long, highly repetitive sequence, referred to as 'HpaII-repeats', arranged in 'scrambled' arrays together with occasional less highly repetitive elements. These DNA segments contain over one-half of all the repetitive sequences and foldback sequences in the *Physarum* genome. Surrounding segments of less highly methylated DNA are represented as dotted lines. The figure is compiled from data of Whittaker and Hardman (1980), Whittaker et al. (1981), Gerrie et al. (1983), and Peoples et al. (1983, 1985).

(b) Structure and "scrambling" of HpaI-repeats in M+ DNA segments: The HpaII-repeat is approximately 8.6 kb in length and contains long terminal repeats (LTRs) of 277 bp. The "consensus" restriction map of genomic copies of the HpaII-repeat is shown. The LTR elements contain short, terminal repeats TGTTGG...CTAACA and putative transcriptional control signals similar to those recognized as common features of eukaryotic transposable genetic elements. Symbols are restriction sites for: ■ BamHI; □ EcoRI; ○ HindII; ▲ HindIII; and ● MspI.

Scrambled clusters of HpaII-repeats are believed to result from transposition-like events that lead to insertion of the element into resident copies of its own sequences. 1) Multiple insertions lead to scrambling of the structure of copies of the repeat that are already present, generating long sequence tracts comprised almost exclusively of rearranged HpaII-repeats. 2) Insertion of mutually inverted copies of the HpaII-repeat (I') leads to arrangements with the capacity to form foldback structures, when the DNA is denatured and allowed to "snap-back" (horizontal arrows and solid/broken lines indicate inverted repeats). HpaII-repeats probably account for about one-half of the repetitive component (15-20% of the genome) and one-half of the mass of foldback DNA. Data are compiled from Peoples and Hardman (1983), Peoples et al. (1985), and Pearston et al. (1985).

it could be demonstrated that these elements are not transcribed in *Physarum* plasmodia. Hence, if these elements do evolve by a retrotranscription mechanism, it remains to be established at what developmental stage the necessary RNA intermediates are generated.

Remarkably, all the transposition target sites so far investigated are located within the same repetitive element. As elaborated by Pearston and summarized in Fig. 3-2, targeted transposition of this element can explain many of the previously confusing properties of M+ DNA segments: the long ($>$ 20-50 kb) clusters of this repeated sequence which constitute M+ DNA segments (see above); the "scrambling" of the sequences that would be a consequence of self-insertion (see Fig. 3-2b); the origin of "foldback DNA" structures (by transposition in inverse orientation into resident copies of the repetitive element, illustrated in Fig. 3-2b).

Many questions remain unresolved concerning the properties and possible biological function of this transposon-like sequence: Does the sequence necessarily always insert into its own sequence? What factors are responsible for targeted insertion? Why are the sequences selectively methylated? At what stage in *Physarum* development are the postulated RNA transposition intermediates generated? Can other, possibly less abundant/more useful, transposable elements be isolated that may pave the way for the development of transformation vectors in *Physarum*, by analogy with work carried out in *Drosophila*? Some of these questions were also addressed in other sessions of the Workshop (Chapter 2).

CHROMATIN PROTEINS

Histone Proteins

Bradbury provided a short but comprehensive review of the current status of chromatin research in *Physarum*. Some of the most definitive experiments on cell cycle-dependent histone modifications have been made in the *Physarum* system.

Physarum chromatin has a well-characterized nucleosome structure (Johnson et al., 1976). The nucleosome core particle has recently been isolated and shown to have a structure, at low resolution, identical with nucleosome core particles from higher organisms (J. P. Baldwin and G. Stone, personal communication). Matthews and his collaborators have shown that *Physarum* chromatin contains a full complement of histones (Mende et al., 1983); H3 and H4 have properties of sequence and composition in keeping with their rigid conservation; H1, H2A, and H2B are diverged from their mammalian homologs with H1 displaying the greatest divergence. Bradbury briefly summarized the amino acid sequence information for histones H1, H2A, and H2B from a number of sources. Comparisons reveal both species and tissue-specific variation with sequence differences located in the unfolded, flexible amino and carboxy terminal domains.

Foote further elaborated some of the structural differences between Physarum histone H1 and its mammalian counterparts. She purified the protein in milligram amounts to a high degree of purity (> 95%), enabling biophysical and biochemical characterization (Foote, 1985). Gel exclusion chromatography demonstrated that Physarum H1 is considerably larger than the calf thymus protein (37 K compared with 21.5 K). Neutron-scattering studies in free solution both under acidic (pH 2.0) and neutral (pH 7.0) conditions showed the protein to behave as a disordered polypeptide that is considerably more elongated than other H1 proteins. Nuclear magnetic resonance and circular dichroism studies of H1 histone revealed additional information concerning the inability of the protein to fold below pH 7.0, titration of the seven histidine residues being required (Cary et al., 1985). Trimethylated lysine residues were identified in the non-folded domains. Further information on folding, derived from limited proteolysis and the application of other techniques of physical analysis, leads to a structural model very similar to that envisaged for other H1 histones (e.g., calf thymus H1)--that of a central globular domain with unfolded amino and extended carboxy terminal regions.

Bradbury also reviewed the work from his own and other laboratories on the various known reversible covalent modifications of the histone proteins that might be expected to modify chromatin and nucleosome structure. Ever since the discovery of such modifications, there has been widespread interest in the possibility of their involvement as important controlling factors in DNA replication, gene transcription, and chromosome condensation. The major postsynthetic modifications are acetylation, ubiquitination, and phosphorylation of the core histones and phosphorylation of the very lysine-rich histones. These modifications can be further categorized according to the levels at which they act. The relatively low levels of acetylation (thought to be involved in the control of replication and transcription) and ubiquitination (thought to be connected with transcriptionally competent chromatin) indicate that these modifications probably involve a small subcomponent of chromatin. In contrast, all the histone H1 and H3 are reversibly modified by phosphorylation at metaphase, and these much more extensive modifications are probably associated in some way with chromatin and chromosome condensation.

Czupryn introduced results of her recent work on correlations between changes in transcriptional activity and chromatin structure during the plasmodium/spherule transition induced by starvation of Physarum microplasmodia (Czupryn et al., 1985). The rate of RNA synthesis (measured in vitro using endogenous RNA polymerase-II) decreased after 12 hr starvation to 30% of its initial level. This was accompanied by a comparable decrease in the amount of chromatin that could be solubilized by DNase 1 digestion, an increase in the content of histone H1 and several additional non-histone proteins in the solubilized chromatin fraction, and changes in the properties of the nucleosomes. Structural reorganization of chromatin thus correlates with repression of transcription during spherulation, but the primary causes of these structural changes are still not known. Could they be induced by phosphorylation of histone H1^{o} (see below)?

Loidl summarized studies of posttranslational histone modifications during the cell cycle and during spherulation. The level of acetylated H4 histone has been shown to fluctuate over the cell cycle in Physarum plasmodia with a maximum in S-phase (Loidl et al., 1983). This correlates well with the time at which newly synthesized H4 histone is translated from H4-mRNA, whose synthesis is initiated during the previous cell cycle in G2 phase (see following section; Wilhelm et al., 1984). These results, in contrast to other studies (Chahal et al., 1980), are interpreted to suggest a lack of correlation between histone acetylation and RNA transcription (Loidl et al., 1983).

Histone synthesis was measured after induction of the plasmodium/spherule transition and was found to decrease to zero after 12 hr. This was followed by a 3- to 5-hr resumed period of synthesis, 35-45 hr after induction, which correlated with the time at which mature spherules were first observed. Nonviable spherules were formed when the period of late histone synthesis was prevented by treatment with cycloheximide. The level of histone acetylation was monitored during spherulation and was shown to be constant, despite the depression of RNA synthesis that is known to follow induction of spherulation. It was demonstrated that considerable incorporation of [^{3}H]-labelled acetate preceded the period of late histone synthesis, although acetate incorporation ceased completely within the initial 12 hr of spherulation. Loidl suggests that these results argue against a simple correlation between levels of histone acetylation and RNA transcription, but rather support the idea that histone acetylation plays an important role in the replacement of preexisting histones by newly synthesized histones. Differentiation-specific histone biosynthesis may be required to accommodate the chromatin rearrangements necessary to complete the differentiation program successfully (P. Loidl and P. Gröbner, personal communication).

Bradbury made the important point that much of the work to date has involved correlations of histone modifications with function, and that more definitive information could perhaps be obtained by the design of experiments to demonstrate cause-and-effect. He provided an example of such an approach: conditional mutants in the pathway leading to phosphorylation and dephosphorylation of histones, for instance, could be used to define the role of phosphorylation in chromosome condensation. Organisms such as Physarum that are amenable to genetic analysis should be useful for such studies.

A more rigorous approach on more traditional lines might be more appropriate for other problems such as defining the role of the less extensive, but possibly more specific, modifications such as acetylation and ubiquitination. For instance, if acetylation is a general requirement of "active" chromatin, then cloned cDNA probes for those genes, which display differential expression during the Physarum cell cycle or life cycle, may provide the means to test this hypothesis. Allfrey's group have already provided evidence that 80-90% of the H3 histone in active rDNA chromatin is in the tri- and tetra-acetylated forms (B. G. Allfrey, personal communication). There should, therefore, be a profitable opportunity in the near future for cross-fertilization of expertise between those groups interested in chromatin structure and others involved in isolating cDNAs for developmentally regulated mRNAs for other purposes (Chapter 24). A similar approach might be adopted for studies of ubiquitination. Studies in Drosophila were referred to (Levinger and Varshavsky, 1982), claiming that 50% of the nucleosomes on the heat-shock 70 K protein gene contain ubiquitinated histone H2A (uH2A), compared with an overall level in total chromatin of 5%. This implies that a distinct family of genes may be marked for histone ubiquitination. Bradbury's group (Mueller et al., 1985) has shown that ubiquitination of H2A and H2B histones persists throughout the cell cycle except for a brief period immediately prior to metaphase until early in anaphase. This possibly reflects a requirement for particular chromatin regions to be in a continuous state of ubiquitination, but that they are deubiquitinated at metaphase probably to allow for their correct packaging into metaphase chromosomes.

New histone subtypes are still being discovered. Bradbury's group has identified a very lysine-rich histone subtype, $H1^{o}$, in Physarum (Yasuda et al., 1985). In mammalian cells $H1^{o}$ is found in mitotically inactive cells, in cells undergoing terminal differentiation, and in the late stages of embryonic development. Much higher levels of histone $H1^{o}$ are present in Physarum compared with animal cells. The ratio of $H1^{o}$ to H1 doubles when plasmodia are converted to sclerotia and, moreover, the protein becomes highly phosphorylated. As with other histone modifications associated with changes in transcriptional activity, the application of appropriate cDNA probes for genes that display differential expression in these different developmental stages should be useful for future studies related to the function of $H1^{o}$.

Histone Genes and Their Expression during the Cell Cycle

The synchronous mitotic cycle of the Physarum plasmodium provides a unique opportunity to study a number of important questions related to chromatin assembly; for example, how the production of new histone proteins is coordinated with the process of DNA replication during the cell cycle. Studies so far in Physarum have been confined largely to investigations of the temporal expression of the histone H4 genes, carried out primarily by F.X. and M.L. Wilhelm with their collaborators.

Wilhelm presented a concise summary of current knowledge of the structure and expression of histone H4 genes, of which there are probably only two in Physarum. The Wilhelms' work on the H4 genes remains the only report of a cloned, fully characterized gene sequence from Physarum, other than the rRNA genes (Wilhelm and Wilhelm, 1984; see also Chapter 23). The Wilhelms' group has encountered difficulties in attempting to isolate the other putative H4 allele. In other parts of the Workshop (Chapter 22), it was the general feeling that this is probably related in some way to the organization of gene sequences in Physarum, perhaps partly owing to a high density of foldback sequences or repetitive sequences. It is still not known with certainty whether the gene isolated is a functional genetic unit or a pseudogene, since it contains a short (86 bp) insertion that interrupts the H4 protein coding region. The inserted DNA segment has some characteristics of a transposable element. It is, therefore, not yet clear whether this intervening sequence is a "normal" intron.

As pointed out by F. X. Wilhelm, the biosynthesis of histone proteins in eukaryotic cells occurs periodically during the cell cycle and is generally believed to be coupled to the synthesis of DNA during S-phase. Regulation is thought to occur mainly at the transcriptional level, but a number of studies indicate that post-transcriptional controls may also be important (Chapters 5 and 7). The availability of a hybridization probe for histone H4 mRNA has made it possible to make a rigorous investigation of H4 gene expression in Physarum during the cell cycle (Wilhelm et al., 1984). Northern blots of mRNA obtained at different stages in the mitotic cycle, hybridized with a histone H4 gene probe, showed that H4 gene transcription is initiated in the second half of G2 phase, but that it is not translated. During the first third of the following S phase, the level of H4 mRNA is the highest, and H4 histone is synthesized. The level of H4 mRNA decreases dramatically approaching G2 phase; around the middle of G2 phase, the level of H4 mRNA is lowest, and no newly synthesized histones can be detected. A number of interesting questions are raised by this study related to posttranscriptional control of H4 expression. Size differences in the H4-specific mRNA band were observed in Northern blot experiments during the cell cycle. These may be indicative of subtle changes taking place in the structure of H4 mRNA that may influence its function (Chapter 5).

The results of a similar, independent study were also reported by Braun (V. Kung and R. Braun, personal communication). They correlated well with the above data. With both homologous and heterologous DNA probes, similar tenfold cyclical variation in the amounts of H4 mRNA during the cell cycle was demonstrated, whereas levels of actin mRNA remained approximately constant.

The main conclusions from this work are that both transcriptional and translational controls operate to modulate the time and level of expression of histone H4 genes during the cell cycle; H4 mRNA is synthesized in the G2-phase of the preceding cycle and is utilized to produce H4 protein during the successive round of DNA replication, when presumably newly made H4 molecules are assembled into daughter chromatin.

THE NUCLEAR MATRIX

The nuclear matrix is considered to be a nonchromatin structure of interphase nuclei. It is envisaged to have several possible roles that are difficult to understand without invoking an image of the matrix as a scaffold-like structure (Wanka et al., 1982; Berezney, 1984). During this Workshop emphasis was placed on the possible function of the matrix in the organization of nuclear DNA during replication and mitosis. Bradbury summarized data from other systems supporting the "loop model" of chromatin organization.

Various methods have been employed for the isolation of matrix structures. In general, these are based on three treatments: 1) removal of the nuclear envelope by treatment with nonionic detergents; 2) extraction of the chromatin proteins by high salt treatment (usually 2M NaCl); and 3) removal of the nucleic acids by digestion with DNase and RNase. Matrix-DNA complexes can be obtained by omitting DNase treatment. These were referred to as "nucleoids" during the discussion.

During the Workshop, Wanka, Waterborg, and Opstelten presented their data on the possible structure and functions of the matrix at various stages, and these are collated in the following sections.

Ultrastructural Analysis

Wanka gave a brief account of the artificial ultrastructural changes that can result from various treatments and fixations, making it clear that final conclusions on the in vivo physical structure of the matrix should be drawn with considerable caution, although the presence of a peripheral nuclear laminar protein layer (the "lamina") is not disputed by even the most ardent critics of this ultrastructural work. The major part of the nuclear space appears to be occupied by a fibrous network connected to the nuclear lamina. Residual nucleolar matrix structures apparently consist of less well-structured, electron-dense material. The fibrous part of the matrix in Physarum nuclei becomes separated from the lamina in prophase and displaced toward the center of the nuclear space, whereas the residual nucleolar mass is dispersed along the nuclear periphery (Bekers et al., 1981). At subsequent phases the "2M-NaCl-resistant" fibers become more condensed and are displaced to those sites where chromosomes are normally observed. In contrast to cells that display open mitosis, where the lamina is degraded at the end of prophase, the lamina in Physarum persists at least until late anaphase but then undergoes the morphological changes required to accommodate mitosis.

Protein Composition of the Matrix

Opstelten reviewed the available data on the nature of matrix proteins. Physarum matrix prepared by extraction with 2M NaCl consists primarily of two major proteins with relative molecular weights of 23 and 37K, together with a greater number of less predominant polypeptides (Fig. 3-3 and Mitchelson et al., 1979). In other organisms it has been argued that the matrix may be an artifact resulting from the precipitation of nucleoplasmic or chromatin proteins by the use of high NaCl concentrations. In attempts to counter this argument, the protein compositions of residual nuclear structures prepared by different treatments have been compared. SDS-PAGE patterns of polypeptides from chromatin treated with DNase at low ionic strength (Fig. 3-3a) are very similar to those of residual protein structures prepared with 2M NaCl (Fig. 3-3e), except that DNase-treated chromatin retains small amounts of actin and some high molecular weight proteins removed by salt treatment. Treatments with staphylococcal nuclease or 2 mg/ml of dextran sulfate under low-ionic strength conditions also remove the histone proteins (Fig. 3-3b and d, respectively). Thus, there is no indication of artifactual precipitation of chromatin proteins or of other nucleoplasmic proteins normally soluble at low salt concentrations in matrix preparations prepared by treatment of nuclei with 2M NaCl. Proteins resembling those of the nuclear lamina of mammalian cells are not detected in matrix preparations for reasons that are still not understood.

Few changes are observed in the protein composition of the matrix during the mitotic cycle. A minor 52 K polypeptide appears gradually during S phase and is most prominent during mitosis.

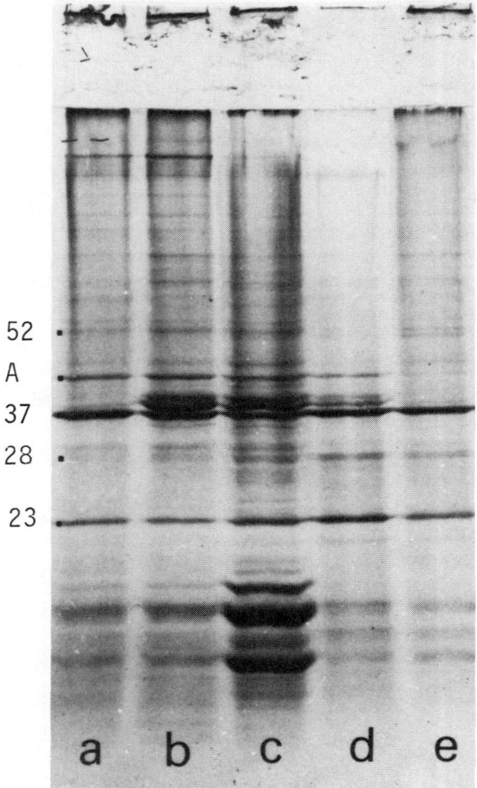

Fig. 3-3. Protein composition of residual nuclear structures. Nuclear matrices were prepared (a) by digestion of isolated nuclei for 60 min with 20 units/ml of DNase I at 37°C; (b) by digestion with 100 units/ml of staphylococcal nuclease at 0°C. Proteins of untreated nuclei are shown in (c). Nucleoids were prepared by extraction with (d) 0.2 mg/ml of dextran sulfate or (e) 2M NaCl. Proteins were dissolved in sample buffer containing 6M urea, 3% sodium dodecyl sulfate, and 10 mM dithiothreitol, and separated by electrophoresis on SDS-polyacrylamide gels (Photograph courtesy of J. Eygensteyn).

Structural and Functional Domains

Both Wanka and Waterborg described in greater depth the current situation regarding attachment of nuclear DNA to the matrix and its possible function in the organization of DNA during replication. Matrix prepared without DNase treatment can retain virtually the complete complement of nuclear DNA. No covalent attachment seems to be involved, since matrix proteins and DNA are dissociated by treatment with 2M NaCl/9M urea and can be separated either by sucrose gradient centrifugation or by chromatography on hydroxyapatite (Fig. 3-4). These data agree with results obtained with mammalian cells and are considered to be of general significance (Dijkwel et al., 1979; Van der Velden et al., 1984).

Wanka and his colleagues were able to exploit the unique property of natural mitotic synchrony and predictability of nuclear events in Physarum macroplasmodia to great effect in experiments designed to study the attachment of newly replicated DNA to the matrix during the cell cycle (Aelin et al., 1983). DNA was labeled at specific stages during the mitotic cycle; origins and termination sites of replication were labeled selectively by incorporating radioactively labeled precursors either at the beginning or at the

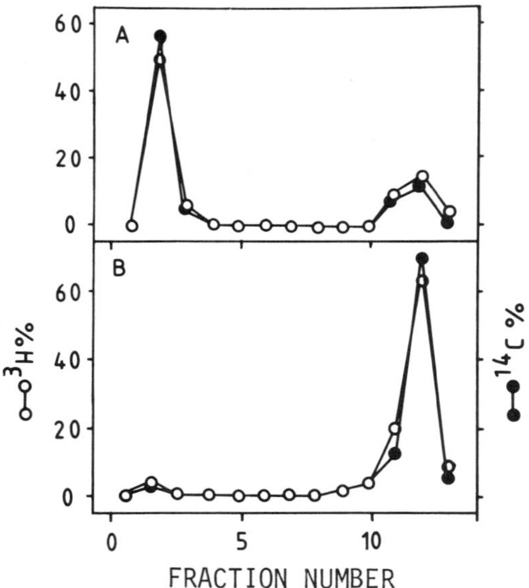

Fig. 3-4. Dissociation of nuclear matrix protein:DNA complexes by urea/NaCl. Nucleoids were prepared by treating isolated nuclei (labeled with ^{3}H-leucine and ^{14}C-thymidine) with 2M NaCl. The material was then carefully resuspended in 50 mM Tris-HCl buffer, pH 7.2, containing either (1) 2M NaCl or (2) 9M urea/2M NaCl. Samples were layered onto a 15-40% sucrose gradient containing 2M NaCl and centrifuged fo 1 hr at 25,000 rpm. Sedimentation is from right to left.

end of S phase. Nucleoids were then prepared after various "chase" periods, and unattached DNA was progressively removed from the matrix by digestion with DNase. Enrichment of the residual matrix-associated DNA for radioactive label was thus used as a measure of its position relative to the DNA:matrix attachment sites. The general conclusions from such analyses of nucleoids revealed that 1) DNA is attached to the matrix at sites at, or immediately adjacent to, the origins of replication during the entire nuclear cycle; and 2) during replication, evidence supports the notion that additional binding sites are generated near the DNA replication forks, from which replicated DNA is progressively displaced.

Additional experiments were described (Waterborg and Shall, 1985; J. H. Waterborg, personal communication) to counter the assertion that particular treatments, e.g., high salt concentrations, might lead to artifactual dislocation of the normal in vivo sites of DNA attachment. This possibility was referred to as "slippage". A number of changes in the preparation of nucleoids were outlined. Nuclei were isolated in the presence of 0.1M NaCl; magnesium ions were used to replace calcium ions; nucleoids were prepared in the presence of 10% polyethylene glycol, followed by the use of restriction nucleases rather than DNase to remove unattached DNA. In all cases DNA attachment appeared to be permanent and stable.

The interest in the organization of DNA into functional units of replication, "replicons", is long-standing and originates primarily from attempts to explain two aspects of the replication process. First, how is the unwinding of the DNA double helix controlled during replication so that resulting daughter DNA molecules can segregate effectively at mitosis? Second, how is the apparently chaotic mass of chromatin rearranged efficiently into chromosomes with reproducibly specific morphologies during mitosis?

The general elements for the organization of replication domains were outlined in a model that is supported by the available data and is gaining increasingly wide acceptance (Dingman, 1974; Dijkwel et al., 1979; Wanka et al., 1982). The model implies that each chromatin fiber is repeatedly bound by consecutive origins of replication to the fibrous protein matrix, as shown diagramatically in Fig. 3-5a. Initiation of DNA replication is accompanied by a duplication of the origin binding sites on the protein scaffold (Fig. 3-5c1 and c2). With progression of replication, the loop of the chromatin fiber is reeled through the replication binding site (Fig. 3-5c3). When the replication of a loop is completed, the termination site becomes detached from the backbone. This type of organizational arrangement could serve to generate two untangled daughter chromatin fibers that remain aligned by virtue of their permanent attachment to the protein scaffold (Fig. 3-5c4 and c5). In order to understand the function of the fibrous protein backbones in the formation of the chromosomes, one must assume that they are

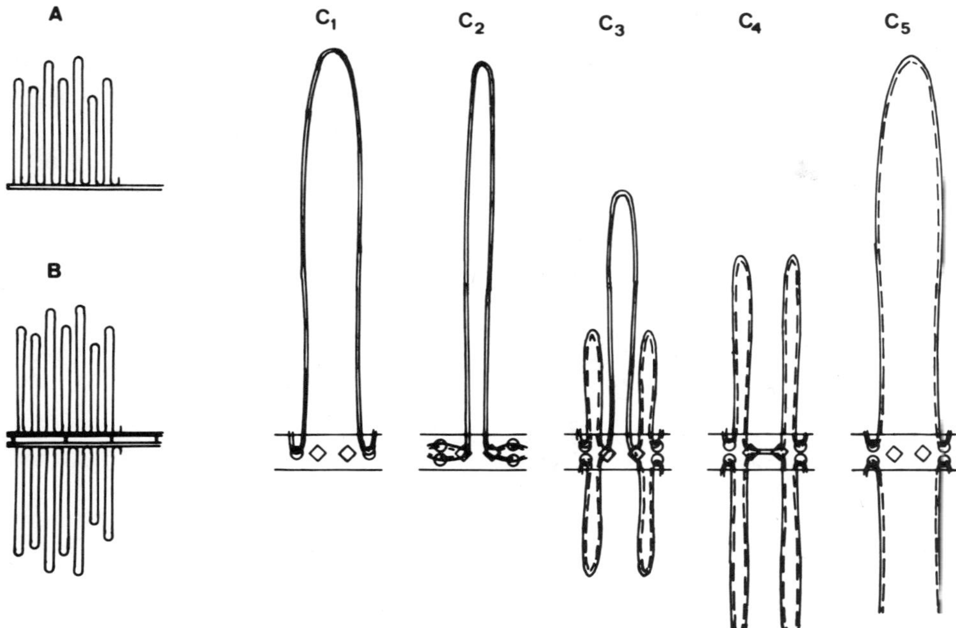

Fig. 3-5. Diagram illustrating a model for the organization of chromatin fibers in Physarum nuclei during interphase. (A) Part of a chromatin fiber attached by consecutive origin regions to a protein backbone in G1-phase. (B) The same structure after duplication of the protein backbone (late G2-phase); C_1 represents a single chromatin loop (double line) between two origins attached to the origin binding sites (circles) on the protein backbone. C_2 shows the situation after initiation of DNA replication: origins and origin sites are duplicated; replication forks have become bound to the replication binding sites (squares). In C_3 the DNA loop is reeled through the replication binding sites, giving rise to four loops of newly replicated chromatin (represented by interruptions in one line). In C_4 replication is completed. Finally, the termination site is detached from the replication binding site, resulting in the generation of two daughter chromatin loops bound to the backbone at the origin binding sites (C_5).

duplicated at some time between DNA replication and mitosis, but that newly formed backbones remain joined to each other along their entire length (Fig. 3-5, B). At present it is not clear how this duplication process is achieved.

Further Aspects of a Chromatin Loop Model

Bradbury referred to the increasing evidence from other systems in support of a "chromatin loop" or "domain" model for eukaryotic chromosomes (Laemmli et al., 1978; Marsden and Laemmli, 1979). In *Drosophila*, these loops are thought to be in the range 10-80 kb in size, with an average of 40-45 kb. This model is in many respects similar to that outlined in Fig. 3-5. The base of each DNA loop is attached to major matrix proteins, and evidence suggests that one of these is a protein of molecular weight 172,000 daltons that cross-reacts with antibodies to topoisomerase-II (Earnshaw and Heck, 1985; Earnshaw et al., 1985). This is consonant with the importance of DNA supercoiling in the generation of transcriptionally active chromatin. Bradbury pointed out that established properties of active chromatin, such as DNase and S1 nuclease hypersensitivity, are probably influenced by DNA supercoiling. Factors affecting the level of DNA supercoiling in a closed DNA loop in chromatin include: the action of topisomerase I and II; coiling of DNA in the nucleosome; and coiling of nucleosomes into higher-order chromatin structures. It will be important on several fronts to relate the work of Wanka and his colleagues, by use of the *Physarum* system, to the studies carried out by other groups working on other organisms. For example, where do topoisomerases fit into the picture drawn in Fig. 3-5? Is topoisomerase activity associated with any of the protein components of the matrix presented in Fig. 3-3? Antibodies raised against specific matrix proteins could be used as biochemical tools in a number of different ways: they could provide a means to study and characterize matrix preparation in vitro, and, in combination with appropriate labeling techniques, they could be used as a new approach to confirm and extend current knowledge of the ultrastructure of the matrix, both in situ and after various treatments. Such approaches may help to answer critics who remain to be convinced that artifacts do not arise from some treatments used to prepare nuclear matrix.

In the future there will be unique opportunities with the *Physarum* system to exploit the wealth of information available on histone variants and on periodic modifications of chromatin proteins to add detail to the chromatin loop model. It will also be of interest to determine whether a relation exists between the periodic organization of DNA sequences that appears to be a feature of the *Physarum* genome, e.g., in the arrangement of hypermethylated and undermethylated domains (Peoples et al., 1985) and the functional "domains" of replication and transcription.

The evidence that has accumulated for the periodic attachment of DNA to the nuclear matrix, both in *Physarum* and other systems (McCready et al., 1980; Aelin et al., 1983), leads to the suggestion that such attachment sites might be located at specific sequences. If this is the case, it should be possible to use molecular cloning techniques for their isolation and characterization. Some preliminary work along these lines has been carried out in mouse cells (Goldberg et al., 1983). With a different approach, cloned segments derived either from the histone gene cluster or the hsp70 heat-shock gene in *Drosophila melanogaster* have been used as hybridization probes to isolated matrix DNA, in both cases providing evidence for matrix attachment sites being located in close proximity to 5' regulatory elements of transcription (Mirkovitch et al., 1984). Similar experiments demonstrating differential attachment of genes to the nuclear protein scaffold have been carried out by others for the globin gene loci and transcriptionally active viral genes (Cook and Brazell, 1980; Cook et al., 1982). It should be possible to exploit the availability of homologous gene-specific probes for analogous experiments in *Physarum*.

REFERENCES

Aelin, J. M. A., Opstelten, R. J. G., and Wanka, F., 1983, Organization of DNA replication in *Physarum polycephalum*: attachment of origins of replication and replication forks to the nuclear matrix, *Nucl. Acids Res.*, 11:1181.

Bekers, A. G. M., Gijzen, H. J., Taalman, R. J. F. M., and Wanka, F., 1981, Ultrastructure of the nuclear matrix from *Physarum polycephalum* during the mitotic cycle, *Ultrastruct. Res.*, 75:352.

Berezney, R., 1984, Organization and function of the nuclear matrix, *in*: "Chromosomal Non-Histone Proteins", L. S. Hnilica, ed., CRC Press, Boca Raton.

Bergold, P. J., Campbell, G. R., Littau, V. C., and Johnson, E. M., 1983, Sequence and hairpin structure of an inverted repeat series at termini of the *Physarum* extrachromosomal rDNA molecule, *Cell*, 32:1287.

Blum, B., Seebeck, T., Braun, R., Ferris, P., and Vogt, V. M., 1983, Localization and sequence around the initiation sites of rDNA transcription in *Physarum polycephalum*. *Nucl. Acids Res.*, 11:8519.

Cary, P. D., Carpenter, C. G., and Foote, A. M., 1985, Physical studies by NMR and circular dichroism determining three structurally different domains in *Physarum polycephalum* histone H1, *Eur. J. Biochem.*, 151:579.

Chahal, S. S., Matthews, H. R., and Bradbury, E. M., 1980, Acetylation of histone H4 and its role in chromatin structure and function, *Nature*, 287:76.

Cheung, M. K., Drivas, D. T., Littau, V. C., and Johnson, E. M., 1981, Protein tightly bound near the rDNA termini of the *Physarum* extrachromosomal rDNA molecule, *J. Cell. Biol.*, 91:309.

Christensen, M. E., LeStourgeon, W. M., Jamrich, M., Howard, G. C., Serunian, L. A., Silver, L. M., and Elgin, S. C. R., 1981,

Distribution studies on polytene chromosomes using antibodies directed against hnRNP, J. Cell. Biol., 90:24.

Cook, P. R., and Brazell, I. A., 1980, Mapping sequences in loops of nuclear DNA by their progressive detachment from the nuclear cage, Nucl. Acids Res., 8:2895.

Cook, P. R., Lang, J., Hayday, A., Lania, L., Fried, M., Chiswell, J., and Wyke, J. A., 1982, Active viral genes in transformed cells lie close to the nuclear cage, EMBO J., 1:447.

Czupryn, M., Fronk, J., and Kazimierz, T., 1985, Chromatin reorganization during early differentiation of *Physarum polycephalum*, Biochim. Biophys. Acta, in press.

Dijkwel, P. A., Mullenders, L. H. F., and Wanka, F., 1979, Analysis of the attachment of replicating DNA to the nuclear matrix in mammalian interphase nuclei, Nucl. Acids Res., 6:219.

Dingman, C. W., 1974, Bidirectional chromosome replication: some topological considerations, J. Theoret. Biol., 43:187.

Earnshaw, W. C., Halligan, B., Cooke, C. A., Heck, M. M. S., and Liu, L. F., 1985, Topoisomerase-II is a structural component of mitotic chromosome scaffolds, J. Cell Biol., 100:1706.

Earnshaw, W. C., and Heck, M. S., 1985, Localization of Topoisomerase-II in mitotic chromosomes, J. Cell Biol., 100:1716.

Ferris, P. J., 1985, Nucleotide sequence of the central nontranscribed spacer region of *Physarum polycephalum* rDNA, Gene, in press.

Ferris, P. J., and Vogt, V. M., 1982, Structure of the central spacer region of extrachromosomal rDNA in *Physarum polycephalum*, J. Mol. Biol., 159:359.

Ferris, P. J., Vogt, V. M., and Truitt, C. L., 1983, Inheritance of extrachromosomal rDNA in *Physarum polycephalum*, Mol. Cell. Biol., 3:635.

Foote, A. M., 1985, Deuteration studies and histone H1 from *Physarum polycephalum*. Ph.D. Thesis, CNAA (Portsmouth, U.K.).

Gerrie, L. M., Humphreys, J., Peoples, O. P., and Hardman, N., 1983, Sequence organization in nuclear DNA from *Physarum polycephalum*: arrangement of highly-repeated sequences, Biochim. Biophys. Acta, 741:214.

Goldberg, G. I., Collier, I., and Cassel, A., 1983, Specific DNA sequences associated with the nuclear matrix in synchronized mouse 3T3 cells, Proc. Natl. Acad. Sci., U.S.A., 80:6887.

Hardman, N., Jack, P. L., Brown, A. J. P., and McLachlan, A., 1979, Distribution of inverted repeat sequences in nuclear DNA from *Physarum polycephalum*, Eur. J. Biochem., 94:179.

Hardman, N., Jack, P. L., Fergie, R. C., and Gerrie, L. M., 1980, Sequence organization in nuclear DNA from *Physarum polycephalum*: interspersion of repetitive and single-copy sequences, Eur. J. Biochem., 103:247.

Hattori, M., Ljljana, A., and Sakaki, Y., 1984, Direct repeats surrounding the ribosomal RNA genes of *Physarum polycephalum*, Nucl. Acids Res., 12:2047.

Jack, P. L., and Hardman, N., 1980, Sequence organization in nuclear DNA from Physarum polycephalum: physical properties of fold-back sequences, Biochem. J., 187:105.

Johnson, E. M., 1980, A family of inverted repeat sequences and specific single strand gaps at the termini of the Physarum rDNA palindrome, Cell, 22:875.

Johnson, E. M., Littau, V. C., Allfrey, V. G., Bradbury, E. M., and Matthews, H. R., 1976, The subunit structure of chromatin from Physarum polycephalum, Nucl. Acids Res., 3:3313.

Judelson, H. S., and Vogt, V. M., 1982, Accessibility of ribosomal genes to trimethyl psoralen in nuclei of Physarum polycephalum, Mol. Cell. Biol., 2:211.

Kukita, T., Sakaki, Y., Nomiyama, H., Otsuka, T., Kuhara, S., and Sagaki, Y., 1981, Structure around the 3' terminus of the 26S ribosomal RNA gene of Physarum polycephalum, Gene, 16:309.

Kunzler, P., 1985, The linear extrachromosomal DNA of Physarum polycephalum replicates and is maintained under nonselective conditions in two different lower eukaryotes, Nucl. Acids Res., 13:1855.

Kunzler, P., Pauli, U., and Braun, R., 1984, Regions in the ribosomal minichromosome of Physarum polycephalum are protected from restriction endonucleases; protection is insensitive to high salt in the G-phase and sensitive in the M-phase of the cell cycle, J. Mol. Biol., 179:651.

Laemmli, U. K., Cheng, S. M., Adolph, K. W., Paulson, J. R., Brown, J. A., and Baumbach, W. R., 1978, Metaphase chromosome structure: the role of nonhistone proteins, Cold Spring Harbor Symp. Quant. Biol., 42:351.

Levinger, L., and Varshavsky, A., 1982, Selective arrangement of ubiquinated and D1 protein-containing nucleosomes within the Drosophila genome, Cell, 28:375.

Loidl, P., Loidl, A., Puschendorf, B., and Grobner, P., 1983, Lack of correlation between histone H4 acetylation and transcription during the Physarum cell cycle, Nature, 305:446.

Marsden, M. P. F., and Laemmli, U. K., 1979, Metaphase chromosome structures: evidence for a radial loop model, Cell, 17:849.

McCready, S. J., Godwin, J., Mason, D. W., Brazell, I. A., and Cook, P. R., 1980, DNA is replicated at the nuclear cage, J. Cell Sci., 46:365.

Mende, L. M., Waterborg, J. H., Mueller, R. D., and Matthews, H. R., 1983, Isolation, identification and characterization of histones from plasmodia of the true slime mold Physarum polycephalum using extraction with guanidine hydrochloride, Biochemistry, 22:38.

Mirkovitch, J., Mirault, M.-E., and Laemmli, U. K., 1984, Organization of the higher-order chromatin loop: specific DNA attachment sites on nuclear scaffold, Cell, 39:223.

Mitchelson, K. R., Bekers, A. G. M., and Wanka, F., 1979, Isolation of a residual protein structure from nuclei of the myxomycete Physarum polycephalum, J. Cell Sci., 39:247.

Mueller, R. D., Yasuda, H., Hatch, C. L., Bonner, W. M., and Bradbury, E. M., 1985, Identification of ubiquinated histones 2A and 2B in Physarum polycephalum, J. Biol. Chem., 260: 5147.

Ochs, R. L., Spohn, W. H., and Lischwe, M. A., 1984, Autoimmune serum from a patient with scleroderma recognizes a new protein of the nucleolus: fibrillarin, J. Cell. Biol. 99:12a.

Otsuka, T., Nomiyama, H., Yoshida, H., Kukita, T., Kuhara, S., and Sakaki, Y., 1983, Complete nucleotide sequence of the 26S rDNA gene of Physarum polycephalum: its significance in gene evolution, Proc. Nat. Acad. Sci., U.S.A., 80:3163.

Pearston, D. H., Gordon, M., and Hardman, N., 1985, Transposon-like properties of the major, long repetitive sequence family in the genome of Physarum polycephalum. EMBO J., in press.

Peoples, O. P., and Hardman, N., 1983, An abundant family of methylated sequences dominates the genome of Physarum polycephalum, Nucl. Acids Res., 11:7777.

Peoples, O. P., Robinson, A. C., Whittaker, P. A., and Hardman, N., 1983, Sequence organization in nuclear DNA from Physarum polycephalum: genomic organization of DNA segments containing foldback sequences, Biochim. Biophys. Acta, 741:204.

Peoples, O. P., Whittaker, P. A., Pearston, D. H., and Hardman, N., 1985, Structural organization of a hypermethylated nuclear DNA component in Physarum polycephalum, J. Gen. Microbiol., 131:1157.

Prior, C. P., Cantor, C. R., Johnson, E. M., Littau, V. C., and Allfrey, V. G., 1983, Reversible changes in nucleosome structure: histone H3 accessibility in transcriptionally-active and inactive states of rDNA chromatin, Cell, 34:1033.

Van der Velden, H. M. W., van Willigen, G., Wetzels, R. H. W., and Wanka, F., 1984, Attachment of origins of replication to the nuclear matrix and the chromosomal scaffold, FEBS Lett., 171:13.

Wanka, F., Peick, A. C. M., Bekers, A. G. M., and Muellenders, L. H. F., 1982, The organization of replicating DNA on the nuclear matrix, in: "The Nuclear Envelope and Nuclear Matrix", G. Maul, ed., Alan R. Liss Inc., New York.

Waterborg, J. H., and Shall, S., 1985, The organization of replicons, in: "The Cell Division Cycle in Plants", Soc. Exp. Biol. Seminar Series, vol. 26, J. A. Bryant and D. Francis, eds.

Whittaker, P. A., and Hardman, N., 1980, Methylation of nuclear DNA in Physarum polycephalum, Biochem. J., 191:859.

Whittaker, P. A., McLachlan, A., and Hardman, N., 1981, Sequence organization in nuclear DNA from Physarum polycephalum: methylation of repetitive sequences, Nucl. Acids Res., 9:801.

Wilhelm, M. L., Toublan, B., Jalouzot, R., and Wilhelm, F. X., 1984, The histone H4 gene is transcribed in S-phase but also in late G2-phase in Physarum polycephalum, EMBO J., 3:2659.

Wilhelm, M. L., and Wilhelm, F. X., 1984, A transposon-like DNA fragment interrupts a Physarum polycephalum histone H4 gene, FEBS Lett., 168:249.

Yasuda, H., Mueller, G., Logan, K. A., and Bradbury, E. M., 1985, Chromatin structure and histone modification through mitosis in the plasmodium of *Physarum polycephalum*, *in*: "Molecular Regulation of Nuclear Events in Mitosis and Meiosis," R. A. Schlegel, M. S. Halleck, and P. N. Rao, eds., Academic Press, New York, in press.

Chapter 4: TEMPORAL ORDER OF REPLICATION AND GENE EXPRESSION IN *PHYSARUM POLYCEPHALUM*

Gerard Pierron

Laboratory of Biology and Ultrastructure of the Nucleus
Institute of Scientific Research on Cancer
Villejuif, FRANCE

SCOPE

The modern methods of molecular biology have led to major discoveries on the fine structure of eukaryotic genes and their expression in various biological situations. Comparatively, the progress in the definition of the replication units of eukaryotic DNA has been very slow; one has to deal with a periodic phenomenon restricted to a short period of the cell cycle. Moreover, eukaryotic cellular genomes are replicated as a series of several thousand replicons, so the replication of a specific DNA fragment is a short-lived event, which cannot be analyzed in nonsynchronously growing cultures of cells.

As a consequence, a relation between replication and transcription activities, although suspected for some time, has never been studied in detail. Considering that the overwhelming majority of the genes are not rearranged during the development of the organisms, it has been tempting to link the *dynamics of gene expression* (turning "on" or "off" of a gene) to a *dynamic process* (the temporal order of replication). Purely speculative hypotheses have flourished (Stambrook and Flickinger, 1970; Wilkins, 1976; Weintraub et al., 1978; Sauer, 1978; Smithies, 1982; Taylor, 1984; Goldman et al., 1984). They associate the specific expression of a gene in different cell lineages with a specific program of replication. By a change of the timing of replication and/or by selection of a subset of origins of replication within a chromosomal domain, differential gene expression could be generated from identical genomes.

For a test of these speculations, a few questions need to be addressed: Are the eukaryotic cellular genomes made of definite replicons characterized by temporally regulated origins of replication? Is there a fixed temporal order of replication in a given differentiated stage? What is a temporal unit? a replicon? a cluster of replicons? How are the genes and the temporal units of replication distributed in the genome?

These questions cannot be addressed with the model systems currently used for DNA replication studies, which, like *E. coli*, T4 phage, or SV40 virus, are single replicon genomes. On the other hand, it is clear that a model system for DNA replication studies should be devoted to the discovery of origins of replication even in the case of the complex eukaryotic genome.

In the following, I would like to argue that *Physarum* is a suitable system for the analysis of the temporal order of replication within specific chromosomal domains and to evaluate any relation with transcriptional activities. Moreover, the natural synchrony of S phase in *Physarum* should allow an original approach toward the characterization of replication origins.

WHAT MAKES *PHYSARUM* NATURALLY SYNCHRONOUS?

In one phase of their complex life cycle, the myxomycetes develop a particular form of organization -- the plasmodium (Rusch, 1980). A plasmodium is a multinucleated cell that results from the multiplication of the nuclei by intranuclear mitosis. It is, therefore, by definition, somewhat different from a syncytium, which is a multinucleated cell produced by the fusion of cells. In fact, *Physarum* exhibits both of these properties, which in the laboratory are used in conjunction in order to obtain large numbers of synchronous nuclei. Deposited on a filter paper, small plasmodia of *Physarum*, grown in shaken liquid cultures, fuse spontaneously into a single macro-plasmodium (Daniel and Baldwin, 1964). Commonly, one can obtain more than 10^8 nuclei in a gigantic cell (diameter of 5-10 cm) in which vigorous cytoplasmic streaming maintains a high degree of homogeneity. Being in a common cytoplasm, these nuclei behave synchronously and divide every 10 hr.

THE MITOTIC CYCLE OF *PHYSARUM*

Autoradiographic and cytological studies have demonstrated a tight coupling between mitosis and S phase in the plasmodium of *Physarum* (Nygaard et al., 1960; Braun et al., 1965). The observation of the synchronous mitosis defines the chronology of the cell cycle as the mitosis is followed by a 3-hr S phase and then by a 6- to 7-hr G2 phase, which leads to the next mitosis (Mohberg and Rusch, 1969). The absence of a G1 phase is a remarkable feature of the mitotic cycle of *Physarum*. The onset of DNA replication in

Physarum is thus morphologically defined by the observation of telophase. This is extremely convenient for DNA replication studies.

The degree of synchrony within a plasmodium has been estimated by flow cytometry (Kubbies and Pierron, 1983). Measurements of the fluorescence of Hoechst-stained nuclei, isolated at various time points after mitosis, provided a convincing confirmation of the natural synchrony of mitosis and S phase in Physarum (Fig. 4-1). G2 phase nuclei have a uniform DNA content and are consequently resolved as a narrow and symmetrical peak. At least 99% of the 10^8 to 10^9 nuclei of a plasmodium are cycling. They divide in synchrony, as evidenced by a shift of DNA fluorescence from 4C to 2C, which takes place in less than 5 min about every 10 hr.

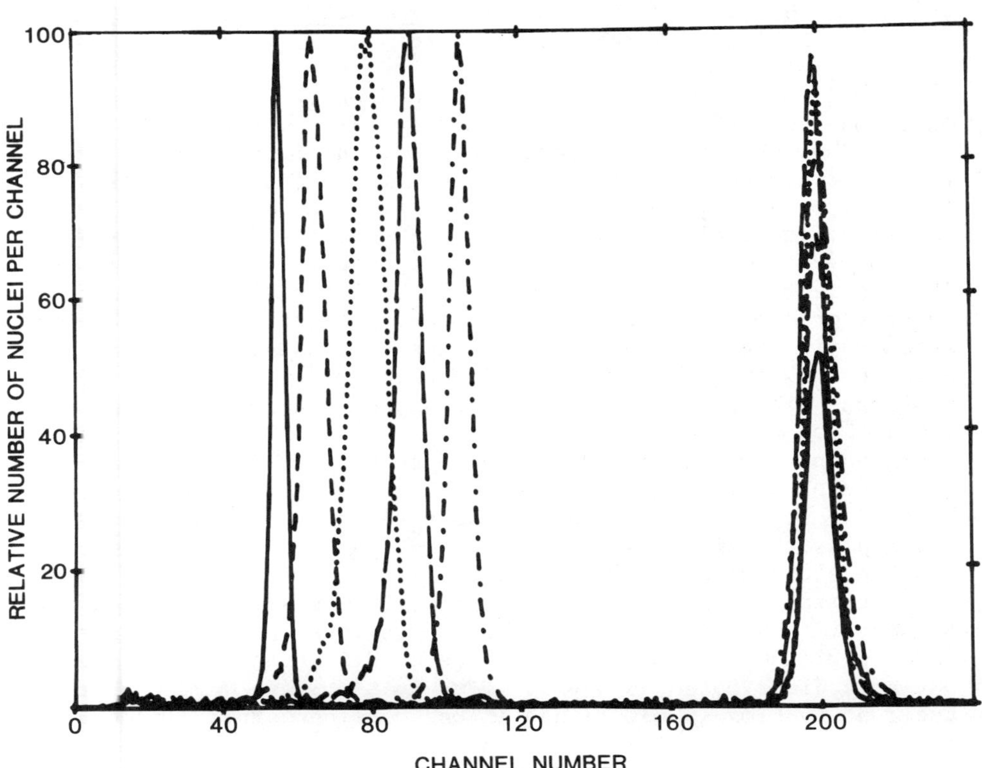

Fig. 4-1. Flow-cytometric analysis of the natural synchrony of Physarum plasmodial nuclei over the cell cycle. Internal standard chicken erythrocyte histograms overlap at channel 200. Left-hand peaks represent Hoechst-stained nuclei isolated at 5 min, 30 min, 1 hr, 2 hr, and 7 hr after the onset of S phase (Kubbies and Pierron, 1983).

Furthermore, the high degree of synchrony is maintained throughout S phase. All nuclei initiate DNA replication within minutes after telophase. The rate of DNA synthesis is high for the first 90 min of S phase, at which time about 75% of the genome has been replicated, and then decreases from 5000 to 1200 kb/min/nucleus, so that it takes another 90 min to replicate the last quarter of the genome (Kubbies and Pierron, 1983). This is not true for the ribosomal DNA genes of *Physarum*, which, as palindromic extrachromosomal elements, are replicated throughout the cycle, except perhaps for the first hour of the S phase (Vogt and Braun, 1977). Their pattern of replication will not be discussed here.

THE MECHANISMS OF DNA REPLICATION IN *PHYSARUM*

The main features of the pattern of replication of the eukaryotic cellular genomes are at work in *Physarum*. The genome is replicated as a series of subchromosomal replication units. In a replicon, the synthesis is bidirectional from the functional origin, with approximately equal rates at both forks (Funderud et al., 1978a). Okazaki fragments of 200 bp (Funderud and Haugli, 1975) primed by short oligoribonucleotides (Waqar and Huberman, 1975) provide evidence for a discontinuous synthesis, at least in the 3'-5' direction. Measurements of the replicon size by alkaline sucrose gradient centrifugation (Funderud et al., 1978b), electron microscopic studies, or autoradiograms of pulse-labeled DNA (Funderud et al., 1979) consistently demonstrate varying lengths of the replicons with a mean size of about 35 kb. The progression of the replication forks has been evaluated at 1.2 kb/min/replicon in independent investigations (Funderud et al., 1978a, 1979; Hunt and Vogelstein, 1981). Thus, it takes about 30 min to synthesize a mean-size replicon. When one knows the size of the genome (C value = 0.3 pg), the kinetics of DNA replication (5000 kb/min/nucleus), and the elongation rate of the forks (1.2 kb/min/replicon), one can estimate that at most 4000 of the 18,000 replicons are active at one time. This determines what is obvious from the electron microscopic data: the replicons are of different sizes and are variably distributed at any one time in S phase. Nevertheless, their activation is not random if one considers that (1) they are activated only once in S phase as multi-fork structures have never been observed despite extensive electron microscopic observations of replicating DNA or chromatin; (2) they are activated in clusters of 2-8 adjacent relicons; this clustering has no known biological significance (Funderud et al., 1978b, 1979); and (3) there exists a definite temporal order of replication, i.e., the DNA sequences replicated early in one S phase are replicated early in the following S phase.

TEMPORAL ORDER OF GENOME REPLICATION IN PHYSARUM

This has been demonstrated by sequential labeling of the early replicated DNA with a radioactive precursor in one S phase, followed by a bromodeoxyuridine (BUDR) labeling in the next S phase, and comparison of the distribution of the heavy-light and radioactive fractions of the genome in cesium chloride gradients (Braun et al., 1965; Braun and Wili, 1969). Other indirect evidence, such as the differential buoyant density of the early- and late-replicating DNA (Braun and Ruedi-Wili, 1971), the interaction of Hoechst dye with the early- and late-replicating chromatin (Kubbies and Pierron, 1983), and the renaturation kinetics of DNA labeled in early S phase (Fouquet and Sauer, 1975), suggests that the GC-rich nonrepetitive DNA is preferentially replicated in early S. Finally, an invariant order of replication of specific DNA sequences, the four unlinked actin gene loci of *Physarum*, has been demonstrated recently (Pierron et al., 1984). The members of this multigene family had been identified previously by the meiotic assortment of restriction fragments hybridizing to actin-cloned genes (Schedl and Dove, 1982; see Chapter 2). For the first time, the timing of replication of these specific DNA fragments has been established by complementary approaches. One involves the isolation of newly replicated DNA after in vivo bromosubstitution and demonstration by "Southern hybridization" that three of the four loci are replicated during the first 20 min of S phase. The other approach takes advantage of this information. Since the members of the multigene family, which are displayed in a single DNA preparation, do not replicate simultaneously, we determined the *relative* intensity of the hybridization signals from DNAs extracted at different times in S and G2 phase. As expected, we found a transient increase of the hybridization bands on "Southern blots" as the respective genes were replicated (gene dosage determination, Fig. 4-2). The temporal resolution of this second approach is limited only by the degree of synchrony of S phase in the plasmodium. On the basis of different analytic principles, these experiments have shown an invariant order of replication of the four actin gene loci; three have replicated at 8-10 min of S phase, the fourth being late, replicating at 80-90 min of S phase (Pierron et al., 1984).

CHRONOLOGY OF DNA REPLICATION AND GENE EXPRESSION IN PHYSARUM

The patterns of replication of the actin-gene family of *Phyarum* and of the murine alpha globin gene family present some similarity. In the mouse, the two adult alpha globin genes and the unlinked intron-containing pseudogene (ψ4) are replicated early, whereas the intron-less pseudogene (ψ5) is late-replicating (Calza et al., 1984). Therefore, in both cases, nonrepetitive sequences have been found to be late replicating. It is further demonstrated that homologous sequences in different chromosomal locations can

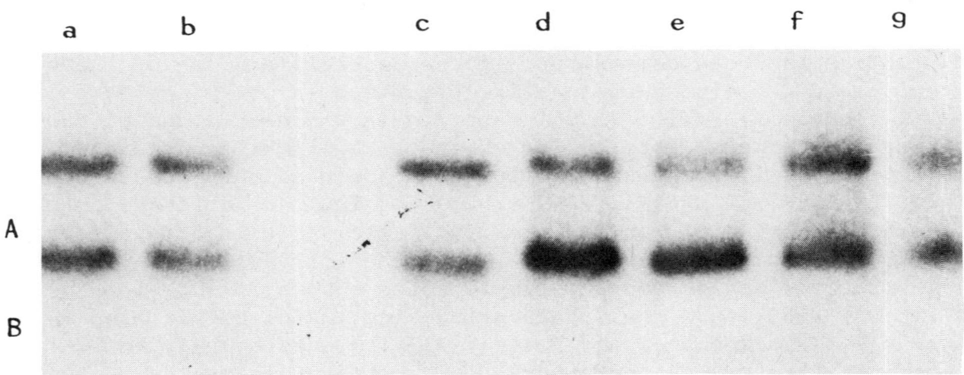

Fig. 4-2. Gene-dosage determination of 2 actin gene-containing Hind III fragments of Physarum. (A = 17.4 kb ardA2 allele, B = 13.6 kb ardB1 allele). Total DNA from two independent G2-phase preparations and from three time points in S phase was purified, digested with Hind III, and hybridized under standard conditions. Scanning of the autoradiograph and integration of the peak areas establish that the relative intensity of the hybridization bands, constant in G2 phase (lanes a,b,f,g) at an A/B ratio of 86±6%, is unchanged at +8 min of S phase (lane c, 94%), but varies significantly at +30 min of S (lane d, 58%) and at +45 min (lane e, 52%). Their findings suggest replication of B between +8 and +30 min and replication of A after +45 min. Results obtained by density shift of BUDR-substituted DNA had shown replication of B between 0 and 20 min of S and replication of A after 40 min (from Pierron et al., 1984).

replicate at very different times. It is, therefore, of interest to compare the expression and the structure of the various actin-genes of Physarum and to determine whether they, like murine alpha globin genes and human arginosuccinate synthetase genes (Goldman et al., 1984), represent another case of gene expression apparently restricted to the early replicating members of a multigene family. The sequencing of one early- and one late-replicating actin gene recently cloned (Monteiro and Cox, personal communication; Nader et al., 1985; see Chapters 22 and 23), should shed some light on this problem.

The functional importance of the early-replicating DNA has been recognized for some time in Physarum. Specific inhibitors of DNA replication have drastic effects on the nuclear morphology and on the transcriptional activity when they inhibit DNA replication in early S-phase and have none of these effects when applied in late S or in G2 phase (Rao and Gontcharoff, 1969; Fouquet et al.,

1975; Pierron and Sauer, 1980). This was the rationale for the hypothesis of "replication-transcription coupling," which links the order of expression of the genes during S phase to the sequential replication of the genome and, therefore, the program of transcription to the chronology of DNA replication (Sauer, 1978). This hypothesis has been substantiated by the observation of nascent replicons undergoing active transcription (Fig. 4-3) on Miller spreads of early S-phase chromatin (Pierron et al., 1982). This was direct evidence for a rapid activation of newly replicated genes. Moreover, this was a direct visualization of a set of genes that are both early-replicated and transcriptionally competent. Finally, a striking feature of these replicon-associated transcription units (Fig. 4-3) is their location, almost without exception,

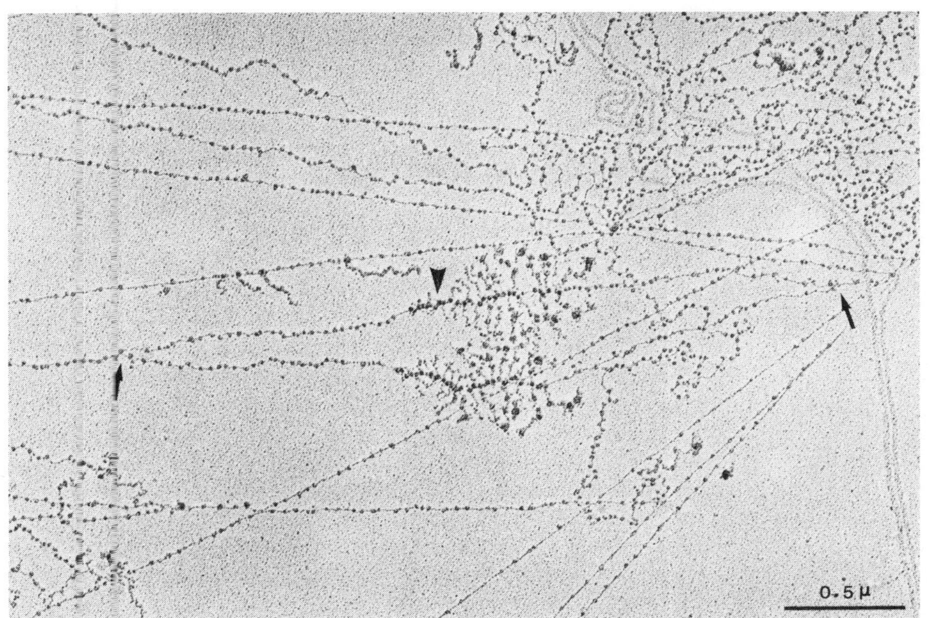

Fig. 4-3. Replicon-associated transcription units in *Physarum*. Plasmodia were harvested in early S phase (15 min after telophase) and processed for electron microscopic spread preparations. Arrows indicate replication forks. Arrowhead marks the presumptive location of the origin of replication, assuming a bidirectional synchronous fork-rate movement (from Pierron et al., 1982).

at the center of the replication bubbles; this feature strongly suggests that origins of replication (we now have more than 100 examples) might be contained within the transcription units.

The electron microscopic pictures do not establish the complete sequence of events at these specific loci. They clearly demonstrate a post-replicative transcription but do not allow prediction on the transcriptional status of the gene before its replication. Starting from G2 phase, these genes are experiencing in less than 30 min the mitotic chromatin condensation, their replication, and transcription. Thus, the period between the condensed state of the chromatin and the initiation of replication at the locus must be very short. The analysis of RNA levels from specific genes throughout the cell cycle has shown by Northern blotting a rather constant level for the actin mRNA and for two uncharacterized mRNAs for which cDNA probes were obtained (Schedl et al., 1984). On the other hand, the alpha- and beta-tubulin mRNAs (Schedl et al., 1984), as well as the histone H4 mRNA (Wilhelm et al., 1984), the latter coded for by an early-replicating gene (Jalouzot et al., 1985), are showing periodic accumulation, peaking in late G2 phase (see Chapter 5). Thus, so far, with the paucity of the sample in mind, not a single gene has been found to be activated by its replication as postulated by the replication-transcription coupling hypothesis (Sauer, 1978). An alternative interpretation, in which a transient early S phase transcription potentiates origins of replication cannot be formally excluded. However, the presence of origins of replication activated in early S-phase within early S-phase active genes (Fig. 4-3) renders the search for eukaryotic origins of replication even more urgent.

PROSPECTS

The replication of the genome is in itself a fundamental process that deserves to be studied in detail. A possible role of the temporal order of replication in the regulation of gene expression reinforces the necessity to define specific replicons and their origins. The most elaborate approach has been performed in yeast where Autonomously Replicating Sequences (ARS) have been found that confer on transforming DNA molecules the ability to replicate autonomously, i.e., without being integrated in the genome (Stinchcomb et al., 1979). Nevertheless, there is no direct biochemical evidence that DNA sequences from yeast, which have an ARS function on plasmids, are origins of replication in the chromosome (Mechali and Kearsey, 1984). One possible way to determine this would be to study the relative timing of replication of an ARS sequence and of its flanking sequences. Being an origin, the ARS sequence should replicate earlier. This has not been achieved in yeast or in *Physarum*, but it may be technically feasible in the slime mold.

First, ARS sequences defined by their action in yeast have been isolated from the Physarum genome (Gorman et al., 1981). Second, the natural synchrony has permitted the unambiguous establishment of some aspects of the kinetics of replication. As seen earlier, the composite photograph of Physarum replicons depicted a 30-kb-long unit synthesized in about 30 min. Third, we have devised a method having such a temporal resolution (Pierron et al., 1984). As pointed out above, by gene dosage determination, on the basis of the quantitation of relative intensities of hybridization signals throughout S phase (Fig. 4-3), one can detect the replication of actin Hind III restriction fragments within the first 8-10 min of S phase. The size of these fragments ranges from 4.2 to 13.6 kb. This suggests that one can indeed establish the chronology of replication of restriction fragments contained within a replicon of Physarum and therefore study the function of an ARS sequence integrated in the chromosome. In more general terms, the possibility of mapping specific replicons would greatly improve our understanding of S phase and help to document any link between replication and transcription units.

ACKNOWLEDGMENTS

I would like to thank Drs. E. Puvion and A. M. De Recondo (Villejuif) for showing interest in my work, and Professor H. W. Sauer (Texas A & M University) for encouragement and fruitful discussions.

REFERENCES

Braun, R., Mittermayer, C., and Rusch, H. P., 1965, Sequential temporal replication of DNA in Physarum polycephalum, Proc. Natl. Acad. Sci., U.S.A., 53:924.

Braun, R., and Ruedi-Wili, H., 1971, Early replicating DNA of Physarum is denser than late replicating DNA, Experientia, 27:1412.

Braun, R., and Wili, H., 1969, Time sequence of DNA replication in Physarum, Biochim. Biophys. Acta, 174:246.

Calza, R. E., Eckhardt, L. A., Delguidice, T., and Schildkraut, C. L., 1984, Changes in gene position are accompanied by a change in time of replication, Cell, 36:689.

Daniel, J. W., and Baldwin, H. H., 1964, Methods of culture for plasmodial myxomycetes, in: "Methods in Cell Physiology", Vol. 1, p. 9, D. M. Prescott, ed., Academic Press, New York.

Fouquet, H., Bohme, R., Wick, R., Sauer, H. W., and Scheller, K., 1975, Some evidence for replication-transcription coupling in Physarum polycephalum, J. Cell Sci., 18:27.

Fouquet, H., and Sauer, H. W., 1975, Variable redundancy in RNA transcripts isolated in S and G2 phase of the cell cycle of Physarum, Nature, London, 255:253.

Funderud, S., Andreassen, R., and Haugli, F., 1978a, DNA replication in Physarum polycephalum: bidirectional replication of DNA within replicons, Nucl. Acids Res., 5:713.

Funderud, S., Andreassen, R., and Haugli, F., 1978b, Size distribution and maturation of newly replicated DNA through the S and G2 phases of Physarum polycephalum, Cell, 15:1519.

Funderud, S., Andreassen, R., and Haugli, F., 1979, DNA replication in Physarum polycephalum: electron microscopic and autoradiographic analysis of replicating DNA from defined stages of the S period, Nucl. Acids Res., 6:1417.

Funderud, S., and Haugli, F., 1975, DNA replication in Physarum polycephalum: characterization of replication products in vivo, Nucl. Acids Res., 2:1381.

Goldman, M. A., Holmquist, G. P., Gray, M. C., Caston, L. A., and Nag, A., 1984, Replication timing of genes and middle repetitive sequences, Science, 224:686.

Gorman, J. A., Dove, W. F., and Warren, N., 1981, Isolation of Physarum DNA segments that support autonomous replication in yeast, Mol. Gen. Genet., 183:306.

Hunt, B., and Vogelstein, B., 1981, Association of newly replicated DNA with the nuclear matrix of Physarum polycephalum, Nucl. Acids Res., 9:348.

Jalouzot, R., Toublan, B., Wilhelm, M. L., and Wilhelm, F. X., 1985, Replication timing of the H4 histone genes in Physarum polycephalum, Proc. Natl. Acad. Sci., U.S.A., 82:6475.

Kubbies, M., and Pierron, G., 1983, Mitotic cell cycle control in Physarum, Exp. Cell Res., 149:57.

Mechali, M., and Kearsey, S., 1984, Lack of specific sequence requirement for DNA replication in Xenopus eggs compared with high sequence specificity in yeast, Cell, 38:55.

Mohberg, J., and Rusch, H. P., 1969, Growth of large plasmodia of the myxomycete Physarum polycephalum, J. Bacteriol., 97:1411.

Nader, W. F., Edlind, T. D., Huettermann, A., and Sauer, H. W., 1985, Cloning of Physarum actin sequences in an exonucleus-deficient bacterial host, Proc. Natl. Acad. Sci., U.S.A., 82:2698.

Nygaard, O. F., Guttes, S., and Rusch, H. P., 1960, Nucleic acid metabolism in a slime mold with synchronous mitosis, Biochim. Biophys. Acta, 38:298.

Pierron, G., Durica, D. S., and Sauer, H. W., 1984, Invariant temporal order of replication of the four actin gene loci during the naturally synchronous mitotic cycles of Physarum polycephalum, Proc. Natl. Acad. Sci., U.S.A., 81:6393.

Pierron, G., and Sauer, H. W., 1980, More evidence for replication transcription coupling in Physarum polycephalum, J. Cell Sci., 41:105.

Pierron, G., Sauer, H. W., Toublan, B., and Jalouzot, R., 1982, Physical relationship between replicons and transcription units in Physarum polycephalum, Eur. J. Cell Biol., 29:104.

Rao, B., and Gontcharoff, M., 1969, Functionality of newly synthesized DNA as related to RNA synthesis during mitotic cycle in Physarum polycephalum, Exp. Cell Res., 56:269.

Rusch, H. P., 1980, The search. in: "Growth and Differentiation in Physarum polycephalum", p. 1, W. F. Dove and H. P. Rusch, eds., Princeton University Press, Princeton.

Sauer, H. W., 1978, Regulation of gene expression in the cell cycle of Physarum polycephalum, in: "Cell Cycle Regulation", p. 149, J. Jeter, I. L. Cameron, G. M. Padilla, and A. M. Zimmermann, eds., Academic Press, New York.

Schedl, T., Burland, T. G., Gull, K., and Dove, W. F., 1984, Cell cycle regulation of tubulin RNA level, tubulin protein synthesis, and assembly of microtubules in Physarum, J. Cell Biol., 99:155.

Schedl, T., and Dove, W. F., 1982, Mendelian analysis of the organization of actin sequences in Physarum polycephalum, J. Mol. Biol., 160:41.

Smithies, O., 1982, The control of globin and other eukaryotic genes, J. Cell Physiol., Suppl. 1:137.

Stambrook, P. J., and Flickinger, R. A., 1970, Changes in chromosomal DNA replication patterns in developing frog embryos, J. Exp. Zool., 174:101.

Stinchcomb, D. T., Struhl, K., and Davis, R. W. 1979, Isolation and characterization of a yeast chromosomal replicator, Nature 282:39.

Taylor, J. H., 1984, Origins of replication and gene regulation, Mol. Cell. Biochem., 61:99.

Vogt, V. M., and Braun, R., 1977, The replication of ribosomal DNA in Physarum polycephalum, Eur. J. Biochem., 80:557.

Waqar, M. A., and Huberman, J. A., 1975, Covalent linkage between RNA and nascent DNA in the slime mold Physarum polycephalum, Biochim. Biophys. Acta, 383:410.

Wilhelm, M. L., Toublan, B., Jalouzot, R., and Wilhelm, F. X., 1984, Histone H4 gene is transcribed in S phase but also in late G2 phase in Physarum polycephalum, EMBO J., 3:2659.

Wilkins, A. S., 1976, Replicative patterning and determination, Differentiation, 5:15.

Weintraub, H., Flint, S. J., Leffak, I. M., Groudine, M., and Grainger, R. M., 1978, The generation and propagation of variegated chromosome structures, Cold Spring Harbor Symp. Quant. Biol., 42:401.

Chapter 5: THE *PHYSARUM* CELL CYCLE

Thomas G. Laffler[1] and John J. Tyson[2]

[1]Department of Microbiology-Immunology
Northwestern University School of Medicine
Chicago, IL USA

[2]Department of Biology
Virginia Polytechnic Institute and State University
Blacksburg, VA USA

With contributions from: L. D. Barnes, R. Braun, T. Burland, J. Carrino, R. A. Cox, J. W. Daniel, H. Evans, R. Exenberger, G. Garcia-Herdugo, L. Green, P. Loidl, A. M. MacNicol, H. R. Matthews, J. M. Mitchison, E. C. A. Paul, H. Sauer, J. H. Waterborg, F. X. Wilhelm, and M. Wright

INTRODUCTION

Various aspects of the synchronous nuclear division cycle in *Physarum polycephalum* have been reviewed in detail recently (Holt, 1980; Tyson, 1982; Schedl et al., 1984b). In this chapter we shall concentrate on research published since 1982, particularly on reports discussed at the Workshop.

Physarum has been an important experimental organism for studies of the cell cycle, because in the multinucleate plasmodial stage of the life cycle there is natural synchrony of the stages of mitosis and of the onset and progression of DNA synthesis. At a closer level, other markers of progress through the cell cycle are also naturally synchronous, e.g., the replication of certain genes, the synthesis of a number of periodic enzymes, and the transition points for the action of various chemical and physical agents. Cell division itself, of course, is absent in syncytial plasmodia, but, curiously enough, cell division is one of the more expendable features of the "cell division cycle" (Pringle and Hartwell, 1982).

Another characteristic and exceedingly useful feature of *Physarum* as an experimental organism is the spontaneous fusion of macroplasmodia (of appropriate *fus* genotypes) to produce plasmodia with two or more distinct populations of nuclei. This feature can often be used advantageously in studying genetic and physiological problems.

At present the major limitation in using *Physarum* as a model organism for cell cycle studies is a lack of mutants defective in cell cycle functions, analogous to the *cdc* and *wee* mutations in yeast cells. The reason for this limitation seems to be primarily a lack of extended effort in isolating and characterizing *cdc*-type mutations in the amoebal stage of the life cycle. Several people have reported success in isolating plasmodial cell cycle mutants (Burland and Dee, 1979, 1980; Laffler et al., 1979), but none of these mutants have been studied in detail. Continued search for and characterization of cell cycle mutants should be a high priority for the future. Expression of cell cycle mutations in plasmodia is, of course, the goal, but a thorough study of the cell cycle in *Physarum* amoebae is also desirable. Next to nothing is known about the amoebal cell cycle, although such knowledge would be useful in comparing the *Physarum* cell cycle with those of yeasts, ciliates, and mammalian cells in culture.

PERIODIC PROTEIN SYNTHESIS AND MODIFICATION

As molecular biologists, we expect that controls of gene expression play an important role in regulation of the cell cycle. Control of gene expression can be studied on many levels, e.g., replication timing and replication-transcription coupling (see Chapter 4), mRNA synthesis and processing, translation and posttranslational modification of proteins, and activation and inhibition of enzyme activity. This section reviews recent results on protein synthesis and modification obtained by two-dimensional polyacrylamide gel electrophoresis (PAGE) and by biochemical characterization of some specific enzymes. A detailed review of the more thoroughly studied cases of histone and tubulin gene expression is given below in a separate section.

Two-dimensional PAGE Surveys

A new and promising approach to studying periodic polypeptides -- proteins that change significantly in either their intracellular concentration or their rate of synthesis during the cell cycle -- is provided by two-dimensional gel electrophoresis. Hundreds of distinct polypeptide chains can be resolved and quantified by this technique. With the help of sophisticated digital image processing routines, these spots can be surveyed throughout the cell cycle for evidence of periodic protein synthesis or accumulation. The results of studies of bacteria (Lutkenhaus et al., 1979), yeast (Elliott and McLaughlin, 1978; Lorincz et al., 1982), and mammalian cells (Bravo and Celis, 1980) all agree that virtually all (ca 99%) of the 500-or-so most abundant proteins resolved by two-dimensional PAGE show little variation in either level or synthesis over the cell cycle.

These surveys have inherent technical limitations. All use two-dimensional gel systems that do not resolve basic proteins (in particular, histones which are generally synthesized in S phase only), and the majority of proteins cannot be observed because their relative abundance lies below the limit of detection. Therefore, we are left with the unanswered question of whether these analyses focus on "housekeeping" polypeptides, missing many more interesting but abundant regulatory proteins, or whether there are only a few genes (ca 1%) that are periodically expressed. Furthermore, as far as regulatory enzymes are concerned, changes in activity, which may be undetectable by this technique, may be more important than changes in abundance, synthesis, or degradation. Another limitation of the studies just mentioned is that some form of selection or induction synchrony must be carried out on naturally asynchronous cultures, and it is unclear to what extent lack of perfect synchrony may obscure small but significant and detectable changes in protein synthesis and/or accumulation.

By virtue of its natural mitotic synchrony, Physarum is uniquely adapted for these studies. Visual inspection of autoradiographs of two-dimensional gels by Laffler et al. (1981) and Turnock et al. (1981) showed only two proteins (α and β tubulin) to vary significantly in rate of synthesis at different stages in the cell cycle. A recent survey by Pahlic and Tyson (1983a, 1985), using digital image processing of silver-stained gels, uncovered only five spots (of 250 low- to moderate-abundance proteins) that showed reproducible, significant (greater than threefold), cell cycle-dependent fluctuations. These surveys reinforce the conclusion, derived from studies of synchronized cultures, that in general only 1-2% of the gel spots show significant changes during the cell cycle.

Gröbner and Loidl (1985a) have recently reported an interesting variation of this approach. From antiserum raised against proteins of S phase extracts of Physarum, antibodies were purified and attached to protein-A-Sepharose CL-4B. Immunoadsorption of a G2 phase extract with these anti-S-antibodies decreased the 700 predominant proteins seen on silver-stained two-dimensional gels to only 20 spots. However, Gröbner and Loidl observed no qualitative difference in protein pattern of two-dimensional gels between S phase and G_2-phase extracts either before or after immunoadsorption with anti-S antibodies. Nonetheless, by a bioassay, Loidl and Gröbner (1982) were able to detect in G2 phase extracts a mitosis-stimulating activity, which could be enriched tenfold by immunoadsorption of G2 extracts to anti-S antibodies. If the mitosis-stimulating activity is due to one or more proteins, their abundances were evidently below the level of detection by silver-stained gels.

Although the initial results of two-dimensional gel studies may be disappointing to some, the full power of the technique has not yet been exploited. Neither Laffler and his colleagues nor Turnock and his colleagues used digital image processing to analyze their gels for significant, reproducible changes that may be overlooked by visual comparison of gels. On the other hand, Pahlic and Tyson (1985) were limited in the conclusions they could draw by the nonstoichiometric nature of the silver stain they used to quantify protein accumulation. There has yet to be carried out a thorough, computer-assisted survey of protein synthesis and accumulation patterns, quantified by autoradiography, on a naturally synchronous cell cycle. Furthermore, it is possible to resolve and quantify basic proteins by nonequilbrium isoelectric focusing, and it should be possible to fractionate plasmodia into nuclear, microsomal, and cytosolic components for a closer look at protein distributions and translocations. Enrichment by antibody binding, as pioneered by Loidl and Gröbner, is also worth pursuing.

Cell cycle-regulated gene expression should also be approachable at the RNA level. Stage-specific messenger RNA populations ought to be detectable by RNA:DNA hybridization techniques. A careful analysis of hybridization kinetics can outline the number and abundance of these cell cycle-regulated transcripts. Methods exist for cDNA cloning of differentially regulated messengers. For example, Hedrick et al. (1984) have used subtraction hybridization to clone a cDNA specific to the T cell beta-chain, and Cox and Monteiro (see Chapter 23) have described a competition method for identifying clones of sequences specific for certain stages of the life cycle of _Physarum_. Similar approaches should allow the cloning of cell cycle-regulated transcript sequences from _Physarum_.

Specific Enzymes

Thymidine kinase (TK) and thymidylate synthetase are both synthesized at a constant basal rate throughout the cell cycle, whereas additional periodic synthesis starts before the onset of DNA synthesis and lasts until mid-S phase (Wright and Tollon, 1979a; Gröbner and Loidl, 1983). The periodic synthesis of these two dTMP-synthesizing enzymes thus coincides with the phase of increased requirement for dTMP. TK activity may also be controlled by cell cycle-dependent, posttranslational phosphorylation of the protein. Fluctuations in TK activity provide a fine marker for progress through the normal cell cycle, but the existence of TK-deficient mutants makes it clear that this enzyme is not essential for progress through the cycle.

It has been known for some time that TK exists in five isoelectric variants whose relative abundances change during the cell cycle (Gröbner and Sachsenmaier, 1976; Gröbner, 1979; Pahlic and Tyson, 1983b), but further analysis of these variants was limited by their extreme lability. Recently, however, Gröbner and Loidl (1984) have accomplished an 8000-fold purification of TK by one-step affinity chromatography on a dThy-Sepharose column. With the purified enzyme they could prove that the acidic variants are phosphorylated forms of the enzyme. The phosphorylated variants, which are present mainly during S phase, are one-third as sensitive to the metabolic inhibitor (dTTP) as the dephosphorylated variants.

ADP-ribosyltransferase, in the hands of Gröbner and Loidl (1985b), exhibits a typical peak enzyme pattern, with a maximum in S phase, when enzyme activity in isolated nuclei is measured under conditions of substrate saturation (exogenous histones as substrate and DNA as stimulator). This activity peak presumably represents availability of the enzyme itself and is presumably attributable to an increase in rate of synthesis commencing just before mitosis. Interestingly, unstimulated enzyme activity (measured in isolated nuclei without exogenous histones and DNA), when expressed as activity per macroplasmodium, exhibits a typical step-enzyme pattern with a doubling of activity in S phase of each cell cycle. The step in unstimulated activity presumably reflects the availability of endogenous substrate, e.g., chromosomal proteins. On the other hand, H. Evans (personal communication) finds a peak in ADP-ribosylation of endogenous nuclear proteins in late G2. In these studies, nuclei were incubated with ^{32}P-NAD, solubilized, and all nuclear proteins were then separated by SDS-PAGE. Evans located eight ^{32}P-labeled bands at apparent molecular weights: 16, 20, 23, 39, 58 (doublet), 84, and 100K. The 39K protein showed a sharp increase in poly ADP-ribosylation in early prophase.

Histone acetylation during the cell cycle is a controversial subject. In an early report, Chahal et al. (1980) observed that the acetate content of histone H4 fluctuates in a biphasic manner, with maxima in S phase and late G2 phase, and they pointed out the similarity of this pattern to the biphasic rate of RNA synthesis reported by Mittermayer et al. (1964). However, this correlation between histone acetate content and RNA synthesis has been questioned on both fronts. On the one hand, Loidl et al. (1983) have questioned the reality of the second (late G2) peak in highly acetylated histone H4, and on the other hand, the biphasic pattern in rate of RNA transcription, which was established by pulse-labeling with ^{3}H-uridine, has not been confirmed by other methods. In particular, Hall and Turnock (1976) found, using a sensitive technique of isotope dilution, that rRNA is synthesized continuously and exponentially throughout the cell cycle except for a brief pause at mitosis. This suggests that the biphasic pattern observed by Mittermayer et al. may reflect fluctuations in the uridine pool rather than fluctuations in the rate of RNA synthesis. Furthermore, several groups have reported that mRNA synthesis occurs primarily during S phase (Fouquet and Braun, 1974; Pierron and Sauer, 1980).

The difficulty in resolving this controversy stems from differences in techniques used in different laboratories, and from the possibility that patterns of histone acetylation and of RNA transcription may vary with growth conditions. To establish a link between histone acetylation and transcription in *Physarum* it is imperative that one laboratory measure both properties under identical growth conditions and, preferably, by more than one method.

The study of histone acetylation in *Physarum* has been improved by the determination of the complete amino acid sequence of histones H4 (Wilhelm and Wilhelm, 1984), and the identification of acetylation sites in histones H3 and H4 (Waterborg and Matthews, 1983a).

Recent data on histone acetate turnover (as distinct from acetate content) during the cell cycle describe separate patterns of acetate turnover for all four core histones. The patterns are associated with transcription and replication (Waterborg and Matthews, 1983b, 1984). The specificity of these patterns of acetate turnover suggests a reason for the lack of success in demonstrating effects of histone acetylation in vitro where histones are acetylated nonspecifically. Recent data indicate that the specificity involves a specific subset of the in vivo acetylation sites (Pesis and Matthews, personal communication).

DNA polymerase, which one might expect to be a peak-enzyme in S phase, has been established by Cox and his coworkers to be a step-enzyme in G2. In accordance with unpublished work by Cox that DNA polymerase protein increases stepwise in G2 phase, MacNicol (personal communication) finds that DNA polymerase activity of whole plasmodia also increases stepwise in G2. [Earlier studies (Brewer and Rusch, 1966), which showed DNA polymerase to be a peak enzyme in S phase, were carried out on isolated nuclei and may reflect variations in binding of enzyme to nuclei rather than variations in enzyme protein or activity.] The DNA polymerase assay used by MacNicol involves SDS-PAGE of crude cell lysates, removal of SDS, and in situ measurement of DNA polymerase activity by autoradiography. Three bands, of apparent molecular weights 75K, 112K, and 130K, show polymerase activity and, when Western blots of such gels are probed with DNA polymerase antibodies, only those bands showing polymerase activity are detected by polymerase-specific antibodies. Furthermore, when G2 phase poly(A)$^{+}$-mRNA is translated in vitro and the protein products are immunoprecipitated by DNA polymerase antibodies, DNA polymerase polypeptides can be detected. Efforts to clone the gene for *Physarum* DNA polymerase are currently in progress (MacNicol and Cox, personal communication).

Other Specific Cases

Myc-related protein, as defined by cross reaction with antisera against a v-myc fusion protein, has been reported by Sauer and coworkers (personal communication; Shipley et al., 1985). The cross reacting species in Physarum is a single protein of about 70K as determined by Western blots of polyacrylamide gels. This protein is absent at metaphase and increases to a maximum level in late G2 phase.

The MC29 v-myc probe cross reacts under stringent hybridization conditions with poly(A)$^{+}$-RNA from all stages of the cell cycle, and it detects six positive clones among the equivalent of six genomes in a Physarum EMBL 3 library (Sauer, personal communication). Green (personal communication) also finds cross-reacting sequences when probing genomic DNA with a 3'-region subclone of v-myc DNA. Sauer and coworkers have found that one of their clones contains a Bam H1 fragment, which hybridizes to a similar fragment genomic Southern blots. While suggestive, cross hybridization data are insufficient to prove rigorously the existence of a myc protooncogene in Physarum. Sequence data are needed.

Because the c-myc gene has been shown to be active in cycling mammalian cells, and because an activated form of the myc gene has been implicated in the immortalization of some cancer cells, it is possible that an homologous gene could be expressed during the growth phase of Physarum. It will be interesting to study its role (if any) in the cell cycle and to find out whether this gene is repressed during the reversible spherulation process and the irreversible sporulation process.

Diadenosine tetraphosphate (Ap_4A) levels during the cell cycle have been analyzed by Barnes (personal communication), in order to test the proposal of Rapaport and Zamecnik (1976) that Ap_4A is a signal dinucleotide coupling the rate of protein synthesis to the initiation of DNA replication. Barnes and his coworkers found that the cellular Ap_4A concentration did not change significantly during traverse of the cell cycle in Physarum macroplasmodia. [In contrast, Weinmann-Dorsch et al. (1984) reported an 8- to 30-fold increase in Ap_4A in Physarum during S phase.] Barnes, however, did find significant changes in Ap_4A levels during oxidative stress (induced by dinitrophenol), in agreement with a recent proposal by Ames and coworkers that Ap_4A and related dinucleotides are signal compounds, or alarmones, for oxidative stress (Bochner et al., 1984).

PERIODIC EXPRESSION OF TUBULIN AND HISTONE GENES

Periodic synthesis of histones during S phase is a general feature of the cell cycle in eukaryotic cells (Maxson et al., 1983), and *Physarum* is no exception (Schofield and Walker, 1982). Large changes in the magnitude of expression of tubulin genes during the cell cycle and life cycle of *Physarum* are now well-documented, whereas changes in tubulin synthesis are only beginning to be studied in other cell cycles.

By pulse-labeling plasmodia with radioactive amino acids, then resolving newly synthesized proteins by two-dimensional PAGE and autoradiography, Laffler et al. (1981) and Turnock et al. (1981) showed that the rate of α- and β-tubulin synthesis is much greater before mitosis than in early S-phase. The differential rate of tubulin protein synthesis increases 40-fold in the 2-3 hr before mitosis, but virtually disappears soon after mitosis (Laffler et al., 1981). All of the tubulin isotypes expressed in the plasmodia appear to be coordinately regulated. By quantifying the total amount of protein in two-dimensional gels by digital integration of the optical density of silver-stained spots, Pahlic and Tyson (1985) found that a (α and β) composite tubulin signal was nearly fourfold more abundant in mitosis and early S phase than in mid-G2. Thus, *Physarum* tubulins define a dramatic peak in synthesis and a modest peak in accumulation.

In mammalian cells only a threefold greater rate of synthesis of tubulins has been observed in M than in S (Bravo and Celis, 1980), far less than in *Physarum*. The periodicity of tubulin synthesis in flagellates such as *Chlamydomonas* (Ares and Howell, 1982) and *Naegleria* (Lai et al., 1979) represents a special case: these cells duplicate their flagella in each cycle. Interestingly, the reported periodicity of tubulin synthesis in *Chlamydomonas* may be due to a synchronization artifact, since periodicity is lost when synchronized cultures are returned to constant growth conditions (Rollins et al., 1983). Thus, the tubulins of *Physarum* remain unique among nonhistone proteins for their highly periodic synthesis within an uncomplicated cell cycle.

Periodic proteins are useful from two perspectives. Their periodicity defines an interesting series of cell cycle events worthy of detailed study. While many protein periodicities might be due to posttranslational effects, we presently know that either the level or rate of synthesis of some proteins peak at characteristic stages of the cell cycle. In some cases, the level of mRNA and even the rate of individual gene transcription has been studied. Our present challenge is to address the molecular mechanisms that are responsible for these periodicities. In addition, periodicities provide useful landmarks for studies of cell cycle regulation. Thus the activation of a gene's transcription in mid-G2 phase can be used to divide G2 phase into two compartments. The effect of experimental manipulations that prolong or shorten the cell cycle by adjusting the length of G2 phase can now be localized to one of these compartments. Both perspectives can be informative, as evidenced by recent studies on the Physarum cell cycle now to be addressed.

Periodicities in Unperturbed Cell Cycles

Periodic expression of tubulin genes. Why is tubulin synthesis strongly periodic in Physarum but weakly periodic in HeLa cells? Physarum's tubulin periodicity may reflect the lack of a cytoskeletal tubulin reserve. Physarum plasmodia lack a microtubular cytoskeleton -- no interphase microtubules have been observed, and microtubular structures are assembled only near mitosis (Daniels and Jarlfors, 1972; Havercroft and Gull, 1983; Schedl et al., 1984a). In contrast, most other eukaryotes have an extensive microtubular cytoskeleton that can provide a reserve of tubulin subunits for spindle assembly (Hyams and Hyams, 1979). Given the periodic demand for tubulin protomers, it is not surprising that there is periodic synthesis. What is surprising is that the level of tubulin protein varies over the cell cycle by up to a factor of four (Carrino and Laffler, 1985; Pahlic and Tyson, 1985): we would expect a stable protein merely to double. While tubulins are generally stable, there is a period in early S phase when they appear to be less stable (Laffler et al., 1981; Carrino and Laffler, 1985). In the 100 min period directly following mitosis, the tubulin level may drop to 20% of the peak level (Carrino and Laffler, 1985), remaining essentially constant until the next cycle of accumulation begins. It could be that, because of this instability, intracellular tubulin concentration drops below a critical threshold, so that spindle assembly can be reinitiated only as protomers accumulate in the following cycle.

Schedl et al. (1984a) showed by molecular hybridization that the level of the α- and β-tubulin mRNAs varied in parallel with tubulin protein synthesis. Fig. 5-1 (Laffler and Carrino, unpublished data) presents a similar Northern analysis. J. Carrino and T. Laffler (1986) have used an assay in which nascent RNA chains initiated in vivo are completed in vitro in isolated nuclei (a nuclear 'run-on' assay). Their data suggest that α-tubulin gene transcription begins in mid-G2 phase and ends at mitosis (Fig. 5-2). Since the period of transcription parallels that of messenger accumulation, they infer that tubulin gene expression is probably regulated by a transcriptional control mechanism. As in vitro experiments may not accurately mirror the in vivo situation, this conclusion needs to be critically tested by careful in vivo labeling experiments.

Possible posttranscriptional controls (see Homeostatic Controls, below) must also be considered. There is a quantitative disparity between the rates of tubulin mRNA synthesis as measured by 'run-on' experiments and the accumulation of tubulin mRNA, as measured by titration against Ppc-α125. In the absence of posttranscrip-

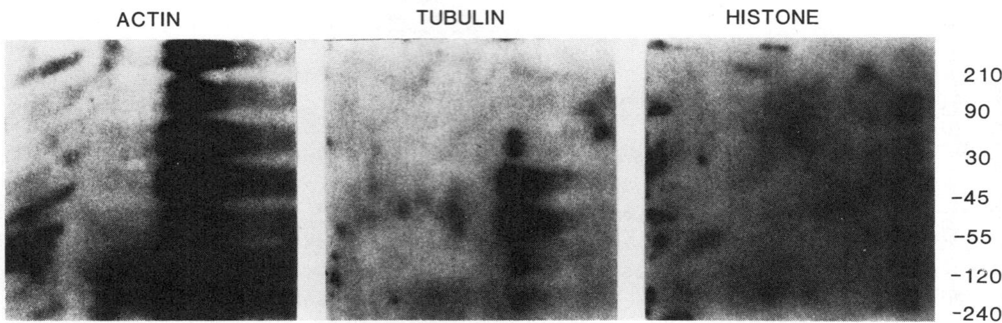

Fig. 5-1. Northern blot analysis of alpha-tubulin, histone H4, and actin RNAs (Laffler and Carrino, unpublished data). Total RNAs were isolated at timepoints relative to the second mitosis after plating. Aliquots (20 μg) were glyoxalated, electrophoresed through 1% agarose gels, and blot-transferred to Gene Screen Plus (New England Nuclear, Boston, MA) membranes. Hybridization was carried out per manufacturer's instructions in 10 ml of hybridization buffer containing 5 x 10^5 cpm/ml nick-translated probe. The following sequences were used as probes: actin, a 2.8 kb HinIII-EcoRI fragment carrying the ardA genomic sequence provided by W. Nader; tubulin, a 1.1 kb EcoR1-Sal1 fragment for the alpha tubulin cDNA clone Ppc-α125 provided by T. Schedl; and histone, a 600 bp HinIII fragment carrying a histone H4 gene sequence, provided by X. Wilhelm. Blots were exposed at -80°C for 4 hr with intensifying screens.

tional controls, the simple gene switch indicated by the 'run-on' data would predict that tubulin mRNA accumulates at an ever-decreasing rate to a steady-state in which synthesis is balanced by degradation. However, Schedl et al. (1984a, b) claim that tubulin mRNA accumulates at an ever-increasing rate (i.e., exponentially) during G2 phase. They present a model in which tubulins act as positive autogenous regulators of transcription, but they must postulate a regulated partitioning of the protomer pool between nucleus and cytoplasm. This model must be reconsidered in light of the 'run-on' data, which suggest that once the genes are activated transcription continues at a constant rate. It could be that a level of posttranscriptional control adjusts the messenger level in response to an increasing rate of withdrawal from the tubulin protomer pool (see Homeostatic Controls, below). The relative rates of mRNA synthesis and turnover would be better addressed by measuring in vivo rates of mRNA synthesis and turnover by pulse-chase analysis, but large and variable precursor pools render this approach extremely difficult (L. Green, personal communication). Whether or not results of 'run-on' experiments adequately reflect the in vivo rate of mRNA synthesis remains to be confirmed.

Periodic expression of histone genes. Using as a probe a *Physarum* histone H4 gene they have cloned, Wilhelm and Wilhelm (1984) found histone H4 mRNA to be present in both late G2 phase and S phase, reporting a decrease in the apparent size of the transcripts following mitosis. Carrino and Laffler (1986, see Fig. 5-1) and Kung and Braun (personal communication) did not see this shift. This discrepancy may not be an artifact. Both polyadenylated and unpolyadenylated forms of histone H4 and mRNA are present in both G2 and S phases (Wilhelm et al., personal communication; Kung and Braun, personal communication). In some G2-phase preparations of poly(A)$^+$ RNA, a minor (ca 600 bp) band can be seen in some preparations along with the major 500 bp band. This minor band becomes more evident when protein synthesis is inhibited with cycloheximide (Kung and Braun, personal communication). This could be the form originally detected by Wilhelm et al. (1984). If this larger species is a precursor, it might contain the 86 bp sequence interrupting the cloned histone H4 coding sequence (Wilhelm and Wilhelm, 1984). However, there appears to be at least one more histone H4 locus that might provide the second species (Jalouzot et al., 1985). It remains to be shown whether both loci are expressed. Alternatively, there may be differences in polyadenylation or in untranslated flanking messenger sequences that could account for the larger species.

Hereford et al. (1981) observed in budding yeast that histone gene transcription begins in late G1 phase, after the completion of START. Transcription is turned off early in S phase. When progress of a synchronized culture through the cycle is blocked in late G1 (by temperature shifting a *cdc7* mutant), histone gene transcription continues, but if the culture is blocked early in S phase (by shifting a *cdc8* mutant) transcription stops. They concluded that completion of an early S phase step was needed to turn off histone gene transcription, and hypothesized that the necessary event was histone gene replication. Presumably, after their replication the histon genes would assume an inactive chromatin structure that would then be reversed at the next mitosis. This model would account for the lack of histone expression in G2 phase and the activation of histone expression in G1 phase observed in yeast.

Recently, histone gene replication has been shown to be an early event in S phase in both mammalian cells (Iqbal et al., 1984) and in *Physarum* (Jalouzot et al., 1985). These data are cited in support of the replication-termination model. However, several observations clearly contradict this model. For instance, when DNA replication is blocked in HeLa cells with aphidicolin in mid-S phase, histone mRNAs disappear, but when the block is released, messengers reaccumulate (Heintz et al., 1983). The reaccumulation suggests that histone genes are transcribed in mammalian cells well beyond the time at which histone genes are replicated. However, it remains possible that the drug treatment leads to atypical mRNA synthesis. In _Physarum_, there are no data to support the turn-off of histone gene transcription in early S phase. Measurements of transcription rates by nuclear 'run-on' assays (Carrino and Laffler, 1986 see Fig. 5-2) suggest that transcription of the histone H4 genes continues at the maximum rate at least 90 min into S phase. On the other hand, the rate of DNA replication peaks in early S phase and then declines until S phase is over. Messenger levels probably are coupled to DNA synthesis by a posttranscriptional mechanism (see below). Therefore, at least in _Physarum_ the replication-termination model does not appear to be correct. We need

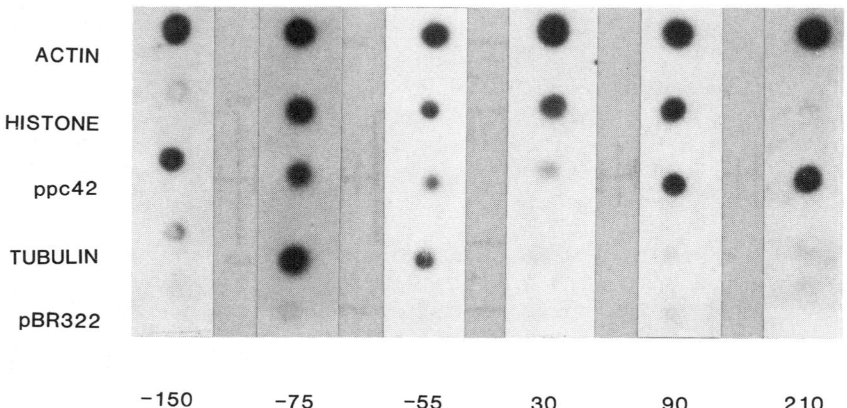

Fig. 5-2. Transcription of alpha-tubulin, histone, Ppc42, and actin genes during the cell cycle (Laffler and Carrino, unpublished data). Ppc42 is a cDNA clone for an mRNA whose level is constitutive over the cell cycle (Scheidl et al., 1984a). Nuclear transcription reactions were performed at 26°C for 30 min, and the labeled RNA product was isolated. Two-microgram aliquots of the indicated plasmid DNAs were denatured with NaOH, and an aliquot was spotted on nitrocellulose filters. Following overnight prehybridization, 10^6 cpm of labeled transcript was added in four parts prehybridization solution/one part 50% dextran sulfate, and hybridization was carried out for 4 days at 42°C. Exposures were for 72 hr at -80°C with intensifying screens.

to consider an appropriate mechanism for turning off histone gene transcription at the end of S phase that does not involve histone gene replication. Perhaps the replication of some other gene at the end of S phase initiates a chain of events that leads to the inactivation of histone gene transcription. A hypothetical model for transcriptional switching has been suggested by Tyson and Sachsenmaier (1979).

Elsewhere in this volume Pierron discusses the apparent coupling of replication and transcription (Chapter 4). The basic observation is that transcription complexes are often observed in the middle of replication bubbles, where the replication origin ought to be located. These results have been interpreted as the activation of gene transcription at the time of gene replication. Presumably, as the transcribed sequences are centered in replicons, their activation would be a result of replication origin function. The transcription experiments in Fig. 5-2 indicate that for none of the four sequences examined is there evidence for the S phase activation of RNA synthesis. While one can argue that the histone and tubulin genes were selected as special cases, neither actin nor Ppc42 show S phase activation. Instead, they are transcribed throughout the cell cycle. Rather, it could be that origin activity is stimulated by active transcription. Thus, transcription may be required for origin function rather than replication origin activity being required for transcription. However, correlation studies cannot discern cause from effect. A possible test of the origin activation hypothesis might be to compare the replication timing of an inducible gene under conditions of induction and repression. A gene that replicates very early in S phase when it is induced but not when it is repressed would provide supporting evidence.

The accumulation of histone mRNA in anticipation of S phase is far more clearly defined in *Physarum* then in yeast (Hereford et al., 1981). Since histone mRNA synthesis begins in mid-G2 phase, but histone protein synthesis is largely restricted to S phase, there must be a mechanism that prevents both the translation and the degradation of the histone mRNAs in late G2 phase. A dowry of histone transcripts may be stored in G2 phase for use early in S phase, at a time when the rate of DNA synthesis and the demand for histone protein are maximal. There are a number of possible storage mechanisms. The transcripts could be retained in the nucleus or stored in the cytoplasm as ribonucleoprotein complexes. Alternatively, the messengers could remain as inactive precursors, with final processing reserved until the beginning of S phase. Over the years, the storage of inactive histone mRNAs in sea urchin oocytes has been explained by polyadenylation, processing, nuclear compartmentalization, and cap methylation mechanisms, to name only a few possibilities (see Caldwell and Emerson, 1985). Despite a long history, the issue is far from resolved. The question of messenger storage may be more accessible in *Physarum*.

The regulation of histone messenger accumulation is proving to be quite complex. Data from the nuclear 'run-on' assays of Laffler and Carrino (1986) suggest that histone H4 gene transcription begins in mid-G2 phase (surprisingly at the same time as α-tubulin transcription begins), and continues at essentially the same level through S phase. Since there is a 4-fold drop in mRNA level with no change in the rate of 'run-on' transcription following mitosis, posttranscriptional regulation seems to play a major role in the control of histone gene expression.

Are the tubulin and histone genes cotriggered? Transcription of both the α-tubulin and histone H4 genes seems to begin at the same time in mid-G2 phase (Carrino and Laffler, 1986; see Fig. 5-2). This is an intriguing result. Why should the major structural proteins of the mitotic spindle be co-regulated with the major chromatin proteins? The timing of the tubulin gene activation fits well with expectations drawn from mRNA and protein synthesis data, but the activation of the histone genes is curious. Indeed, the G2 accumulation of histone mRNAs requires a specific storage mechanism. Certainly, a dowry of histone mRNA permits a high level of histone protein synthesis at the beginning of S phase, but the triggering in mid-G2 phase seems quite early. Rather, it could be that the pathway leading to nuclear division and chromosome replication begins in G2 phase, and the activation of transcription of tubulin and histone genes represents an early event in that pathway.

In both budding and fission yeast, the cell cycle is regulated at two points to maintain a constant cell size at cell division (Nurse, 1975; Fantes, 1977; Beach et al., 1982; Piggot et al., 1982; Pringle and Hartwell, 1982). The first, called START, acts in G1 phase to control entry in S phase, and the second acts in G2 phase to regulate nuclear division. Since in yeast START is the major control point, its cell cycle is said to begin in G1 (Pringle and Hartwell, 1982).

Physarum, however, lacks a G1 phase. Any 'G1' events necessary for the onset of DNA synthesis must, in Physarum, be completed in G2 phase of the previous mitotic cycle. In a sense, mitosis in Physarum has been displaced to the 'G1/S' boundary, and we might regard the Physarum cell cycle as beginning at some point in G2 phase. From this perspective, S phase does not lie at the beginning of the cycle but at the middle or the end. Since chromosome replication closely follows mitosis, events that must be completed in anticipation of either process could be initiated by a common trigger. Thus, there can be a certain economy in co-regulating histone mRNA synthesis (an event triggered in anticipation of chromosome replication) and tubulin mRNA synthesis (an event triggered in anticipation of nuclear division). Further evidence for the importance of a mid-G2 event in the Physarum cycle is provided by Loidl and Gröbner (1982) and others (Oppenheim and Katzir, 1971;

Blessing and Lempp, 1978), who observe a maximum in mitosis-stimulating activity about 2 hr before metaphase.

There is an apparent inconsistency between the mid-G2 transition point for tubulin and histone mRNA synthesis claimed by Laffler and Carrino (1986) and the exponential increase in tubulin messenger level reported by Schedl et al. (1984a). With an exponential accumulation curve, a transition point cannot be determined. However, the 'run-on' experiments suggest that the tubulin and histone gene transcription rate shows a dramatic step in mid-G2 phase. Clearly, there is something that is radically different about the transcription of these genes in nuclei isolated in early G2 phase as compared with those isolated in mid-G2 phase. Final proof will depend on well-executed in vivo labeling experiments.

The messenger accumulation data need a more critical appraisal. The data of Schedl et al. (1984b) are presented without confidence intervals. Given that the early-G2 phase level of tubulin mRNA is two orders of magnitude less than the peak level at mitosis, it is questionable whether these data bear significance prior to mid-G_2 phase. Although the exponential accumulation of tubulin mRNA in late G2 phase is quite evident, the results may not be inconsistent with a mid-G2 transition point if turnover of tubulin RNA is slow in G2 phase.

Since the rate of replication falls off continuously during S phase, it is difficult to define a clear endpoint to S phase. Primarily as a matter of convenience, we consider S phase to be finished approximately 3 hr after mitosis, by which time the rate of incorporation of ^{3}H-thymidine has dropped to G2 levels. However, alkaline sucrose sedimentation studies indicate that ligation of the newly replicated DNA into mature strands is not completed until at least 90 min past the conventional end of S phase (Funderud et al., 1978). Helmut Sauer (Chapter 1) has suggested that the transition observed by Laffler and Carrino (1986) could perhaps be determined by the end of this ligation period.

Homeostatic Controls

While the transcription of the tubulin and histone genes appears to be initiated by a cell cycle trigger in G2 phase, there is a clear necessity for a fine tuning mechanism to match the supply of tubulin and histone proteins to changes in demand. There are precedents for homeostatic control over both histone (Maxson et al., 1983) and tubulin synthesis (Ben Ze'ev et al., 1979).

Histone synthesis. Nuclear 'run-on' assays (Fig. 5-2) indicate that the rate of histone H4 transcription remains high throughout most of S phase, although the messenger level at 90 min into S phase is only 10% of the level at 30 min (Wilhelm et al., 1984).

Therefore, degradation of histone mRNA must become increasingly important as S phase proceeds; that is, homeostatic control (to keep the level of histone mRNA in general agreement with the decreasing activity of DNA replication in late S phase) seems to be exerted at a posttranscriptional level (namely, mRNA degradation). However, since this conclusion is based on a comparison of *in vivo* messenger levels and *in vitro* transcription rates, its veracity is limited by the credibility of 'run-on' activity as a metric of the true rate of transcription. Significantly, Osley and Hereford (1981) have shown in yeast that when the dosage of the H2A/H2B genes is doubled, the messenger level is unchanged although the transcription rate doubles, concluding that there must be homeostatic regulation of messenger degradation.

Histone protein synthesis appears to be coupled with DNA replication. When replication is blocked with different concentrations of hydroxyurea (HU), histone synthesis is reduced to the same extent as DNA synthesis (Kelley et al., 1983). Kung and Braun (personal communication) have now extended this type of experiment to the RNA level. Using RNA blot hybridization, they find that, whereas G2 phase application of HU has no effect, the level of histone H4 mRNA drops to a low value when HU is applied in S phase. Conversely, supernormal levels (a four-fold excess) accumulate when cycloheximide is used to inhibit protein synthesis during S phase. Since DNA replication stops shortly after cycloheximide is added, these data suggest that replication and messenger level are not directly coupled. Rather, both results are consistent with a homeostatic mechanism that couples supply and demand for histone proteins. This mechanism could regulate mRNA levels by modulating either transcription or turnover of the mRNA. However, as addressed above, transcription assays suggest that messenger turnover is modulated. The same assays can be used to address this point in HU-inhibited plasmodia.

Tubulin synthesis. Experiments with colchicine-treated mammalian cells strongly argue for a homeostatic coupling of tubulin mRNA levels and microtubule assembly (Ben Ze'ev et al., 1979; Cleveland et al., 1981, 1983). To date, there has been relatively little work in this area with *Physarum*. Interest has centered on cell cycle regulation where the salient features are the triggering of gene expression in mid-G2 phase and the initiation of spindle assembly shortly before mitosis. These events are clearly interrelated. If there is a dependent link between these events, there are two possible chains of causality. According to the popular view, tubulin gene transcription is directly triggered as a cell cycle event. Messengers accumulate, are translated to make tubulin proteins, and the tubulin concentration increases. A threshold concentration may need to be exceeded before the assembly of spindle microtubules can be initiated. In this scenario, the major controlling event is the cycle-specific activation of transcription.

In the alternate model, tubulin gene transcription is under homeostatic control. The driving force is the assembly of microtubules. As the soluble tubulin pool becomes diminished, tubulin gene transcription is activated. To account for the transcription data, this would need to begin in mid-G2 phase. Since microtubules are not seen until shortly before mitosis, these initial assemblies could not readily be detected by traditional methods. Nonetheless, withdrawal from the tubulin protomer pool would be sufficient to stimulate tubulin gene expression. Transcription of the tubulin genes would then not be controlled by a direct timing mechanism. The lack of observable G2 phase microtubules and the rapid, complete activation of tubulin gene transcription argue against the alternative model.

There are supporting data to be considered as well. Eon-Gerhardt et al. (1981) found that either methylbenzimidazole carbamate (MBC) or griseofulvin delayed both mitosis and the premitotic increase in thymidine kinase activity to the same extent. They concluded that microtubule assembly must precede the increase in thymidine kinase activity. However, their conclusions relied on enzyme assays which might be affected by the drugs used, and drug-resistant controls that could rule out nonspecific drug effects were not yet available. With the availability of benomyl-resistant mutants as controls (Burland et al., 1984) and the ability to monitor directly the synthesis, accumulation, and translation of tubulin mRNAs, this question can be given further attention.

A third possibility is that the triggers of transcription and of spindle assembly are not interdependent but lie on parallel pathways. In this case, the initiation of spindle assembly need not be dependent on tubulin gene expression. However, under normal conditions both pathways would be coordinated by a master timer. Heat-shock experiments that delay mitosis but do not alter the timing of tubulin expression can be interpreted in this light (Carrino and Laffler, 1985; see below), but can also be interpreted in other ways.

There are some interesting new observations that relate to spindle microtubule assembly during the cell cycle. E.C.A. Paul (personal communication) has followed by two-dimensional gel analysis the appearance of tubulins in nuclei isolated under conditions that preserve microtubular structures, and also followed spindle assemblies visualized by indirect immunofluorescence. The tubulin content of these nuclei probably represents subunits sequestered in microtubules; there are no data to indicate that monomeric tubulin is retained in these nuclei. Two-dimensional gel analysis indicates that all four of the plasmodial tubulin isotypes are found in the spindle in the same ratio as they are found in intact plasmodia. There does not appear to be preferential utilization of any isotype. Tubulins that are labeled before one mitosis can be used in the spindle of the following mitosis. (This does not imply, of course, that tubulin protein is completely stable; only that some tubulin protomers survive for more than one cycle.) Whether the same subunits were used in two successive spindles or the subunits used in the second cycle were not used in the first cycle could not be determined. Polymerized tubulin rapidly disappears from nuclei following nuclear division. Both results are consistent with the interpretation that tubulin gene transcription is activated well before microtubular structures become evident.

While the major assembly of microtubules in the nucleus begins less than 30 min before metaphase, a new tubulin-containing structure appears approximately 60 min prior to metaphase. This dot-like structure, which is located on the periphery of the nucleolus, presages an active microtubule-organizing center, as microtubules radiate from this structure at later stages in mitosis (Schedl et al., 1984b).

Periodicities in Perturbed Cell Cycles

Heat-shock experiments. Tubulin synthesis has been followed in plasmodia where mitosis has been delayed by a heat shock (Carrino and Laffler, 1985). Two discrete peaks of tubulin synthesis are observed. The first peak occurs at the same time as that of the single, premitotic peak in controls, and the second peak directly precedes the delayed mitosis. Tubulin protein levels follow a related course. There is an initial accumulation accompanying the first peak, followed by a virtual disappearance of tubulin. A second peak of tubulin accumulates prior to the delayed mitosis. This second peak of accumulation is higher than the first peak. In both delayed and in control plasmodia, the tubulin level drops by 75-80% after mitosis. These results suggest that plasmodia are committed to a scheduled burst of tubulin synthesis prior to the heat shock, but they are delayed in entering mitosis. The period of tubulin degradation that normally follows mitosis occurs on the precommitted schedule. Thus, the initial burst of tubulin synthesis, the initial peak of tubulin accumulation, and the normally post-mitotic period of tubulin instability all continue on the old schedule, while mitosis is rescheduled to occur at a later time. Mitosis follows the reinitiation of tubulin synthesis on a delayed schedule. Thus, the events observed result from an admixture of events determined on the old and new schedules. The lingering questions of which effects are manifest at transcriptional, post-transcriptional, and translational levels can now be resolved by RNA-blot and 'run-on' assays.

Wright and Ducommon (personal communication) also observe two peaks in tubulin protein synthesis following a shift from normal growth temperature to a higher, non-lethal temperature. Incidentally, the curious admixture of events determined on old and new schedules was first observed by Wright and Tollon (1979b) in terms of peaks of activity of thymidine kinase after a temperature shift.

The second peak of tubulin protein synthesis is curious. Since tubulin protein virtually disappears after the first peak, mitosis cannot occur until the tubulin pool is restored. The question is how cell cycle regulation copes with this problem. Tubulin protein accumulation is driven by two components: synthesis and degradation. The programmed decline in tubulin protein level following the first peak of synthesis and accumulation was either determined prior to or was unaffected by the heat shock. Since most of the tubulins that are degraded are synthesized after the heat shock, the turnover of heat-denatured tubulins cannot explain this result. Is the tubulin clock reset or can we invoke homeostatic controls? On the one hand, if the pathway were retriggered, we would expect to see two peaks in tubulin and histone gene transcription. On the other hand, heat shocks may delay mitosis by delaying spindle assembly. Then the second cycle of tubulin expression could be a homeostatic response to delayed withdrawal of protomers from the tubulin pool. We would not expect a parallel effect on histone gene expression. If these homeostatic controls act at the level of tubulin mRNA degradation, we would predict that, following a heat shock, there will be a single broad peak of transcription (from mid-G2 phase to the delayed mitosis) and two peaks of mRNA accumulation. Clearly, two peaks of transcription would not be consistent with a degradative homeostatic control model. This is currently being examined (J. Carrino, personal communication).

Models of Cell Cycle Control

Size control. Much evidence suggests that the DNA-division cycle is coordinated to cell growth as to maintain a constant cell size at cell division or a constant nucleocytoplasmic ratio at the onset of DNA synthesis (reviewed in John, 1981; Nurse and Streiblova, 1984; Donnan et al., 1985). The evidence for size control is especially strong in *Physarum* (Tyson, 1982). Since *Physarum* lacks a G1 phase, the point (or points) at which size control is exerted must lie somewhere in G2 phase. But where? The initiation of transcription of tubulin and histone genes is an attractive candidate for a size control point, but cell fusion experiments (e.g., Tyson and Sachsenmaier, 1978; Loidl and Sachsenmaier, 1982) indicate that the timing of mitosis is regulated by a diffusible substance that triggers nuclear division very late in G2 phase (approx. 45 min before metaphase). Cycloheximide-pulse experiments (Scheffey and Wille, 1978; Tyson et al., 1979) confirm the conclusion from fusion studies that irreversible commitment to mitosis is a very late step in the cycle.

There seem to be (at least) two options for explaining these data. The traditional assumption has been that the size-control checkpoint occurs just prior to mitosis. From this perspective, those events necessary for nuclear division and a new round of DNA synthesis (events such as the initiation of synthesis of tubulin and histone) are completed in G2 according to some prescribed schedule (i.e., a dependent sequence of events connected to the completion of DNA replication) and then, when all preparations are complete, the final question is asked: has the protein-to-DNA ratio reached the critical value of, roughly, 100-to-1? If the answer is yes, then mitosis and a new round of DNA synthesis are triggered. If the answer is no, then the machinery remains poised until the final requirement is met.

This model of a late trigger accounts nicely (in qualitative terms) for the timing of and the protein-to-DNA ratio at mitoses subsequent to heat shocks, cycloheximide pulses, and radiation treatment, and for the ability to contract G2 phase down to a minimum duration. Furthermore, one particular version of the late-trigger paradigm, the nuclear sites titration model of Sachsenmaier et al. (1972), can be pushed to give quantitative details of fusion and perturbation experiments (Sudbery and Grant, 1975; Tyson and Sachsenmaier, 1978; Tyson et al., 1979). In the nuclear sites titration model, an activator of mitosis is assumed to accumulate as the plasmodium grows (the number of activator molecules is proportional to total protein content, i.e., total activator concentration is <u>constant</u>) and to bind to nuclear sites (whose number is proportional to total DNA content). Only after all nuclear sites have been 'titrated' by activator (i.e., only after a specific protein-to-DNA ratio has been reached) does the free activator concentration in the cytoplasm increase to a level sufficient to trigger mitosis and DNA synthesis.

In the late-trigger paradigm, early and mid-G2 events are necessary but not sufficient for the occurrence of mitosis. If the initiation of tubulin synthesis is one of these preparatory events, with tubulin protomers waiting in reserve until a late trigger is pulled, why does one observe two peaks of tubulin accumulation when mitosis is delayed by heat shock? Why is tubulin degraded after the heat shock and resynthesized later in preparation for the delayed mitosis? A possible explanation is homeostatic control: because assembly of microtubules is delayed by heat shock, the tubulin protomer pool swells, inducing tubulin degradation and inhibiting tubulin mRNA synthesis. Then the process of preparation for mitosis must be started over again (at least in part). In the nuclear sites titration model, nuclear sites are assumed to be emptied by heat treatment and to be refilled after return to normal temperature. It could be that the initiation of tubulin and histone synthesis is triggered by the filling of sites (for instance, when nuclear sites are half-filled).

A second option for explaining regulation of the Physarum cell cycle places the size-control checkpoint in early or mid-G2 phase. The 'early-trigger' paradigm provides a more natural account of the two peaks of tubulin synthesis and of thymidine kinase activity after heat shock. To account for the results of fusion experiments and cycloheximide-pulse treatments, one must imagine that the early trigger initiates the accumulation of a substance that later triggers mitosis directly.

Thus, it seems that, in either option, there must be at least two triggers in G2 phase, but it is not clear whether one or the other is the primary size-control checkpoint. Now that we can divide G2 phase into early and late portions with respect to initiation of transcription of tubulin and histone genes, it may be possible to address this question directly.

Autonomous mitotic oscillator. Size control is not the only possible mechanism for regulating the cell cycle in Physarum. Events of the DNA-division cycle could be timed by a mitotic oscillator that cycles independently of the events it times (i.e., independently of nuclear division and chromosome replication). This hypothesis, as originally framed for Physarum by Kauffman and his coworkers (Kauffman and Wille, 1975; Wille et al., 1977), was flawed in two respects. First, it did not couple the mitotic oscillator to cell growth, whereas the period of the nuclear division cycle must be identical to the mass doubling time of cell growth in order for a plasmodium to maintain its nucleocytoplasmic ratio within reasonable bounds. Secondly, it required that the oscillation involve relatively smooth changes in all components, whereas fusion experiments show clearly that a major abrupt change in the regulatory mechanism occurs just prior to metaphase (Tyson and Sachsenmaier, 1978; Loidl and Sachsenmaier, 1982). These two flaws can be remedied easily, but at present there are no experimental results on the Physarum cell cycle that would compel one to reconsider the possibility of an autonomous mitotic oscillator.

Nonetheless, for the cell cycle in early embryos, there is growing evidence (Hara et al., 1980; Newport and Kirschner, 1984) for oscillatory processes, with period close to the normal intermitotic time, that continue in the absence of periodic nuclear division and DNA replication (e.g., in enucleated cells). Perhaps this is reason enough to reconsider autonomous oscillations in Physarum. A good test would be to block mitosis, preferably by a temperature-sensitive cell-cycle mutation, and look for periodic bursts of tubulin and histone gene transcription at intervals thereafter (Mitchison, personal communication). Another test would be to block mitosis temporarily, say by shifting a temperature-sensitive cell-cycle mutant to a non-permissive temperature for various lengths of time and then returning the plasmodium to a permissive temperature, and to determine the timing of subsequent mitosis.

Size control would predict mitotic times correlated to the protein-to-DNA ratio after the block, whereas an autonomous oscillator would predict mitotic times correlated to mitoses in an unshifted control culture (assuming the oscillator continues to run normally during the block).

THE TIMING OF NUCLEAR DIVISION

The 1960's and 1970's saw many studies of the changes in the timing of synchronous nuclear divisions in macroplasmodia following various perturbations of the cell cycle: plasmodial fusion, heat shocks, ionizing irradiation, or treatment with drugs such as cycloheximide and actinomycin. These studies were very fruitful in probing the responses of the "black box" that controls the timing of mitosis (for review, see Tyson, 1982 or 1984); however, they have not been pursued much lately.

Exenberger et al. (1985) have looked at mitotic delays induced in macroplasmodia by X-irradiation. As for treatments with cycloheximide, heat shock, and UV irradiation (Sudbery and Grant, 1975; Tyson et al., 1979), the first post-X-irradiation mitosis is delayed and the protein-to-DNA ratio at mitosis is abnormally large. The next cycle is shortened so that the protein-to-DNA ratio is normal at the second postirradiation mitosis. The same pattern, of an initial delayed mitosis at an increased protein/DNA ratio followed by a shortened second cycle, is observed after adding butyrate to the growth medium (Loidl et al., 1982). These results provide further support for the notion that the timing of nuclear division is regulated by a system sensitive to the nucleocytoplasmic ratio.

The mechanism of induction of mitotic delay by ionizing radiation is not well-known, but Daniel (personal communication) argues that the target for radiation-induced mitotic delay is a membrane receptor complex or component modifying adenylate cyclase. Compatible with this view are the rapidly induced transient increase in cellular cAMP content observed following gamma radiation and a subsequent transient increase in cellular cGMP content during the ensuing mitotic recovery. Both cyclic nucleotide responses are appropriately modified in the presence of caffeine, which decreases radiation-induced mitotic delay (Daniel and Oleinick, 1984).

Caffeine treatment following X-irradiation partially reverts the immediate antimitotic effect on the postirradiation mitosis, but the second postirradiation cycle is prolonged by caffeine treatment (Garcia-Herdugo, personal communication). Does caffeine, therefore, merely delay the antimitotic effects of ionizing radiation (rather then abolishing them)? Since caffeine induces a number of biochemical responses in treated cells (Daniel and Oleinick, 1984), long-term exposure to caffeine may delay the second postirradiation cycle for reasons unrelated to the interplay of caffeine and radiation. Determination of protein-to-DNA ratios during the first and second postirradiation cycles in caffeine-treated plasmodia should prove informative.

Future usefulness of cell cycle-perturbation experiments is most likely to be in conjunction with molecular dissection of the control system, as in the experiments of Laffler and Carrino measuring tubulin synthesis after heat shock. In this regard; (1) a variety of perturbing agents should be used in addition to heat shock (e.g., UV irradiation and cycloheximide pulse treatment); (2) effects should be followed past the first to the second posttreatment mitosis; (3) low growth temperatures should be employed to lengthen the cell cycle and avoid complications introduced by the minimal duration of G2; and (4) total DNA and protein contents should be monitored to keep track of the gross nucleocytoplasmic ratio.

CONCLUSION

The future of cell cycle research in Physarum has never been brighter. There has been rapid progress in studies of both histone and tubulin gene expression. The transcription of these genes could be triggered by the same cell cycle event. These systems are beginning to tell an interesting story about cell cycle regulation. While we have generally been taught to think about biological systems as being continuously variable and to cast a wary eye on simple on-off mechanisms as oversimplifications, the cell cycle may be teaching us a new lesson. If the in vitro 'run-on' data correctly reflect the in vivo reality, tubulin and histone gene transcription may have only two states: on and off. The homeostatic 'fine tuning' that balances supply with demand and provides for continuous variation in messenger level may operate at a posttranscriptional level by modulating the rate of mRNA turnover. However, to be convincing, we must develop the ability to measure accurately messenger synthesis and turnover rates in intact plasmodia. The cell cycle regulation of gene expression might be more complicated than our simplest models, yet be far simpler than we have previously dared to believe.

There is a large catalog of well-documented periodic enzymes in the Physarum cell cycle. Fluctuations in enzyme activity for a few of these enzymes is known to be correlated with periodic synthesis of new protein. The cause of periodic enzyme synthesis in any of these cases is unknown, and the purpose for such periodicities (if there is one) is equally vague. There does seem to be a disproportionately large number of periodic proteins whose synthesis peaks in late G2 phase, and the simplest assumption is that they are all regulated by a common control system associated as well with the timing of mitosis and the onset of DNA synthesis. This optimistic scenario, of course, needs to be ruthlessly tested and refined over the next several years.

It is not known with any certainty whether periodic synthesis of any enzyme in Physarum persists in the absence of nuclear division and DNA synthesis. The answer to this question is crucial to the debate over regulation of the cell cycle: is periodic enzyme synthesis inseparably connected to periodic changes in the nucleo-cytoplasmic ratio caused by the interplay of cell growth and periodic DNA synthesis, or is it an autonomous control system that can continue to run in the absence of periodic nuclear division and DNA synthesis? In fission yeast there is evidence that periodicities in CO_2 production and thymidine diphosphokinase activity continue for more than one generation time after the DNA/division cycle has been blocked by expression of cdc-mutant phenotype (Mitchison, Creanor, and Novak, personal communication). In enucleated eggs, cortical contractions continue with a period close to the normal interdivision time (Hara et al., 1980), and maturation-promoting factor continues periodic fluctuations as well (Newport and Kirschner, 1984). These observations suggest autonomous control systems, at least in yeast cells and early embryos. But what about Physarum? Do such autonomous oscillators exist? How would we look for them? Enucleation is impractical, to say the least, and treatment with 5'-fluorodeoxyuridine is not workable because DNA synthesis continues at a low but significant rate even in the presence of high concentrations of this drug. Temperature-sensitive cell cycle mutants (which aren't "leaky") would provide the best test material for this question, but at present we have too few practical candidates in Physarum, and none of these candidates has been carefully studied.

Two-dimensional gel electrophoresis of plasmodial proteins has proven to be a powerful and informative tool in studying cell cycle-dependent changes in protein synthesis and accumulation, but much remains to be done. A definitive survey of two-dimensional gels by digital image-processing techniques should be carried out, and there are numerous possibilities for studying individual proteins once they have been located as spots on a gel. Tubulin is a good example of what can be done. Other proteins that may warrant further study are DNA polymerase, thymidine kinase, thymidylate synthetase, and calmodulin.

It is easy to block or delay the onset of nuclear division by a variety of chemical and physical agents, e.g., ultraviolet irradiation, X irradiation, heat shock, cycloheximide pulses, FdUrd, etc. The timing of nuclear division after exposure to such agents can be used to test various theoretical notions about the coordination of overall cell growth and nuclear division. This interplay of theory and experiment has identified one model -- the nuclear sites-titration model -- as giving the most parsimonious and believable explanation of all the data. In fission yeast there is evidence that a similar control system operates: yeast geneticists have identified two regulatory genes (weel and wee2 = cdc2) that seem to correspond roughly to the nuclear sites and titrating agent of the Physarum model (Tyson, 1982).

The weakness of Physarum as a test case for cell cycle control is the lack of good mutants defective in progress through the cell cycle. Cell cycle mutants define cell cycle-specific functions. In addition, a mutant plasmodium at the restrictive temperature provides an ideal assay system for the mutant-defined macromolecule. Mutant rescue experiments can be accomplished by introducing wild-type gene products by plasmodial fusion or by injection of subcellular fractions into living plasmodia. Laffler and Okada (personal communication) demonstrated that the temperature-sensitive cell cycle mutant MA67 (Laffler et al., 1979) becomes temperature-resistant when fused with wild-type plasmodia at a nuclear ratio of 20 mutants: 1 wild-type. Injection of protein or mRNA fractions could similarly be used in rescue experiments with the added dividend of providing an assay for purification of the active component.

The potential of this approach has yet to be realized. More effort needs to be invested in temperature-sensitive mutant isolation. As most mutant-defined functions appear to be shared between plasmodia and amoebae (Burland and Dee, 1979, 1980), direct screening or enrichment with myxamoebae is in order. Laffler (personal communication) has proposed a BrdUrd-UV suicide enrichment strategy for isolating thermosensitive growth mutants in myxamoebae, but success has been limited by the poor plating efficiency of current axenic strains. Improved axenic strains currently under development in Dee's laboratory (see Chapter 18) might render this approach more feasible. Matthews (personal communication) is actively pursuing an approach to select amoebal mutants with a conditional block in G2 or mitosis, using a fluorescence-activated cell sorter. There is nothing intrinsic to Physarum to prevent success in mutant isolation if a sufficient number of clever scientists are willing to apply themselves to this problem.

There should be a determined effort to isolate cdc-type and wee-type mutants in Physarum amoebae and to study their effects on both the amoebal cell division cycle and the plasmodial nuclear division cycle. In particular, we should be on the lookout for Physarum genes that complement the cdc28 (START) mutation in budding yeast and the cdc2 ("wee") mutation in fission yeast as well as temperature-sensitive mutant analogs to the yeast START mutants. If we could demonstrate a functionally homologous cell cycle control gene in Physarum, we would be in an excellent position to tie together the phenomenological picture of size control in Physarum with the corresponding genetic/molecular picture emerging for yeast cells.

REFERENCES

Ares, M., and Howell, S., 1982, Cell cycle stage-specific accumulation of mRNAs encoding tubulin and other polypeptides in Chlamydomonas, Proc. Natl. Acad. Sci., U.S.A., 79:5577.

Beach, D., Durkacz, B., and Nurse, P., 1982, Functionally homologous cell cycle control genes in fission and budding yeast, Nature, 300:706.

Ben Ze'ev, A., Farmer, S., and Penman, S., 1979, Mechanisms of regulating tubulin synthesis in cultured mammalian cells, Cell, 17:319.

Blessing, J., and Lempp, H., 1978, An immunological approach to the isolation of factors with mitotic activity from the plasmodial stage of the myxomycete Physarum polycephalum, Exp. Cell Res., 113:435.

Bochner, B. R., Lee, P. C., Wilson, S. W., Cutler, C. W., and Ames, B. N., 1984, AppppA and related adenylated nucleotides are synthesized as a consequence of oxidative stress, Cell, 37:225.

Bravo, R., and Celis, J., 1980, A search for differential polypeptide synthesis throughout the cell cycle of HeLa cells, J. Cell Biol., 84:795.

Brewer, E. N., and Rusch, H. P., 1966, Control of DNA replication: effect of spermine on DNA polymerase activity in nuclei isolated from Physarum polycephalum, Biochem. Biophys. Res. Comm., 25: 579.

Burland, T. G., and Dee, J., 1979, Temperature-sensitive mutants of Physarum polycephalum -- expression of mutations in amoebae and plasmodia, Genet. Res. Camb., 34:33.

Burland, T. G., and Dee, J., 1980, Isolation of cell-cycle mutants of Physarum polycephalum, Mol. Gen. Genet., 179:43.

Burland, T., Schedl, T., Gull, K., and Dove, W., 1984, Genetic analysis of resistance to benzimidazoles in Physarum: differential expression of beta-tubulin genes, Genetics, 108:123.

Caldwell, D., and Emerson, C., 1985, The role of cap methylation in the translational activation of stored maternal histone mRNA in sea urchin embryos, Cell, 42:691.

Carrino, J., and Laffler, T., 1985, The effect of heat shock on the cell cycle regulation of tubulin expression in Physarum polycephalum, J. Cell Biol., 100:642.

Carrino, J., and Laffler, T., 1986, Transcription of the alpha-tubulin and histone H4 genes begins at the same point in the Physarum cell cycle, J. Cell Biol., in press.

Chahal, S. S., Matthews, H. R., and Bradbury, E. M., 1980, Acetylation of histone H4 and its role in chromatin structure and function, Nature, 287:76.

Cleveland, D., Lopata, M., Sherline, P., and Kirschner, M., 1981, Unpolymerized tubulin modulates the level of tubulin mRNAs, Cell, 25:537.

Cleveland, D., Pittenger, M., and Feramisco, J., 1983, Elevation of tubulin levels by microinjection suppresses new tubulin synthesis, Nature, 305:738.

Daniel, J. W., and Jarlfors, U., 1972, Plasmodial ultrastructure of the myxomycete Physarum polycephalum, Tissue and Cell, 4:15.

Daniel, J. W., and Oleinick, N. C., 1984, Cyclic nucleotide responses and radiation-induced mitotic delay in Physarum polycephalum, Radiat. Res., 97:341.

Donnan, L., Carvill, E. P., Gilliland, T. J., and John, P. C. L., 1985, The cell cycles of Chlamydomonas and Chlorella, New Phytol., 99:1.

Elliot, S., and McLaughlin, C., 1978, Rate of macromolecular synthesis through the cell cycle of the yeast Saccharomyces cerevisiae, Proc. Natl. Acad. Sci., U.S.A., 75:4384.

Eon-Gerhardt, R., Tollon, Y., Chraibi, R., and Wright, M., 1981, Regulation of thymidine kinase synthesis during the cell cycle of Physarum polycephalum: the effects of two microtubule inhibitors, Cytobios, 32:47.

Exenberger, R., Garcia-Herdugo, G., and Sachsenmaier, W., 1985, Perturbation by X-rays and spontaneous readjustment of the nuclear cycle in Physarum polycephalum, Eur. J. Cell Biol., 38(Suppl.9):8.

Fantes, P., 1977, Control of cell size and cycle time in Schizosaccharomyces pombe, J. Cell Sci., 24:51.

Fouquet, H., and Braun, R., 1974, Differential RNA synthesis in the mitotic cycle of Physarum polycephalum, FEBS Lett., 38:184.

Funderud, S., Andreassen, R., and Haugli, F., 1978, Size distribution and maturation of newly replicated DNA through the S and G_2 phases of Physarum polycephalum. Cell 15:1519.

Gröbner, P., 1979, Thymidine kinase enzyme variants in Physarum polycephalum, J. Biochem., 86:1595.

Gröbner, P., and Loidl, P., 1983, Response of the dTMP-synthesizing enzymes to differentiation processes in Physarum polycephalum, Exp. Cell Res., 144:385.

Gröbner, P., and Loidl, P., 1984, Thymidine kinase: A novel affinity chromatography of the enzyme and its regulation by phosphorylation in Physarum polycephalum, J. Biol. Chem., 259:8012.

Gröbner, P., and Loidl, P., 1985a, An immunological approach to enrich a mitotic stimulator and to reveal G2 phase-specific proteins in Physarum polycephalum, J. Cell Biol., 100:1930.

Gröbner, P., and Loidl, P., 1985b, ADP-ribosyltransferase in isolated nuclei during the cell cycle of Physarum polycephalum, Biochem. J., 232:21.

Gröbner, P., and Sachsenmaier, W., 1976, Thymidine kinase enzyme variants in Physarum polycephalum; change of pattern during the synchronous mitotic cycle, FEBS Lett., 71:181.

Hall, L., and Turnock, G., 1976, Synthesis of ribosomal RNA during the mitotic cycle in the slime mold Physarum polycephalum, Eur. J. Biochem., 62:471.

Hara, K., Tydeman, P., and Kirschner, M. W., 1980, A cytoplasmic clock with the same period as the division cycle in Xenopus eggs, Proc. Natl. Acad. Sci, U.S.A., 77:462.

Havercroft, J., and Gull, K., 1983, Demonstration of different patterns of microtubule organization in Physarum polycephalum myxamoebae and plasmodia using immunofluorescence microscopy, Eur. J. Cell Biol., 32:67.

Hedrick, S. M., Cohen, D. I., Nielsen, E. A., and Davis, M. M., 1984, The isolation of cDNA clones encoding T cell-specific membrane associated proteins, Nature, 308:148.

Heintz, N., Sive, H., and Roeder, R., 1983, Regulation of human histone gene expression: kinetics of accumulation and changes in the rate of synthesis and in the half-lives of the individual histone mRNAs during the HeLa cell cycle, Mol. Cell Biol., 3:539.

Hereford, L., Osley, M., Ludwig, J., and McLaughlin, C., 1981, Cell-cycle regulation of yeast histone mRNA, Cell, 24:367.

Holt, C. E., 1980, The nuclear replication cycle in Physarum polycephalum, in: "Growth and Differentiation in Physarum polycephalum," W. F. Dove and H. P. Rusch, eds., Princeton Univ. Press, Princeton, NJ.

Hyams, K., and Hyams, J., 1979, "Microtubules," Academic Press, London.

Iqbal, M., Plumb, M., Stein, J., Stein, G., and Shildkraut, C., 1984, Coordinate replication of members of the multigene family of core histone genes, Proc. Natl. Acad. Sci., U.S.A., 81:7723.

Jalouzot, R., Toublan, B., Wilhelm, M., and Wilhelm, F. X., 1985, Replication timing of the H4 histone genes in Physarum polycephalum, Proc. Natl. Acad. Sci. U.S.A., 82:6475.

John, P. C. L., ed., 1981, "The Cell Cycle", Cambridge Univ. Press, London.

Kauffman, S., and Wille, J. J., 1976, The mitotic oscillator in Physarum polycephalum, J. Theor. Biol., 55:47.

Kelley, P., Schofield, P., and Walker, I., 1983, Histone gene expression in Physarum polycephalum II: Coupling of histone and DNA synthesis, FEBS Lett., 161:79.

Laffler, T., Chang, M.-T., and Dove, W., 1981, Periodic synthesis of microtubular proteins in the cell cycle of Physarum, Proc. Natl. Acad. Sci., U.S.A., 78:5000.

Laffler, T., Wilkins, A., Selvig, S., Warren, N., Kleinschmidt, A., and Dove, W., 1979, Temperature sensitive mutants of Physarum: viability, growth, and the nuclear replication, J. Bacteriol., 138:499.

Lai, E., Walsh, C., Wardell, D., and Fulton, C., 1979, Programmed appearance of translatable tubulin mRNA during cell differentiation in Naegleria, Cell, 17:867.

Loidl, P., and Gröbner, P., 1982, Acceleration of mitosis induced by mitotic stimulators of Physarum polycephalum, Exp. Cell Res., 137:469.

Loidl, P., Gröbner, P., Csordas, A., and Puschendorff, B., 1982, Cell-cycle-dependent effects of sodium-n-butyrate in Physarum polycephalum, J. Cell Sci., 41:105.

Loidl, P., Loidl, A., Puschendorf, B., and Gröbner, P., 1983, Lack of correlation between histone H4 acetylation and transcription during the Physarum cell cycle, Nature, 305:446.

Loidl, P., and Sachsenmaier, W., 1982, Control of mitotic synchrony in Physarum polycephalum: Phase shifting by fusion contradicts a limit cycle oscillator model, Eur. J. Cell Biol., 28:175.

Lorincz, A. T., Miller, M. J., Xuong, N. H., and Geiduschek, E. P., 1982, Identification of proteins whose synthesis is modulated during the cell cycle of Saccharomyces cerevisiae, Mol. Cell Biol., 2:1532.

Lutkenhaus, J. F., Moore, B. A., Masters, M., and Donachie, W. D., 1979, Individual proteins are synthesized continuously through the Escherichia coli cell cycle, J. Bacteriol., 138:352.

Maxson, R., Cohn, R., Dedes, L., and Mohun, T., 1983, Expression and organization of histone genes, Ann. Rev. Genet., 17:239.

Mittermayer, C., Braun, R., and Rusch, H. P., 1964, RNA synthesis in the mitotic cycle of Physarum polycephalum, Biochim. Biophys. Acta, 91:399.

Newport, J. W., and Kirschner, M. W., 1984, Regulation of the cell cycle during early Xenopus development, Cell, 37:731.

Nurse, P., 1975, Genetic control of cell size at cell division in yeast, Nature, 256:547.

Nurse, P., and Streiblova, eds., 1984, "The Microbial Cell Cycle", CRC Press, Boca Raton, FL.

Oppenheim, A., and Katzir, N., 1971. Advancing the onset of mitosis by cell free preparations of Physarum polycephalum, Exp. Cell Res., 68:224.

Osley, M. A., and Hereford, L., 1981, Yeast histone genes show dosage compensation, Cell, 24:377.

Pahlic, M., and Tyson, J. J., 1983a, A cell-cycle-dependent protein of Physarum polycephalum revealed by two-dimensional gel electrophoresis, Biochem. J., 212:245.

Pahlic, M., and Tyson, J. J., 1983b, Identification and changes in activity of five thymidine kinase forms during the cell cycle of Physarum polycephalum, Exp. Cell Res., 144:159.

Pahlic, M., and Tyson, J. J., 1985, Analysis of Physarum proteins throughout the cell cycle by two-dimensional PAGE, Exp. Cell Res. (in press).

Pierron, G., and Sauer, H. W., 1980, More evidence for replication-transcription-coupling in Physarum polycephalum, J. Cell Sci., 41:105.

Piggot, J., Rai, R., and Carter, B., 1982, A bifunctional gene product involved in two phases of the yeast cell cycle, Nature, 298:391.

Pringle, J., and Hartwell, L., 1982, The Saccharomyces cerevisiae cell cycle, in: "Molecular Biology of the Yeast Saccharomyces cerevisiae," J. Strathern, E. Jones, and J. Broach, eds., Cold Spring Harbor Laboratories, Cold Spring Harbor, New York.

Rapaport, E., and Zamecnik, P. C., 1976, Presence of diadenosine 5',5"-P1,P4-tetraphosphate ($A_{p4}A$) in mammalian cells in levels varying widely with proliferative activity of the tissue: a possible positive "pleiotypic activator", Proc. Natl. Acad. Sci., U.S.A., 73:3984.

Rollins, M. J., Harper, J. D. I., and John, P. C. L., 1983, Synthesis of individual proteins, including tubulins and chloroplast membrane proteins, in synchronous cultures of the eukaryote, Chlamydomonas reinhardtii, J. Gen. Microbiol., 129:1899.

Sachsenmaier, W., Remy, V., and Plattner-Schobel, R., 1972, Initiation of synchronous mitosis in Physarum polycephalum: A model of the control of cell division in eukaryotes. Exp. Cell Res., 73:41.

Schedl, T., Burland, T., Dove, W., Roobol, A., Paul, E., Foster, K., and Gull, K., 1984b, Tubulin expression and the cell cycle of Physarum polycephalum, in: "Molecular Biology of the Cytoskeleton," G. G. Borisy, D. W. Cleveland, and D. B. Murphy, eds., Cold Spring Harbor Laboratories, Cold Spring Harbor, New York.

Schedl, T., Burland, T., Gull, K., and Dove, W., 1984a, Cell cycle regulation of tubulin RNA level, tubulin protein synthesis, and assembly of microtubules in Physarum, J. Cell Biol., 99:155.

Scheffey, C., and Wille, J. J., 1978, Cycloheximide-induced mitotic delay in Physarum polycephalum, Exp. Cell Res., 36:124.

Schofield, P. N., and Walker, I. O., 1982, Control of histone gene expression in Physarum polycephalum, I. Protein synthesis during the cell cycle, J. Cell Sci., 57:139.

Shipley, G. L., Christensen, M. E., Nader, W. F., Choi, J. J., Murphy, C. D., Diller, J. D., Martin, R. J., and Sauer, H. W., 1985, Evidence for a myc protooncogene in the mitotic cycle of Physarum polycephalum, J. Cell Biol., 101:79a.

Sudbery, P. E., and Grant, W. D., 1975, The control of mitosis in Physarum polycephalum: the effect of lowering the DNA: mass ratio by UV irradiation, Exp. Cell Res., 95:405.

Turnock, G., Chambers, J., and Birch, B., 1981, Regulation of protein synthesis in the plasmodial phase of Physarum polycephalum, Eur. J. Biochem., 120:529.

Tyson, J. J., 1982, Periodic phenomena in Physarum, in: "Cell Biology of Physarum and Didymium," H. Aldrich and J. Daniel, eds., Academic Press, New York.

Tyson, J. J., 1984, Control nuclear division in Physarum polycephalum, in: "The Microbial Cell Cycle," P. Nurse and E. Streiblova, eds., CRC Press, Boca Raton.

Tyson, J., Garcia-Herdugo, G., and Sachsenmaier, W., 1979, Control of nuclear division in Physarum polycephalum: Comparison of cycloheximide pulse treatment, UV irradiation, and heat shock, Exp. Cell Res., 119:87.

Tyson, J. J., and Sachsenmaier, W., 1978, Is nuclear division in Physarum controlled by a continuous limit cycle oscillator?, J. Theor. Biol., 73:723.

Tyson, J. J., and Sachsenmaier, W., 1979, Derepression as a model for control of the DNA-division cycle in eukaryotes, J. Theor. Biol., 79:275.

Waterborg, J. H., Matthews, H. R., 1983a, Acetylation sites in histone H3 from Physarum polycephalum, FEBS Lett., 162:416.

Waterborg, J. H., and Matthews, H. R., 1983b, Patterns of histone acetylation in the cell cycle of Physarum polycephalum, Biochemistry, 22:1489.

Waterborg, J. H., and Matthews, H. R., 1984, Patterns of histone acetylation in Physarum polycephalum: H2A and H2B acetylation is functionally distinct from H3 and H4 acetylation, Eur. J. Biochem., 142:329.

Weinmann-Dorsch, C., Pierron, G., Wick, R., Sauer, H., and Grummt, F., 1984, High diadenosine tetraphosphate (Ap_4A) level at initiation of S phase in the naturally synchronous mitotic cycle of Physarum polycephalum, Exp. Cell Res., 155:171.

Wilhelm, M., and Wilhelm, F.-X., 1984, A transposon-like DNA fragment interrupts a Physarum polycephalum histone H4 gene, FEBS Lett., 168:249.

Wilhelm, M., Toublan, B., Jalouzot, R., and Wilhelm, F., 1984, Histone H4 is transcribed in S phase but also late in G2 phase in Physarum polycephalum, EMBO Journal, 3:2659.

Wille, J., Scheffey, C., and Kauffman, S., 1977, Novel behavior of the mitotic clock in Physarum, J. Cell Sci., 27:91.

Wright, M., and Tollon, Y., 1979a, Physarum thymidine kinase: a step or peak enzyme depending upon temperature of growth, Eur. J. Biochem., 96:177.

Wright, M., and Tollon, Y., 1979b, Regulation of thymidine kinase synthesis during the cell cycle of Physarum by the heat-sensitive system which triggers mitosis and S phase, Exp. Cell Res., 122:273.

Chapter 6: CELLULAR TRANSFORMATIONS OF MYXAMOEBAE

Roger W. Anderson[a], Jennifer Dee[b], and Keith Gull[c]

[a]Department of Genetics
University of Sheffield
Sheffield, UK

[b]Department of Genetics
University of Leicester
Leicester, UK

[c]Biological Laboratory
University of Kent
Canterbury, UK

with contributions from: A. B. Blindt, R. A. Cox, K. E. Foster, K. Kohama, H. Maruta, M. J. Monteiro, D. J. Pallotta, T. Schreckenbach, G. L. Shipley, G. Sweeney, and M. Wright.

INTRODUCTION

The myxamoebae of Physarum polycephalum are uninucleate, free-living cells, which can be maintained in vegetative growth or induced to follow any one of three different pathways of cellular transformation, the end results of which are cysts, flagellated cells, or plasmodia (see Chapter 2). There are many reasons why this system should provide favorable material for the study of cell differentiation. Our long-term aim is to understand the transformations at all levels, relating the regulation of specific gene activities and the differential synthesis of gene products to the organization of cell structures and the functioning of each cell type.

The most widely studied of the myxamoebal transformations is that which generates plasmodia, often referred to as the "amoebal-plasmodial transition". Reports at the Workshop illustrated a diversity of approaches already being employed, and a major focus of the sessions on Genetics and Differentiation was the question of how these different approaches could be related. Consideration was also given to the transformation of myxamoebae into flagellated cells.

THE AMOEBAL-PLASMODIAL TRANSITION

Desirable Features in a Model System for Differentiation

Schreckenbach speculated on the future of *Physarum* as an organism for studies on cell differentiation and raised the question of whether any of the cellular transformations shown by *Physarum* can truly be considered a model system for cell differentiation. Such a model system should ideally include the following features: (1) a range of developmental mutants available with well-characterized phenotypes; (2) a usable DNA transformation system; and (3) synchronous development in response to a clear, defined inducing signal. Although none of these features is yet fully possessed by the amoebal-plasmodial transition, it will emerge below that in every case there is encouraging progress.

Some Features of the Amoebal-Plasmodial Transition

In heterothallic strains, clones of amoebae derived from single cells normally remain in vegetative growth in nutrient conditions and do not differentiate into plasmodia, because all the cells are of the same mating type. When haploid amoebae carrying different alleles of the mating-type genes meet, they undergo cell and nuclear fusion to form diploid zygotes, which differentiate into diploid plasmodia. In certain conditions, amoebae of an apogamic clone such as CL (see Chapter 2) can differentiate individually into haploid plasmodia; every cell in an apogamic colony is potentially capable of differentiating into a separate plasmodium, given a suitable microenvironment. As an amoeba differentiates apogamically, it ceases cytokinesis, grows to a larger size than that usually attained during vegetative amoebal growth, and undergoes mitosis without cytokinesis to become binucleate (Anderson et al., 1976). Further growth then occurs by repeated synchronous mitoses, as in heterothallic plasmodium formation. In both types of plasmodium formation, the multinucleate cells readily fuse with one another so that ultimately only a few large plasmodia emerge from a culture in which initially many cells differentiated individually.

An important step in plasmodium formation is the attainment of cell commitment, when differentiation becomes irreversible. If cells from apogamic colonies are separated by resuspension and replating on fresh medium, cells committed to plasmodium formation continue to grow and differentiate into isolated, multinucleate plasmodia, whereas cells that have not reached the stage of commitment resume proliferation to give rise to new amoebal colonies. Committed cells appear in a culture only after an initial period of amoebal proliferation, during which the level of an "inducer" produced by the cells rises in the medium. Commitment is dependent upon inducer concentration; thus plasmodium formation is dependent upon the attainment of a critical cell density (Youngman et al., 1977). Commitment in apogamic strains must begin in unfused, uninucleate cells, since a substantial proportion of committed cells is present in suspensions from cultures harvested in the early stages of plasmodium formation, before any multinucleate cells have been formed (Burland et al., 1981).

Unlike heterothallic plasmodium formation, apogamic development in most strains does not occur at the highest temperatures at which amoebal growth is still possible (Adler and Holt, 1977); thus, plasmodium formation in apogamic colonies may be conveniently controlled not only by cell density or inducer concentration, but also by incubation temperature.

Amoebae and plasmodia of an apogamic strain have the same genome and are of the same ploidy, but it has long been clear that they differ substantially in the gene products they express, as evidenced particularly by qualitative and quantitative differences in the proteins synthesized during growth, and by cell type-specific expression of various mutations (see later in this chapter).

The amoebal-plasmodial transition is thus well-suited to the study of changes in gene expression during cell differentiation. It links two vegetative stages, each consisting of a homogeneous cell type, which can be cultured in quantities sufficient for biochemical studies. The transition can be induced or repressed by cultural conditions and is affected by genes in which a number of mutants have been isolated and analyzed (see later). There is a useful degree of natural synchrony in the development of monoxenic cultures with _E. coli_ on agar (Youngman et al., 1977); however, this synchrony is unfortunately not adequate for some purposes, and methods to improve the synchrony of plasmodium formation are considered later in this chapter. Finally, progress was reported at the Workshop toward a DNA transformation system for amoebae (see Chapter 25).

GENETIC CONTROL OF MATING

The _matA_ Locus

For the reasons described above, much of the work on the amoebal-plasmodial transition involves apogamic strains. Nevertheless, it is in work on sexual development that genes regulating plasmodium formation have been studied most successfully. The major determinant of mating compatibility is the matA locus, which may be viewed as a developmental switch that directly or indirectly regulates most of the changes that occur during plasmodium formation. It functions by affecting the behavior of amoebal fusion products; diploid zygotes heterozygous for matA show a very strong tendency to differentiate into plasmodia, whereas homozygotes continue to proliferate as diploid amoebae, with a cell division accompanying each mitosis (Youngman et al., 1981).

As described below, there has been significant progress in the study of the matA locus through conventional genetic techniques. Nevertheless, the nature of the matA gene products remains obscure, and the only clues to their sites and times of action during mating are provided by the observation that interphase nuclear fusion quickly follows amoebal fusion in matings heteroallelic for matA, but that nuclear fusion is apparently delayed until mitosis in matA-homoallelic fusions (Holt and Hüttermann, 1979). The interpretation of this finding is that matA products are probably present before karyogamy in the fusion cell, either on the nuclear membrane or in the cytoplasm (Youngman, 1979).

Available information is clearly insufficient to make the identification of matA gene products more than a remote prospect by conventional biochemical techniques. An alternative approach emerged at the Workshop, however, as a result of the encouraging report by Haugli and Johansen (Chapter 25) concerning progress toward a DNA transformation system for amoebae. If such transformations become achievable with high efficiency, it may be possible to identify genomic clones carrying matA sequences without any prior knowledge concerning the gene products themselves. For example, if amoebae of one strain were transformed with DNA from a strain of different matA type, any transformant expressing both resident and foreign matA sequences might behave like a matA-heterozygous diploid amoeba, in which case it could be selected very efficiently on the basis of its ability to develop into a visible plasmodium against a background of vegetative amoebae. It should be noted, however, that the size of the genomic sequence that would need to be introduced and expressed is unknown and perhaps substantial.

Until transformation experiments become feasible, the main approach to the study of matA function will no doubt continue to be conventional genetic analysis of mutations affecting the matA locus. One type of mutation, known as gad (= greater asexual differentiation), confers on gad mutant amoebae the ability to develop readily into plasmodia without the normal requirement of heterothallic strains for zygote formation (Adler and Holt, 1977). This apogamic type of development is also shown, of course, by strain CL (see Chapter 2), and in all probability the "matAh allele" carried by CL is the result of a gad mutation (gad-h) occurring in a matA2 amoeba (Anderson, 1979).

The association of the gad phenotype with the matA locus (Adler and Holt, 1977; Gorman et al., 1979) would clearly tend to reinforce the idea of matA as some kind of developmental switch, were it not for a report that three out of eleven matA-associated gad mutations were able to recombine with matA at frequencies as high as 12% (Shinnick et al., 1983). Anderson expressed the view, on the basis of repeated failures to reproduce these observations in his own laboratory, that the recombinants apparently detected in the original analysis might have been, for example, aneuploids rather than haploids; thus, all the matA-linked mutations may map at a single locus, which should perhaps be designated gadA and which may be coincident with the matA locus.

Another type of mutation associated with the matA locus is called npf (= no plasmodium formation). These mutations are obtained by subjecting gad strains to a selection for revertants that have lost the ability to develop apogamically (Anderson and Holt, 1981). Most of the npf mutations map at the gad locus, and those that do may be classified into two types, referred to by Anderson as 'class 1' and 'class 2'. The class 1 mutants include the types previously designated npfC and npfE and have all gained the ability to mate with amoebae that carry the matA allele of their heterothallic gad^+ npf^+ progenitor. The class 2 mutants include those previously designated npfB and npfD; they appear, at least superficially, to be "true" revertants in the sense that they have regained the original mating specificity of their heterothallic progenitor. Anderson described new analyses designed to test an earlier hypothesis (Anderson and Holt, 1981) that the class 1 mutants might be effectively heterothallic amoebae expressing new matA alleles (something that extensive efforts have not obtained directly by mutation of wild-type amoebae [Gorman et al., 1979; R. S. Wildin, R. W. Anderson, and C. E. Holt, unpublished]). A set of 36 class 1 mutants derived from two matA3 gad strains was tested; they appeared to be a very uniform group, failing to mate with one another in any combination and showing similar mating patterns in other tests. The implication is that these class 1 mutants do not express novel matA alleles, but lack a distinct, common function required in plasmodium formation.

The class 1 npf mutations and the gad mutations define functions that are associated with matA, but which may be distinct from its mating specificity function. Thus, the matA locus may be considered to be a complex regulatory locus. Further progress toward an understanding of its functions probably requires a molecular approach. It may eventually be possible to clone genomic sequences corresponding to gad and npf^+ functions by complementation. For example, since the gad mutations appear to be dominant

(Shinnick et al., 1983), transformation of wild-type heterothallic amoebae with DNA from a gad mutant strain might yield gad transformants selectable on the basis of their ability to differentiate into plasmodia asexually. In the same way, since the class 1 npf mutations are recessive (Anderson and Youngman, 1985), npf^+ transformants selectable on the basis of plasmodium formation might be generated by transformation of gad npf mutant amoebae with DNA from a wild-type strain.

Other Genes Controlling Mating

In contrast to matA, which determines whether or not plasmodia will form in an amoebal culture, the matB locus affects only the efficiency of plasmodium formation (Dee, 1978; Youngman et al., 1979). Nevertheless, matB is clearly recognizable as a mating-type locus, at least under laboratory conditions, because heteroallelic matings yield substantially larger numbers of plasmodia than do homoallelic ones. By the same criterion, recent experiments by S. Kawano and R. Anderson, discussed informally at the meeting, suggest that the imz locus (Shinnick et al., 1978) must now be considered as a possible third mating-type locus (matC?), although it was not thought to be a mating-type locus when it was first discovered. There is also some evidence that imz may possess at least three alleles. Most genetic analysis in *Physarum polycephalum* is carried out on inbred strains in which plasmodium formation is normally affected only by allelic variations at matA, matB, and imz. It is clear from the patterns of mating between noninbred strains, however, that there must be further loci affecting the efficiency of plasmodium formation in natural isolates (for example, see Honey et al., 1982).

Both matB and imz affect the frequency of recovery of zygotes from mating mixtures, most likely by an effect on the frequency of amoebal fusions. It seems unlikely that either gene has a sufficiently strong phenotypic effect to permit its eventual cloning in the manner suggested above for matA, gad, and npf^+. Nevertheless, since the most likely site of action of the products of both matB and imz is the amoebal cell surface, it is potentially possible to identify these gene products by detecting differences between surface components from amoebae that carry unlike matB and imz alleles. Clearly, it is essential to use isogenic strains in such experiments to reduce or eliminate surface differences unrelated to the genes in question. It is also necessary to use amoebae that are likely to be expressing the matB and imz gene products; in other words, they must be in a state of mating competence. Shipley reported a study satisfying those conditions, carried out at the same time as his group's work to develop monoclonal antibody techniques for the comparison of amoebal and plasmodial surfaces (see later). With whole, fusion-competent amoebae as antigens, 26 hybridomas were obtained that secreted amoebal-specific antibodies.

None of the hybridomas produced an antibody showing allelic specificity for matB or imz, although some strain differences were detected. It is, of course, not surprising that matB and imz antigens were not identified in this preliminary study, since the task of finding such antigens will remain a formidable challenge as long as they are being sought against an overwhelming background of common antigens.

COMPARISON OF AMOEBAE AND PLASMODIA

Background

In order to study the changes that occur during the amoebal-plasmodial transition, it is clearly necessary to describe the endpoints of the transformation, i.e., vegetative amoebae and vegetative plasmodia. These cell types differ visibly in a number of respects, including size, morphology, pigmentation, behavior, and mode of mitosis. But what are the specific differences in gene action and in the synthesis of gene products underlying the visible differences? Information on such underlying differences may be useful in several ways. First, it may contribute to an overall assessment of the extent of difference between amoebae and plasmodia. Second, it may make possible the definition of specific developmental stages in terms of the appearance or disappearance of particular functions or products. Finally, individual differences may become specifically interesting in terms of their regulation or function, or preferably both.

One approach has been to induce temperature-sensitive mutations in each cell type. It is then possible to detect stage-specific functions by identifying mutations that are not expressed in the alternative cell type. To provide valid results, this type of study should be performed with an apogamic strain, so that both cell types are genetically identical, and the cells should be cultured under identical conditions so as to eliminate environmental or nutritional effect unrelated to cell type. Observing these precautions, Burland and Dee (1979) found that at least 67% of mutations identified in plasmodia and at least 51% of mutations identified in amoebae were expressed in both stages. Although this approach gives some maximum estimate of the extent of difference between the two cell types, more detailed analysis of individual mutants may be hampered by the fact that apogamic development in CL and other gad mutants is itself temperature-sensitive; thus, temperature-shift experiments, usually a powerful tool in determining the time of expression of temperature-sensitive mutations, are potentially difficult to carry out.

Another useful contribution to the comparison of amoebae and plasmodia has been the work of Turnock's group, using two-dimensional gels to establish the extent of polypeptide differences. Again with the use of an apogamic strain cultured under identical conditions in both growth phases, 26% of relatively abundant proteins were judged to be possibly phase-specific, with additional proteins showing different rates of synthesis in the two cell types (Turnock et al., 1981). This kind of comparison provides an overview and does not reveal the nature of the changes; for example, the production of new protein species may occur by modulation of gene expression at several levels, including posttranslational modification. There is also the possibility of a lack of co-identity of coordinate two-dimensional gel spots, and the fact that only abundant polypeptide species will be revealed.

Attempts to overcome some of these problems were reported at the Workshop in the form of two major technical approaches: the use of monoclonal antibody techniques to detect specific antigens (mainly proteins or glycoproteins), and the use of cDNA cloning techniques to search for differences in specific mRNA transcripts.

Monoclonal Antibodies

Shipley reported the successful use of monoclonal antibody techniques to identify antigens that show differential expression between plasmodia and amoebae. Shipley and colleagues took the view that cell surface antigens would be most easily identifed by this approach, and they used whole, fusion-competent myxamoebae as the injected antigen. Putative hybridomas were initially identified by use of whole amoebal cells and were then rescreened against amoebal, plasmodial, and bacterial cell homogenates to establish levels of selectivity. Twenty-six hybridomas were detected that produced amoebal-specific antibodies. Further screening was performed by testing the reactivity of antibodies to proteins separated by sodium dodecyl sulfate polyacrylamide gel electrophoresis and then Western blotted onto nitrocellulose paper. This screen revealed that most of the monoclonal antibodies recognized multiple bands. This result suggests, of course, that the antibodies are recognizing an epitope that is a posttranslational modification of many polypeptides, the most likely candidate being the carbohydrate side chains of glycoproteins. This result has been found in many other systems where cell surface-specific markers have been sought by the same route.

Thus, at first sight it would appear that this approach may not recognize differentially expressed protein antigens on the cell surface. Nevertheless, it has the advantage over other comparative studies of the two different cell types in that it produces probes that can be easily used in studies of the timing of events and acquisition of plasmodial characteristics during the transformation of individual cells. Further, in this particular case, this approach may have highlighted a significant difference in the glycosylation pathways of amoebae and plasmodia.

cDNA Clones

Three laboratories reported the use of cDNA techniques to reveal differences in RNA transcripts. Pallotta's group (Chapter 24) produced poly A+ RNA from myxamoebae and plasmodia, copied it into cDNA, and inserted this into pBR322 by poly (dG:dC) tailing. E. coli cDNA libraries were thus produced for both the plasmodium and myxamoebae. These libraries were screened with radiolabeled poly A^+ RNA from amoebae and plasmodia, and 1344 plasmodial clones and 800 amoebal clones were selected. Sixteen percent of the clones were determined to be plasmodial-specific, with three of these clones apparently detecting abundant mRNAs. Similar differential hybridization studies indicated that about 2.5% of the cDNA clones in the amoebal library hybridized to amoebal poly A+ RNA and not to plasmodial poly A+ RNA. Northern blot analysis indicated that seven of these clones represented stage-specific mRNA species; one was not regarded as stage-specific, merely being more abundant in the myxamoebal stage. This group interpreted their analysis as indicating that stage-specific mRNAs do occur and that the plasmodium contains approximately seven times more stage-specific mRNAs than do myxamoebae. Thus, it may be that differentiation of amoebae into plasmodia involves the activation of many plasmodial-specific genes and the inactivation of a few amoebal-specific genes.

Other laboratories also reported the isolation of both amoebal-specific and plasmodial-specific cDNA clones (Chapter 23). Sweeney and Turnock constructed cDNA libraries from plasmodia and amoebae in an M13 vector and found that, of the clones that hybridized to cDNA probes, 4% were plasmodial-specific. Monteiro and Cox isolated amoebal-specific DNA clones by two approaches. Genomic DNA isolated from plasmodia was cloned into an EMBL4 vector; the recombinants were probed with ^{32}P-labeled cDNA copies from poly A+ mRNA isolated from amoebae (CLd-Axe) and plasmodia (M_3cVIII); visual inspection revealed that 50 of the 7000 recombinants screened hybridized more strongly with the amoebal probe than with the plasmodial probe. Detailed analysis of a number of these clones indicated that they hybridized approximately five times more effectively with amoebal RNA than with microplasmodial RNA. In other work, Monteiro and Cox used the phage λgt11 as the vector for the construction of cDNA libraries. Differential screening of these libraries revealed an interesting cDNA clone that hybridized with an mRNA much more abundant in the amoeba than in the plasmodium.

Thus, three different laboratories have used a variety of experimental protocols, yet all have been successful in recognizing stage-specific probes. These studies have been performed against the background difficulty of providing a clear definition of the terms "plasmodial-specific" and "amoebal-specific". The complete lack of a specific RNA transcript (or indeed a particular antigen) is almost impossible to define biochemically; thus, in these initial investigations, operational definitions of acceptable levels have been adopted. Although these definitions are likely to become more precisely defined as the studies progress, it is clear that they are a potential source of some discrepancy between laboratories in the initial stages of this work.

There is no doubt that the acquisition of these cDNA probes is a useful step forward. They give information on the extent of difference between amoebae and plasmodia in the abundance of specific transcripts and are potentially useful markers of developmental stages. More important, however, the detailed analysis of individual clones in terms of whether they represent early or late messages in development, whether all are under the same form of control, and whether regulatory sequences can be recognized should provide useful data on the types of regulatory events occurring in _Physarum_ cell-type switching. Individual probes will also become specifically interesting if the coding regions can be identified in terms of a specific protein, and progress can then be made from these initial stages through to the cell biology of that individual protein.

Cytoskeletal Components

A problem with the approaches described above is, of course, that generally it may not be easy to relate specific antigens or cDNA clones to defined structures of known function. An alternative approach is to begin the analysis with a structure that is already known to be of likely interest; this has proven to be a particularly fortunate choice of approach in the case of microtubules, which are well-defined cellular structures known to be important in cell movement, form, and division.

The switch to the plasmodium has been found to involve extensive restructuring of the microtubule complement of the _Physarum_ amoebal cell. Myxamoebae have cytoplasmic microtubules, centriole microtubules, and microtubules in the open mitotic spindle. However, the plasmodium has no centrioles and no cytoplasmic microtubules (Havercroft and Gull, 1983). Microtubules are present in the plasmodium only in the nucleus in the closed mitotic spindle. Furthermore, the amoebal cell expresses two electrophoretically separable tubulin isotypes, $\alpha 1$ and $\beta 1$, whereas the plasmodium expresses four isotypes, the $\alpha 1$, $\alpha 2$, $\beta 1$, and $\beta 2$ tubulins (Roobol et al., 1983; Burland et al., 1983). Major questions exist as to how expression of this tubulin multigene family is regulated, why different cells express different tubulins, and whether the different tubulin polypeptides have different functional roles.

Studies of tubulin gene structure in Physarum have shown there to be four α tubulin DNA loci, one of which is complex and contains more than one α tubulin sequence, and at least three β tubulin DNA loci (Schedl et al., 1984; see Chapter 2). However, as stated above, only four tubulin polypeptide isotypes (α1, α2, β1, β2) have been shown to arise directly from RNA species by in vitro translation. Thus, it might appear that there are more tubulin DNA sequence loci than there are tubulin polypeptide isotypes. Work was presented at this meeting by Foster to show that in fact a more complex pattern of tubulin isotypes is detectable by the use of monoclonal antibodies, which show some selectivity for different tubulin isotypes. Foster, Birkett, and Gull used a panel of seven well-defined monoclonal antibodies to probe blots of the myxamoebal and plasmodial tubulin isotypes (Birkett et al., 1985).

The myxamoebal α1 tubulin was recognized by a number of monoclonal antibodies as a single spot on two-dimensional gels. Immunoblotting of one-dimensional isoelectric focusing gels revealed that this myxamoebal species is, in fact, composed of three distinct sub-types, all of which focus in this one two-dimensional gel spot. Also, the plasmodial α1-tubulin two-dimensional gel spot has been shown by a similar immunological analysis to contain four subtypes. The plasmodial-specific α2 tubulin appears to be a single species. Thus, it appears that the myxamoeba may in fact be expressing three α1-tubulin isotypes, with the plasmodium expressing four α1-tubulin isotypes and one α2-tubulin isotype. This immunological study has revealed other interesting relationships between these tubulin isotypes. The monoclonal antibody KMP-1 recognizes all of the myxamoebal α1-tubulin isotypes but recognizes only three out of the four α1-tubulin isotypes present in the plasmodium. Also, KMP-1 does not recognize the plasmodial-specific α2-tubulin isotype.

Thus, it appears that there may well be close matching between the numbers of tubulin genes and tubulin polypeptides in Physarum. In such an event, it again becomes important to address why the organism modulates the expression of the tubulin multigene family. The extent of the modulation is now becoming clearer with the recognition of the different tubulin isotypes expressed in the myxamoeba and the plasmodium.

The phenomenon of cell type-dependent expression of components of cytoskeletal proteins may now be extended to include the amoebal and plasmodial myosins. Kohama (see Chapter 10) reported that he and his colleagues had recently purified myosins from both amoebae and plasmodia. They were able to show that both amoebal and plasmodial myosin molecules consisted of one heavy chain and two different light chain polypeptides. Moreover, in both cell types the molecules were of the consensus myosin shape (two-headed and long-tailed).

SDS polyacrylamide gel electrophoresis, peptide mapping, and immunological studies indicated that the amoebal myosin heavy chain differed from the plasmodial myosin heavy chain. By use of similar criteria the 18K light chain was found to be different in amoebae and plasmodia, whereas the 14K light chain appeared identical.

It is, at present, unclear why the plasmodium and myxamoeba might express these different myosins. However, it does serve to indicate that the transition between a myxamoeba and a plasmodium may involve a radical restructuring of the cell's cytoskeletal architecture and protein composition.

ANALYSIS OF CHANGES DURING PLASMODIUM FORMATION

Induction of Synchronous Development

One aim of the cDNA and monoclonal antibody studies with amoebae and plasmodia is to define the amoebal-plasmodial transition in terms of a sequence of changes in gene expression and to relate this sequence to changes in cell locomotion, mitosis, feeding, cell fusion, etc. An important requirement for achieving this aim is better synchrony than is currently available in cultures that are undergoing development. One possible approach would be to use apogamic strains in axenic culture; such cultures might be expected to display good synchrony by virtue of the uniformity with which inducer would presumably be distributed in the culture medium. Nevertheless, there are difficulties with this approach (see Chapter 18). Another interesting possibility would be to add inducer preparations to cultures of apogamic amoebae, either on agar or in liquid medium.

Shipley summarized the present state of knowledge concerning the phenomenon of induction. The existence of an extracellular, diffusible inducer of apogamic development was first suggested by experiments in which sparse cultures of amoebae on Nuclepore filters were placed over denser cultures on agar (Youngman et al., 1977). The dense cultures induced premature development in the sparse cultures, even though the pores of the filters did not permit cell contacts. A similar kind of induction effect has also been shown for certain matings (Pallotta et al., 1979; Youngman, 1979). Shipley and Holt (1982) grew heterothallic amoebal strains separately to various cell densities, mixed them, and assayed their ability to mate immediately. The results showed that mating competence, i.e., the ability to mate when mixed with other competent cells, was cell density-dependent, and this mating competence could be induced through Nuclepore filters.

The results of these transfilter induction experiments might also be consistent with inactivation of a diffusible inhibitor by binding to cells in the dense cultures. The best evidence for the existence of a diffusible inducer comes from the demonstration that cell-free extracts of dense cultures of Physarum amoebae were able to induce premature differentiation in an apogamic strain (Youngman, 1979). The inducing activity was very labile, however, and subsequent work has concentrated on inducing extracts prepared from liquid culture supernatants of Didymium iridis; such cultures have proved to be a fairly convenient source of inducing activity, not only of Didymium mating, but also of sexual and asexual development in Physarum (Youngman, 1979). Shipley reported that extracts from Physarum cultures appeared to have a reciprocal inducing activity for Didymium matings, although this was not easy to demonstrate. Nader et al. (1984) made some progress with the characterization of the Didymium inducer by column chromatography, and suggested that it was a carbohydrate, glycoprotein, or glycolipid of molecular weight about 120K. Nevertheless, purification was unsuccessful because of losses of activity. The purification and characterization of inducer would obviously be a valuable contribution not only because of the possible value of purified inducer in the synchronization of cultures, but also because it would pave the way for an examination of the mechanisms by which inducer interacts with amoebae to trigger development.

Since it is still necessary to use crude extracts in induction experiments, it is not certain that the same inducer (or inducers) is active in both mating and apogamic development, although there is no evidence to suggest otherwise. One mutation that alters inducing activity has been identified. This is the ind-2 mutation, which was first recognized because it caused apogamic amoebae to form plasmodia more slowly and in larger colonies than usual, but which also causes mating to be slow in ind-2 X ind-2 mixtures, and reduces the ability of amoebal strains to induce asexual development through Nuclepore filters (Youngman, 1979).

Enrichment for Committed Cells

Another approach to the problem of synchrony has been taken by Blindt, Dee, and Gull and was described at the Workshop by Gull. This approach depends upon the observation that committed cells do not flagellate when placed in water, whereas myxamoebae that have not yet reached commitment will, of course, still flagellate. Apogamic cultures of amoebae are incubated until committed cells are present, then the mixed populations of committed and uncommitted cells are suspended in water to permit flagellation of essentially all the uncommitted cells. The resulting mixture of committed cells and flagellates is passed down a glass bead column, from which only the flagellates emerge at the bottom. Committed cells are then removed from the beads and will continue their development to plasmodia with reasonable synchrony in semidefined medium; any contaminating amoebal cells cannot grow under these conditions.

Changes in the Microtubule Cytoskeleton

The protocol described above has enabled Blindt and his coworkers to study the changes in the microtubule cytoskeleton during plasmodium formation by use of immunofluorescence microscopy with monoclonal anti-tubulin antibodies. These studies have shown for the first time a clear developmental pathway of cellular organization during the amoebal-plasmodial transition. The committed cell resembles the myxamoeba in that it possesses an extensive network of cytoplasmic microtubules radiating from a cytoplasmic microtubule organizing center (MTOC). Thus, a transforming cell traverses the interphase period after commitment with much the same cytoskeletal arrangement as a vegetative myxamoebal cell. At the first mitosis after commitment, the nucleus divides but the cell does not; the division is clearly seen by these studies to be intranuclear, with the spindle microtubules being nucleated by an intranuclear MTOC. A surprising finding of these studies has been the discovery that the original amoebal cytoplasmic MTOC does not appear to become the plasmodial intranuclear mitotic MTOC but, rather, it continues to nucleate cytoplasmic microtubules over the first few cell cycles after commitment. Immunofluorescence of committed cells in the first mitosis, and of binucleated cells, reveals the persistence of these cytoplasmic MTOCs. The cytoplasmic MTOCs and microtubules disappear over the next few cell cycles, such that by the time the cells are recognizable as small plasmodia containing 16 or 32 nuclei, they no longer contain any cytoplasmic microtubules.

Thus, it appears that the decision to switch from nuclear division, by use of an open spindle, to an intranuclear mitosis is made during the cell cycle that contains the commitment point. The result of this is that myxamoebal cells, which are born by a division involving an open mitosis, go through commitment in the ensuing interphase and then enter a division that involves a closed mitotic spindle and no cytoskeleton. However, although the acquisition of a plasmodial type of division is very rapid, the cell does not quickly disassemble the microtubule architecture of the myxamoeba. This often seems to take a number of cell cycles. Also, analysis of the tubulin isotype composition in transforming cell populations suggests that synthesis of the plasmodial-specific tubulin isotypes is initiated at or soon after commitment, whereas acquisition of the normal stoichiometry of the plasmodial tubulins appears to take a number of cell cycles. These studies have produced a much clearer view of the transition stages of the amoebal-plasmodial transition, and the described developmental landmarks in the microtubule cytoskeleton should serve to provide a baseline for future studies of timings of both gene expression events and acquisition of a plasmodial cell biology during this cell type switch.

Developmental Mutants

Many of the genes involved in the amoebal-plasmodial transition are likely to remain undetected by cDNA probes of the types described earlier in this chapter. Some regulatory genes, for example, may be transcribed only at low level or transiently, and even among the genes that are detected, it may prove difficult to determine functional significance. A complementary approach is to use conventional genetic techniques to isolate developmental mutants of apogamic strains; any mutation that results in loss of the ability to form a plasmodium must, by definition, be functionally significant. Such mutations will identify genes that are dispensable during amoebal proliferation but are required, however briefly, during the process of plasmodium formation or early in the vegetative plasmodium itself.

The first isolation of mutants with defects of plasmodium formation was by Wheals (1973), who screened 5 X 10^4 colonies derived from individual mutagenized amoebae and isolated four mutant strains. One strain carried a mutation now known as npfF1 (originally apt-1), which is unlinked to matA and which blocks development at the stage of enlarged uninucleate cells.

Most subsequent isolations of npf mutants have employed a selective method that should favor the recovery only of mutants that continue to proliferate as amoebae, under conditions that encourage asexual differentiation of gad npf^+ amoebae. The result has generally been the recovery of npf mutations that map in the matA locus. Nevertheless, a few mutations unlinked to matA have been obtained in this way, including npfA1 (Anderson and Dee, 1977) and three new mutations reported at the meeting by R. Anderson. These four mutations are also unlinked to npfF and, with the exception of two of the new mutations, to one another.

R. Anderson also reported recent experiments in which colonies derived from mutagenized gad amoebae were screened essentially as in Wheals' early work. Even with the same, rather inefficient, mutagenesis method employed by Wheals, these preliminary experiments yielded mutants at a frequency of about 6 X 10^{-4}. Of 54 mutants isolated, 16 apparently carried matA-linked, "class 1" mutations (see earlier in this chapter), 28 showed obvious defects in amoebal encystment as well as in plasmodium formation, and 10 showed a variety of defects, including the formation of abnormal multinucleate cells. The next step will be to isolate additional mutants and to define the existing mutants better, both genetically and in terms of the way in which their development is arrested. It may be possible to define the pathways of abnormal development of the mutants with respect to many different gene activities, and the perturbations that the mutations cause in the normal pattern of gene expression may give information about the way in which the expression of certain genes is related to earlier and later events in development. Finally, as discussed earlier, it may become possible to use DNA transformation methods to identify npf$^+$ genomic sequences

THE AMOEBAL-FLAGELLATE TRANSITION

Wright's laboratory has produced an extensive analysis, over a number of years, of the ultrastructural aspects of the amoebal and flagellate microtubule systems. This work is built on the earlier cytological studies of Aldrich, who clearly indicated that important differences existed between the microtubule cytoskeleton of myxamoebae and flagellates (Aldrich, 1968). The most easily discerned difference between these two cell types is the flagellate cell's possession of two flagella. Unlike certain other amoebae that show amoeba-flagellate transitions, the Physarum myxamoeba possesses centrioles that function as the flagellar basal bodies in the flagellate. Wright and his colleagues have provided complete three-dimensional ultrastructural descriptions of the myxamoebal centrioles and associated microtubule arrays, and of the flagellate cytoplasmic microtubule arrays and flagellum configurations (Wright et al., 1979, 1980). More recently, Wright et al. have looked at the maturation of centrioles in the myxamoebal cell cycle, using three-dimensional reconstructions of mitotic amoebae from serial thin sections. Procentrioles are formed in early prophase on both the anterior and posterior centrioles. During the different stages of mitosis, daughter centrioles elongate and acquire anterior satellites, one of the characteristic features of the anterior centrioles. In contrast to the parental posterior centriole, which does not change morphologically during the successive stages of mitosis, the parental anterior centriole loses its morphological characteristics in late prophase and early prometaphase and then acquires the morphological features characteristic of the posterior centrioles. Thus, Wright and his colleagues have suggested the following maturation scheme for centrioles--a procentriole becomes an anterior centriole during the first mitosis and a posterior centriole during the second mitosis. Since the posterior features are maintained during mitosis, the posterior centriole corresponds to the final state of centriole maturation. This is a rather surprising finding, since it is this posterior centriole that gives rise to the short flagellum in the flagellate, whereas it is the anterior centriole that gives rise to the long flagellum. Thus, it appears that it is the "immature" amoebal centriole that gives rise to the active flagellum in the flagellate.

This amoebal-flagellate transition involves the switch from a locomotory system based on actin to one based on microtubules. Adelman's laboratory has produced some very interesting observations on the changes that occur in the cells during this transformation

(Pagh and Adelman, 1982; Pagh et al., 1985). A distinctive motile cytoplasmic domain--the ridge--appears as a transient phenomenon on the amoebo-flagellate. The ridge contains much of the actin of these cells, and Adelman's recent results with the fluorescent probe rhodamine-phalloidin suggest that as the transformation proceeds the actin is redistributed from the pseudopods of the cells into the ridge and the cortical cytoplasm.

The biochemical and molecular events involved in the biogenesis of a flagellum have been most extensively studied in the alga *Chlamydomonas*. The *Physarum* amoebal-flagellate transition provides an interesting alternative system for the study of this phenomenon. It is already clear that one of the most studied events that occurs during the formation of the *Chlamydomonas* flagellum appears also to be involved in the formation of the *Physarum* flagellum axoneme. This is the posttranslational modification of the α-tubulin. In *Chlamydomonas*, Rosenbaum and his colleagues have clearly shown that this posttranslational event is an acetylation of one lysine residue in α-tubulin (L'Hernault and Rosenbaum, 1985a,b). At this Workshop, Maruta presented evidence for the existence of the enzymes that may be involved in orchestrating this reversible modification. An α-tubulin acetyltransferase (TAT) and a deacetylase (TDA) have been isolated from *Chlamydomonas* flagella and cytoplasm respectively. Most of the flagellar TAT activity is associated with the axonemes and can be released by high salt treatment. The cytoplasm of *Chlamydomonas* also shows significant TAT activity, but its precise quantification has been hampered by an apparent TAT inhibitor. A high TDA activity has also been detected in the cytoplasm, but not in the flagella. Although the biochemistry of the *Physarum* flagellum α-tubulin posttranslational modification is not known, it seems very likely that a similar enzyme system might well be present. Knowledge of such a system in *Physarum* would provide some of the first detailed biochemical differences between the amoeba and the flagellate. Certainly at the present time the presence of the characteristic α3 tubulin is a useful biochemical marker for the flagellate cell type (Green and Dove, 1984).

REFERENCES

Adler, P. N., and Holt, C. E., 1977, Mutations increasing asexual plasmodium formation in *Physarum polycephalum*, *Genetics*, 87:401.

Aldrich, H. C., 1968, The development of flagella in swarm cells of the myxomycete *Physarum flavicomum*, *J. Gen. Microbiol.*, 50:217.

Anderson, R. W., 1979, Complementation of amoebal-plasmodial transition mutants in *Physarum polycephalum*, *Genetics*, 91:409.

Anderson, R. W., Cooke, D. J., and Dee, J., 1976, Apogamic development of plasmodia in the Myxomycete *Physarum polycephalum*: a cinematographic analysis, *Protoplasma*, 89:29.

Anderson, R. W., and Dee, J., 1977, Isolation and analysis of amoebal-plasmodial transition mutants in the Myxomycete Physarum polycephalum, Genet. Res., Camb., 29:21.

Anderson, R. W., and Holt, C. E., 1981, Revertants of selfing (gad) mutants in Physarum polycephalum, Dev. Genet., 2:253.

Anderson, R. W., and Youngman, P. J., 1985, Complementation of npf mutations in diploid amoebae of Physarum polycephalum: the basis for a general method of complementation analysis at the amoebal stage, Genet. Res., Camb., 45:21.

Birkett, C. R., Foster, K. E., Johnson, L., and Gull, K., 1985, Use of monoclonal anibodies to analyze the expression of a multi-tubulin family, FEBS Lett., 187:211.

Burland, T. G., Chainey, A. M., Dee, J., and Foxon, J. L., 1981, Analysis of development and growth in a mutant of Physarum polycephalum with defective cytokinesis, Dev. Biol., 85:26.

Burland, T. G., and Dee, J., 1979, Temperature-sensitive mutants of Physarum polycephalum--expression of mutations in amoebae and plasmodia, Genet. Res., Camb., 34:33.

Burland, T. G., Gull, K., Schedl, T., Boston, R. S., and Dove, W. F., 1983, Cell type-dependent expression of tubulins in Physarum, J. Cell Biol., 97:1852.

Dee, J., 1978, A gene unlinked to mating-type affecting crossing between strains of Physarum polycephalum, Genet. Res., Camb., 31:85.

Gorman, J. A., Dove, W. F., and Shaibe, E., 1979, Mutations affecting the initiation of plasmodial development in Physarum polycephalum, Dev. Genet., 1:47.

Green, L., and Dove, W. F., 1984, Tubulin proteins and RNA during the myxamoeba-flagellate transformation of Physarum, Mol. Cell Biol., 4:1706.

Havercroft, J. C., and Gull, K., 1983, Demonstration of different patterns of microtubule organization in Physarum polycephalum myxamoebae and plasmodia using immunofluorescence microscopy, Eur. J. Cell Biol., 32:67.

Holt, C. E., and Hüttermann, A., 1979, Genetic determination of plasmodium formation in Physarum polycephalum (Myxomycetes), Inst. für den Wissenschaftlichen Film, Göttingen, Film B1337.

Honey, N. K., Poulter, R. T. M., and Aston, R. J., 1982, Non-selfing mutants from selfing (Het^-) strains of Physarum polycephalum, Genet. Res., Camb., 39:261.

L'Hernault, S. W., and Rosenbaum, J. L., 1985a, Reversal of the posttranslational modification on Chlamydomonas flagellar α-tubulin occurs during flagellar resorption, J. Cell Biol., 100:457.

L'Hernault, S. W., and Rosenbaum, J. L., 1985b, Chlamydomonas α-tubulin is posttranslationally modified by acetylation on the ϵ-amino group of a lysine, Biochemistry, 24:473.

Nader, W. F., Shipley, G. L., Hüttermann, A., and Holt, C. E., 1984, Analysis of an inducer of the amoebal-plasmodial transition in the Myxomycetes Didymium iridis and Physarum polycephalum, Dev. Biol., 103:504.

Pagh, K., and Adelman, M. R., 1982, Identification of a microfilament-enriched, motile domain in amoeboflagellates of Physarum polycephalum, J. Cell Sci., 54:1.

Pagh, K. I., Vergara, J. A., and Adelman, M. R., 1985, Improved negative staining of microfilament arrangements in detergent-extracted Physarum amoeboflagellates, Exp. Cell Res., 156:287.

Pallotta, D. J., Youngman, P. J., Shinnick, T. M., and Holt, C. E., 1979, Kinetics of mating in Physarum polycephalum, Mycologia, 71:68.

Roobol, A., Wilcox, M., Paul, E. C. A., and Gull, K., 1983, Identification of tubulin isoforms in the plasmodium of Physarum polycephalum by in vitro microtubule assembly, Eur. J. Cell Biol., 33:24.

Schedl, T., Owens, J., Dove, W. F., and Burland, T. G., 1984, Genetics of the tubulin gene families of Physarum, Genetics, 108:143.

Shinnick, T. M., Anderson, R. W., and Holt, C. E., 1983, Map and function of gad mutations in Physarum polycephalum, Genet. Res. Camb., 42:41.

Shinnick, T. M., Pallotta, D. J., Jones-Brown, Y. R., Youngman, P. J., and Holt, C. E., 1978, A gene, imz, affecting the pH sensitivity of zygote formation in Physarum polycephalum, Curr. Microbiol., 1:163.

Shipley, G. L., and Holt, C. E., 1982, Cell fusion competence and its induction in Physarum polycephalum and Didymium iridis, Dev. Biol., 90:110.

Turnock, G., Morris, S. R., and Dee, J., 1981, A comparison of the amoebal and plasmodial phases of the slime mould, Physarum polycephalum, Eur. J. Biochem., 115:533.

Wheals, A. E., 1973, Developmental mutants in a homothallic strain of Physarum polycephalum, Genet. Res., Camb., 21:79.

Wright, M., Mir, L., and Moisand, A., 1980, The structure of the proflagellar apparatus of the amoeba of Physarum polycephalum: relationship to the flagellar apparatus, Protoplasma, 103:69.

Wright, M., Moisand, A., and Mir, L., 1979, The structure of the flagellar apparatus of the swarm cells of Physarum polycephalum, Protoplasma, 100:231.

Youngman, P. J., 1979, Studies concerning the regulation of plasmodium formation in Physarum polycephalum and other Myxomycetes, Ph.D. Thesis, Massachusetts Institute of Technology.

Youngman, P. J., Adler, P. N., Shinnick, T. M., and Holt, C. E., 1977, An extracellular inducer of asexual plasmodium formation in Physarum polycephalum, Proc. Natl. Acad. Sci., U.S.A., 74:1120.

Youngman, P. J., Anderson, R. W., and Holt, C. E., 1981, Two multiallelic mating compatibility loci separately regulate zygote formation and zygote differentiation in the Myxomycete Physarum polycephalum, Genetics, 97:513.

Youngman, P. J., Pallotta, D. J., Hosler, B., Struhl, G., and Holt, C. E., 1979, A new mating compatibility locus in *Physarum polycephalum*, *Genetics*, 91:683.

Chapter 7: GENE EXPRESSION DURING PLASMODIAL DIFFERENTIATION

Thomas Schreckenbach and Anne-K. Werenskiold

Max-Planck Institute for Biochemistry
Munich, FRG

with contributions from: D. Staiger, R. G. Allen, F. Bernier, G. Lemieux, C. Nations, and D. Pallotta

INTRODUCTION

Scope

The intent of this chapter is to define the present position of research on plasmodial differentiation in *Physarum* and to discuss the potential of this model organism to contribute to an understanding of the molecular mechanisms underlying development. The literature accumulated in the last 10 years of *Physarum* research has been reviewed recently by several authors (Gorman and Wilkins, 1980; Raub and Aldrich, 1982; Sauer, 1982); therefore, in this chapter, we will present mainly novel, unpublished data. We will focus on results that, in our opinion, provide a basis to investigate biological differentiation at the gene level, using *Physarum* as a model.

Starving plasmodia exhibit two developmental pathways, both of which represent transitions into dormant structures (sporulation and sclerotization). In the past, the research on plasmodial differentiation has been limited to the descriptive analysis of overall RNA and protein synthesis as well as of changes in enzyme activity and metabolite concentration (for review see Huettermann, 1982). The use of recombinant DNA technology now allows the identification of regulated genes as molecular markers of a differentiating cell.

For this reason, the main emphasis of this chapter is the expression of distinct genes during plasmodial development. As in many other systems, most scientists working on Physarum differentiation have become heavily engaged in studies on gene cloning and expression. Work that demonstrates differential mRNA regulation still is of a descriptive nature, but is aimed at the ultimate goal of understanding regulatory mechanisms, i.e., the identification of molecules participating in the transduction of morphogenetic signals and in the control of gene expression.

Biological Aspects

Physarum plasmodia react to prolonged starvation with the development of dormant structures by one of the alternate pathways of either sporulation or sclerotization (Gorman and Wilkins, 1980). During sclerotization the cytoplasm is cleaved to form clusters of multinucleated cysts (spherules). This differentiation process is fully reversible upon refeeding of the spherule clusters with growth medium. The biochemistry of sclerotization is more easily studied with microplasmodial suspensions and, in this case, is termed spherulation. Spherulation can be induced by a variety of stress signals (see below) and is reversibly inhibited by blue light (Schreckenbach et al., 1981). Thus, spherulation of Physarum microplasmodia can be regarded as a system to study gene regulation during stress-induced encystment and to investigate the role of blue light receptors in differential gene expression.

In contrast to spherulation, fructification in macroplasmodia is a terminal differentiation process leading to the formation of haploid spores. Upon germination, spores liberate amoebae, which will form new plasmodia after sexual fusion. Under laboratory conditions, the induction of sporulation appears to be critical. It requires a defined period of starvation ("competence formation") and, at least in certain strains, the presence of niacin in the medium (Daniel and Rusch, 1962). Fructification in Physarum is induced by illumination (1-3 hr white light; Daniel and Rusch, 1962); induction by heat shock in the dark (e.g., 37°C, 2 hr) is also feasible (Schreckenbach, unpublished). Once the plasmodium is induced, it proceeds autonomously through the differentiation process (Fig. 7-1). Fruiting body morphogenesis starts about 10 hr after induction and is terminated by melanization of the sporangiophore heads at about 18 hr. The meiotic divisions occur several hours after this time point (Laane and Haugli, 1976). Because of its high synchrony, sporulation in Physarum is an attractive model system for the study of photomorphogenesis in a non-plant organism. The crucial questions are (1) What is the nature of the photoreceptor and the signal transduction chain? (2) Which genes are activated or inactivated by light stimulation? and (3) Which factors participate in the regulation of differential gene expression?

PHOTOINDUCED SPORULATION

Photoinduction

Numerous photomorphogenetic processes have been described in plants and microorganisms. Most of these are induced by either red or blue light or by both wavelengths (Mohr, 1972; Senger and Briggs, 1981). However, only one of the photoreceptors involved in these responses, namely phytochrome, the red light receptor in higher plants, has been identified unequivocally (Pratt, 1979). So far blue light-absorbing pigments have been characterized only by action spectroscopy, which gives an indirect characterization of the photoreceptors (Schmidt, 1980). Action spectroscopy of *Physarum nudum* plasmodia has led to the conclusion that the photoreceptor(s) involved in the induction of sporulation in this organism is a blue and red light-absorbing pigment; green light was found to be inhibitory (Rakoczy, 1980). We have recently confirmed these results using plasmodia of the white mutant strain of *Physarum polycephalum* (Schreckenbach et al., 1981). A strong response with blue ultraviolet light below 400 nm and with red light above 580 nm was observed as well; green light was completely inactive (Schreckenbach, unpublished). A characteristic feature of the photoresponse was that the reaction did not obey the Bunsen-Roscow law of reciprocity*. This situation complicates the interpretation of spectroscopic data, and the action spectra obtained do not necessarily match the absorption spectrum of the photoreceptor but are qualitative in nature (Schaefer and Fukshansky, 1984). A direct isolation of the photoreceptor pigment from plasmodial extracts is not feasible because there is no functional assay for this pigment and because there are no photoreceptor mutants useful for the analysis of differential pigment composition.

Wormington and Weaver (1976) have claimed that one of the wild-type plasmodial pigments is involved in the photoinduction process. They reported that illumination in vitro of an extract containing small molecules (molecular weight 500) confers on the extract the ability to induce sporulation when injected into a starved, unilluminated slime mold. The activity comigrated on a thin-layer chromatography plate with one of the four major yellow plasmodial pigments. Unfortunately, these studies have not been confirmed. Earlier studies by the same authors (Wormington et al., 1975) showed that even injection of salt solutions could result in unspecific induction. Therefore, the question remains whether the fraction isolated actually contains a component of the in vivo

---

*The Bunsen-Roscow law states that the action of photochemically active light depends only on the dosage $d = i \times t$, regardless of how the single factors i (intensity) and t (time) are chosen (Schmidt, 1980).

signal transduction chain. The postulated identity of the factor with the plasmodial pigment should be checked with a highly purified preparation of this molecule. Observations in our laboratory and in many other groups (unpublished) indicate that the white mutant strain (Putzer et al., 1984) exhibits an even stronger response to light compared with wild-type plasmodia. For this reason, the involvement of the yellow pigments in the photoreception and signal transduction processes remain doubtful.

The most plausible explanation of all available data is that the photoreceptor is not identical with the yellow plasmodial pigment but is a highly sensitive ultraviolet/red light-absorbing chromophore present at a very low concentration. Possibly, illumination results in the production of a primary metabolic signal, which must accumulate above a certain level in order to trigger fructification. The photochemical process seems to be saturated at about daylight intensity because the induction time cannot be shortened below a certain limit by increasing the light intensity above this level. Photoinduction of sporulation in *Physarum* resembles to some extent the so-called "high irradiance response" in higher plants (Mancinelli, 1980), which is believed to be mediated by phytochrome. However, so far there are no indications for the existence of this pigment in *Physarum* or any other nonphotosynthetic organism.

Signal Transduction

The morphological differentiation of plasmodial veins into fruiting bodies starts about 10 hr after the beginning of illumination with saturating light intensity (Fig. 7-1). Shortly before the onset of morphogenesis, the induction of sporulation-specific mRNAs is observed (e.g., see below--SMP11 mRNA, Fig. 7-5). Thus, under our standard laboratory conditions, a lag phase of about 8 hr exists between induction and presumptive changes in transcriptional activity of major differentiation-specific genes. To our knowledge, there is no report published in the literature dealing with the characterization of components that participate in the photoinduction process or in secondary events of signal transduction. Results on ion fluxes obtained with microplasmodial suspensions (Daniel and Eustache, 1972) are not necessarily related to the sporulation process, because *Physarum* plasmodia exhibit various photobiologically distinct light responses (for a short review see Schreckenbach, 1984).

Because of their giant size, *Physarum* plasmodia are suitable objects for studies requiring microinjection techniques (see Chapters 20 and 21). Therefore, the most promising approach toward an identification of intracellular signal substances would be the continuation of microinjection experiments initiated by Wormington and Weaver (1976, see above). Such studies should be carried out

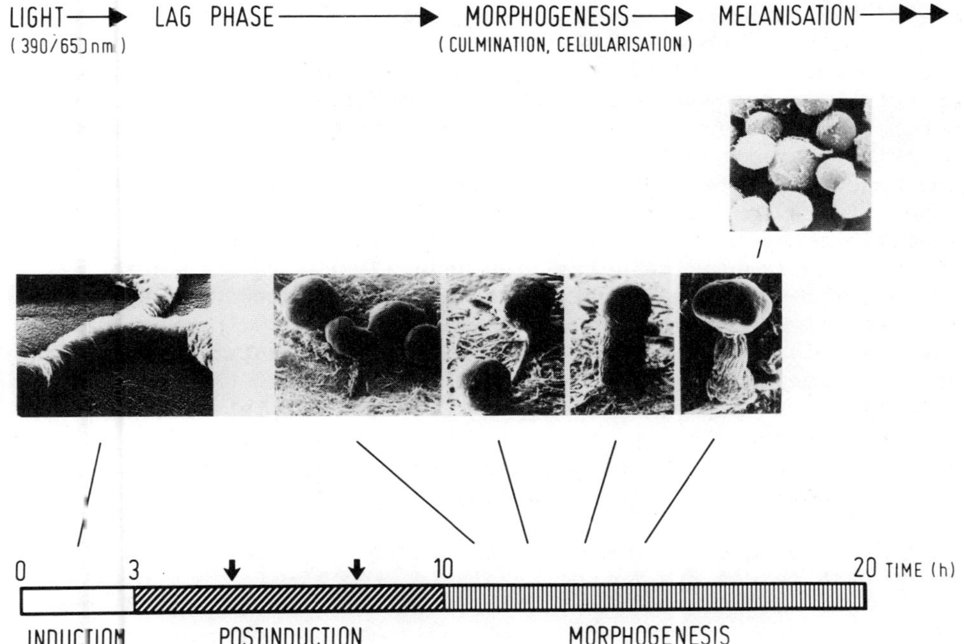

Fig. 7-1. Photoinduced fructification in Physarum. The scanning electron micrographs display plasmodial veins (left), distinct stages of sporangiophore morphogenesis (right), and 15-hr prespore particles (above). The arrow marks the "point of no return" (no inhibition of differentiation by growth medium). For further explanations, see text.

with extracts from induced plasmodia obtained at various time points of development. Microinjection may be performed under green safety light, thereby minimizing photoinduction artifacts. With this type of bioassay, preliminary characterization of an active fraction (e.g., resistance to degrading enzymes) may be carried out, and modern separation techniques like HPLC may be used to fractionate the extracted material.

Developmentally Regulated Genes

Gene products regulated during photoinduced sporulation in Physarum were identified by the following series of experiments. Proteins of sporulating plasmodia were pulse-labeled in vivo with [^{35}S]-methionine, and mRNA preparations from distinct stages of the differentiation process were translated in vitro. The radioactive proteins obtained were separated by two-dimensional gel electrophoresis. Analysis of the autoradiograms revealed that detectable changes in gene expression occurred only at the late, morphogenetic stage of sporangiophore development initiated at 10 hr after photoinduction (Fig. 7-1). Changes in the synthesis of the major regulated protein species correlated with changes in the level of the corresponding translatable mRNAs. These sporulation-specific mRNAs could be classified into various sequentially and transiently induced mRNA families (Fig. 7-2). The induction of sporulation-specific translatable mRNAs was accompanied by an increase of RNA polymerase II activity in isolated nuclei (Werenskiold, 1985); this possibly reflects stimulated transcription of these sequences. The studies further revealed that the mRNA of a prominent 42 K protein disappeared almost completely during fructification. This protein could be purified by affinity chromatography on DNase I Sepharose and thus very likely is identical with plasmodial actin (Werenskiold, 1985; Pahlic, 1985).

From the great number of regulated gene products, four proteins were chosen for further analysis (Fig. 7-3): actin, as the major inhibited plasmodial protein; two tubulins [α- and β-type (Schedl et al., 1984)]; and a sporangiophore morphogenetic protein (SMP11). It is assumed that the two tubulins are components of the mitotic spindle apparatus during a post-illumination mitosis reported by other authors (Guttes et al., 1961). The function of SMP11 is unknown; it represents the most abundant protein induced during fructification and therefore very likely is a prominent structural component of the sporangiophore.

The regulation of the synthesis of these proteins during sporulation correlated with the regulation of the corresponding mRNA sequences as shown by the Northern blot experiments (Figs. 7-4 and 7-5). The 1.6 kb actin mRNA is lost exactly with the onset of sporangiophore morphogenesis (Fig. 7-4, time dependence not shown). The α-tubulin mRNA shows two peaks, at 11 and 17 hr after induction (Fig. 7-5A). The latter peak might reflect synthesis of tubulin mRNA that is deposited in the dormant spore. The SMP11 mRNA appears to represent the earliest response to the light stimulus. It shows a marked increase in concentration at 9 hr after induction and increases further during the morphogenetic period of development (Fig. 7-5B).

None of the Northern blot experiments in Figs. 7-4 and 7-5 showed evidence that a decrease in concentration of the mature mRNA sequences was coupled to the accumulation or disappearance of a high molecular weight RNA precursor. Therefore, it is likely that the expression of the actin, α-tubulin, and SMP11 mRNAs is regulated at the level of mRNA transcription or stability. The analysis of transcriptional activity in vivo is not feasible in

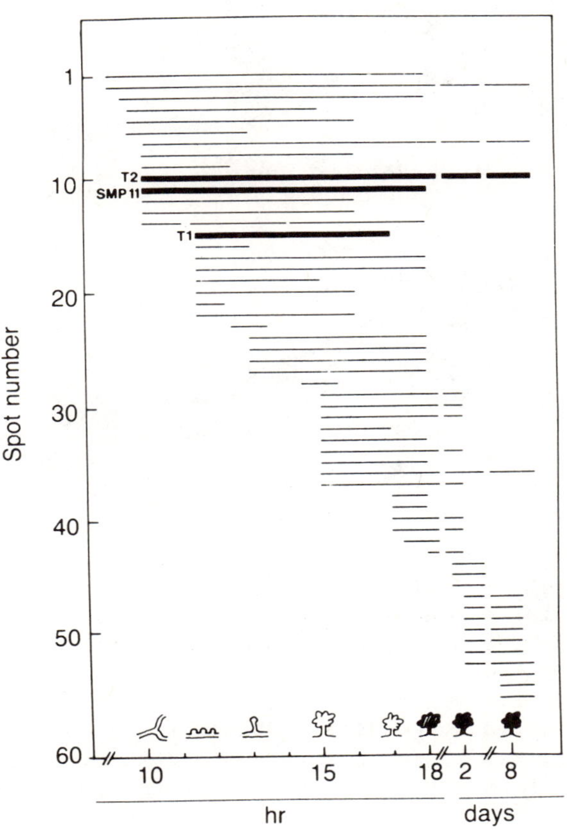

Fig. 7-2. Temporal regulation of translatable sporulation-specific mRNA preparations from various stages of development were translated in vitro, and the proteins were analyzed by two-dimensional gel electrophoresis. The fluorograms were evaluated with respect to the presence (lines) and absence of individual translation products. Times relative to start of the 3-hr illumination (induction) period (Fig. 7-1), and diagramatic representation of developmental stages are given at the bottom, T_2, T_1: α-, β-tubulins (after Putzer et al., 1984).

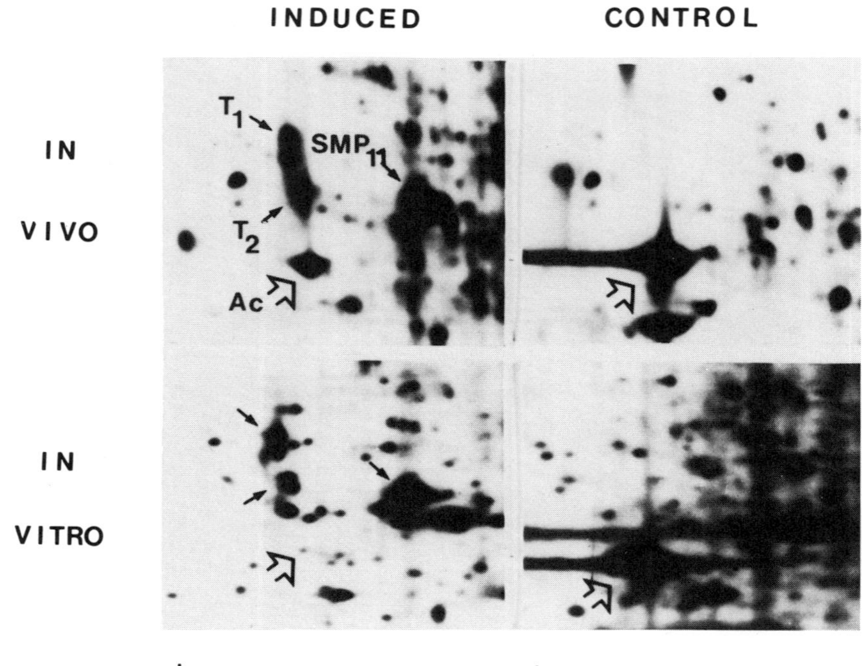

Fig. 7-3. Regulation of actin, tubulin, and SMP11 mRNAs. Plasmodial proteins were labeled in vivo with [^{35}S] methionine during the morphogenetic period of sporulation and in vitro with mRNA isolated at 12 hr of development. Radioactive proteins were separated by two-dimensional gel electrophoresis. Only gel sections around the actin spot (42K) are shown; α- and β-tubulins (T_2, T_1--50 K, 57 K) and sporangiophore protein, SMP11, are marked.

starving Physarum plasmodia because the incorporation of radioactive precursors into RNA is extremely low owing to low uptake and high endogenous precursor pools. Therefore, transcription in isolated nuclei may provide a tool to check for transcriptional control of these genes in sporulating plasmodia. This approach has been used in many other systems (e.g., Berry-Lowe and Meagher, 1985; Piechaczyk et al., 1985) and may be useful for Physarum as well, because actively transcribing nuclei are easily isolated from this organism (see Chapter 19).

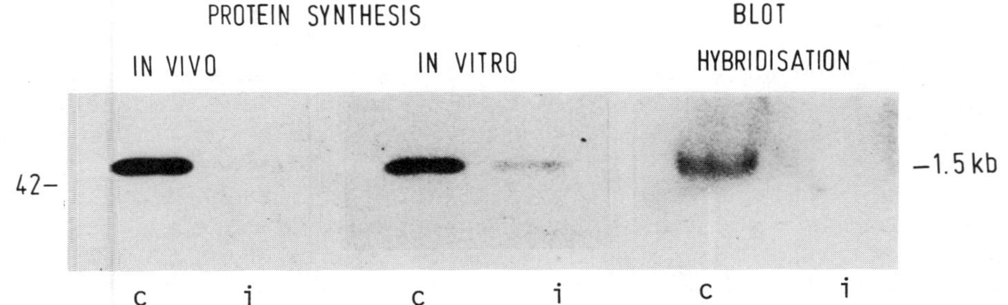

Fig. 7-4. Inhibition of actin synthesis in sporulating plasmodia. Macroplasmodia were labeled with [^{35}S] methionine in vivo during the period of sporangiophore morphogenesis (1, 10-15 hr, see also Fig. 7-1). In vitro translation was carried out with RNA from induced plasmodia (12 hr) and control cells. Actin synthesized in vivo and in vitro was isolated by DNase I affinity chromatography (Zechel, 1980) and subjected to electrophoresis on a 5-20% SDS-polyacrylamide gel. Northern blot hybridization was performed with RNA from induced (15 hr) plasmodia and from control cells, with the entire yeast actin gene (Ng and Abelson, 1980) used as a probe. The bands displayed in the figure were the only bands visible on the three autoradiograms.

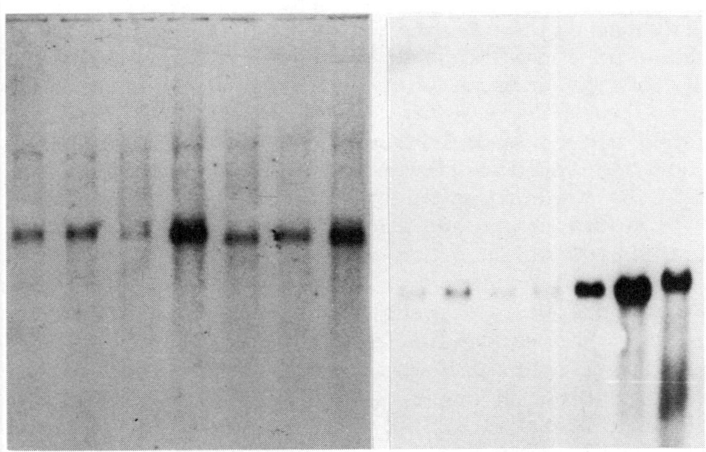

Fig. 7-5. Induction of α-tubulin and SMP11 transcripts. Total RNA was isolated from sporulating plasmodia at the times indicated, size-separated on a denaturing agarose/formaldehyde gel, and transferred to Gene Screen membrane (NEN). Filters were hybridized to radioactively labeled tubulin plasmid Ppcα125 (860 bp SacI/Bgl II insert; Schedl et al., 1984) and radioactively labeled SMP11-SP6 in vitro transcripts (Staiger, 1985).

Perspective

Two important experimental tools are not available so far for *Physarum* research: a transformation system and developmental mutants blocked at distinct stages of differentiation. This severely restricts the analysis of photoreception and signal transduction processes in this organism. The isolation of mutants defective in sporulation is complicated: mutagenesis must be carried out in the amoebal stage with a selfing strain (Haugli et al., 1980; Dee, 1982); each plasmodium derived from an amoebal clone must be maintained on an individual petri dish. Since there is no selection procedure for sporulation mutants, multiple cultures of each plasmodial clone must be screened for photoinduced fructification. This approach appears to be an extremely time-consuming task, which has not been attacked by any of the groups involved in *Physarum* genetics. In contrast, the search for mutants defective in the amoeba-plasmodium transition has been most successful. Here, the genes controlling amoebal fusion have been characterized by genetic analysis (for review see Haugli et al., 1980; Dee, 1982; also see Chapter 6).

In other systems, like *Dictyostelium discoideum* and *Drosophila*, mutant analysis by modern molecular biological techniques has provided important insight into mechanisms of eukaryotic differentiation (Lodish et al., 1982; Gehring, 1985). Experiments aiming toward the construction of a transformation system in *Physarum* have been initiated (see Chapter 25). If these efforts succeed, they will open up a new era in *Physarum* research, allowing the exploitation of the great advantages offered by this model organism.

As stated above, sporulation in *Physarum* represents a model to study photoregulated development in a nonplant system. To our knowledge, it is the only system in which some of the light-regulated gene products have been identified (actin, tubulins) and cloning of their cDNAs or genomic sequences has been initiated. The studies reported here suggest control of gene expression at the posttranscriptional or transcriptional level. The *Physarum* system appears to be especially suitable for further investigation of transcriptional control mechanisms. An approach to identify factors participating in the regulation of specific genes may be transcription in isolated nuclei or chromatin (see Chapter 19). Studies should include fractionation and reconstitution of active chromatin, the search for photoinduced proteins binding to distinct fragments of cloned genes, and mapping of protein-DNA complexes in native chromatin preparations. By a similar approach, factors involved in heat-shock-dependent gene regulation have been characterized (Wu, 1985).

From the genes regulated during Physarum sporulation, only those coding for prominent proteins were chosen for further analysis. In every system studied, these genes are the first to be identified because the corresponding mRNAs are abundant. With the exception of actin and tubulins, none of the regulated gene products has been identified so far, and nothing is known about their function in the differentiating plasmodium. It appears worthwhile to check for sporulation specific changes in gene expression in ALC (amoebae-less life cycle) variants (Adler et al., 1975) because this strain is defective in the re-programming events occurring during spore formation. Further, the developmental regulation of major structural genes might depend on the synthesis of low-abundance, regulatory transcripts, which cannot be identified by the conventional strategy of cDNA cloning and differential screening. For these reasons, the construction of a transformation system is highly desirable for future research on differentiation in Physarum. It might help to answer many of the remaining questions by using integrative gene disruption (Rothstein, 1983; also see Chapters 2 and 17) or in vivo expression of anti-sense mRNA (Kim and Wold, 1985) as experimental tools.

SPHERULATION

Stress Induction

The transition of microplasmodia into spherule clusters is most easily induced by transfer of a shaken culture to a starvation medium. The conversion then proceeds in a roughly synchronous manner and is completed within 30-50 hr, the time depending on the strain used. Besides starvation, mannitol and various stress signals have been found to induce the microplasmodium-spherule transition, e.g., osmotic shock, heavy metal ions, cold temperature, or low pH (Chet and Rusch, 1969). Allen et al. (1985) and Nations et al. (1985) surmise that either oxygen-free radicals or their anti-oxidant defenses might play a causative role in spherulation. Their assumption is based on inhibitor experiments and on the observation that the transition is accompanied by a 46-fold increase in superoxide dismutase activity, an increase in H_2O_2 (i.e., the product of its catalysis), and a significant decline in glutathione concentration, which might be caused by OH-radical-quenching reactions (Fig. 7-6). These authors also observed an increase in CN^--insensitive respiration in plasmodial homogenates (Fig. 7-6) and argue that this could be accounted for by an increase in the activity of metabolic pathways that generate superoxide O_2^- during spherulation. The latter assumption certainly requires a more detailed analysis of respiratory functions in isolated mitochondria.

From the data presented by Nations et al. (1985), a causative role of superoxide O_2-radicals in the spherulation process cannot

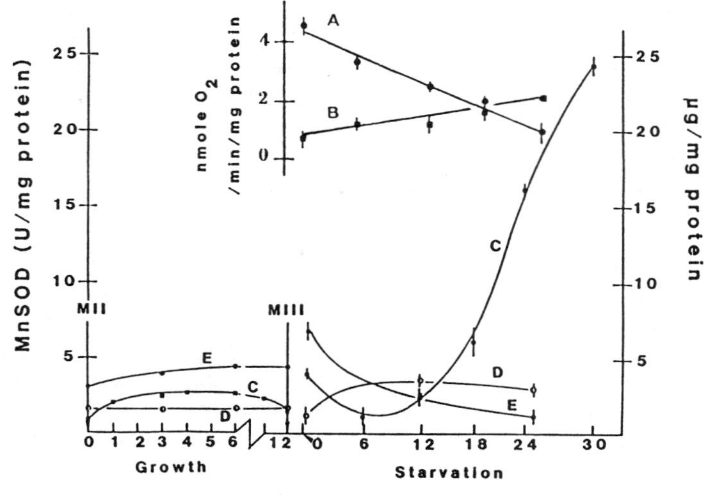

Fig. 7-6. Oxygen metabolism during growth and starvation in macroplasmodia. Oxygen consumption was measured in plasmodial homogenates in pH 3.8 salts medium (A) and in the presence of 1.3 mM KCN in pH 7.1 buffered glucose (B). Superoxide dismutase activity (C), hydrogen peroxide (D), and glutathione concentrations (E) were measured during the growth of plasmodia from the second (MII) through the third (MIII) postfusion mitosis and during starvation-induced differentiation. Oxygen consumption is expressed as nmoles O_2/min/mg protein. Hydrogen peroxide and glutathione concentrations are recorded as µg/mg. (After Nations et al., 1985)

be deduced; the increase in superoxide dismutase activity appears to correlate with spherulation in microplasmodia, but is also observed in macroplasmodia on the second day of starvation, i.e., several days before sclerotization takes place (Fig. 7-6). These data indicate that O_2-radical formation might occur during the period of competence formation (see "Biological Aspects"). Despite the lack of direct evidence for a differentiation-stimulating effect of O_2-radicals, the spherulation system appears to be most suitable to investigate further the pathways leading to the generation and decay of this harmful metabolic intermediate.

Deinduction by Blue Light

When starving microplasmodia are illuminated with blue light, they do not differentiate into spherules but instead form pseudopodia and finally die (Schreckenbach et al., 1981). This inhibition requires a light intensity roughly corresponding to intense sunlight and is fully reversible. Action spectroscopy has indicated that the photoreceptor involved is a blue light receptor similar to the one mediating negative phototaxis in starving macroplasmodia (Haeder and Schreckenbach, 1984). Deinduction of spherulation by blue light is accompanied by an inhibition of the synthesis of all major spherule-specific mRNA species (see below). Therefore, it might be useful not only to investigate blue-light-regulated gene expression, but also to understand the influence of stress signals on differential protein synthesis.

Differential Gene Expression

Changes in protein synthesis and translatable mRNAs. Studies on differential gene expression during spherulation of *Physarum* microplasmodia were initiated by Putzer et al. (1983). These authors separated radioactive proteins obtained by in vivo labeling and by in vitro protein synthesis with mRNA from different time points during spherulation. They deduced from their two-dimensional gel studies that significant changes in protein synthesis and mRNA composition occurred with the onset of formation of morphologically distinct spherule clusters. Some of the proteins induced during the transition comigrated on two-dimensional gels with cell wall proteins of the mature spherule (Fig. 7-7). The concentration of translatable actin mRNA was found to be significantly reduced during spherulation (Fig. 7-7; also see Bernier et al., 1985a). When spherulation was inhibited by blue light, the changes in gene expression were suppressed; this suppression indicated that the new protein and mRNA species were spherulation- and not starvation-specific. Earlier studies on the RNA polymerase II inhibitor α-amanitin had suggested that the spherule-specific changes in gene expression were regulated at the level of transcription (Schreckenbach and Verfuerth, 1982).

In a recent study, Bernier et al. (1985a) investigated in more detail the expression of the spherule-specific proteins and mRNAs. Using two-dimensional gel analysis and a spherule-viability assay, they classified the induced mRNAs according to their occurrence in spherulating cells and in mature spherules. The authors surmise that the former set of proteins and mRNAs might be involved in the formation of viable spherules, whereas the latter might be required for efficient germination of new plasmodia. The major spherulation-specific gene products were found in two size classes, approximately 25-38 K and approximately 14 K. In some cases (14 K protein family), in vitro translation products appeared to be identical with the protein synthesized in vivo. Interestingly, the actin mRNA, which was reduced in spherulating plasmodia, was a prominent component again in poly A^{+} RNA from mature spherules. A similar observation was made regarding actin regulation during photoinduced sporulation (Putzer et al., 1984; see above). It

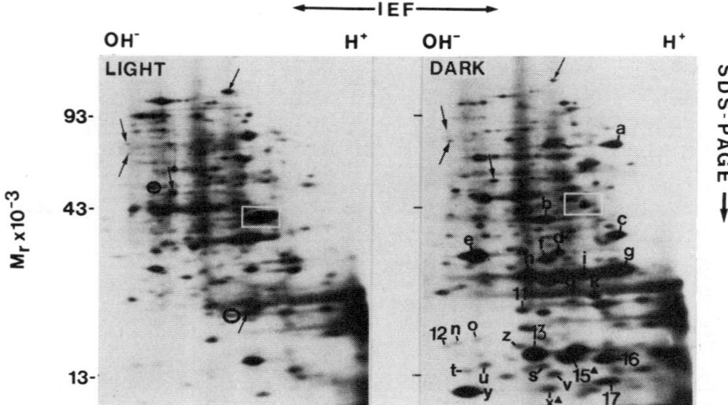

Fig. 7-7. Blue light inhibits the expression of spherule-specific mRNAs. Total RNA was extracted from illuminated starving microplasmodia and from dark-incubated control plasmodia after 25 hr of incubation in starvation medium (0% spherulation in the sample, 80% in the dark control). The RNA was translated in vitro, and the products were analyzed by two-dimensional electrophoresis. Spherule-specific proteins are marked by letters or numbers, the numbers indicating the identity with the corresponding in vivo-labeled material. Cell wall proteins are marked by a triangle; starvation proteins, present in both samples, by an arrow; and light-induced proteins by a circle. Plasmodial actin is enclosed by a rectangle (Putzer et al., 1983).

appears that both spores and spherules accumulate actin mRNA for the germinating amoebae and plasmodia respectively. Using the spherule viability test (Bernier et al., 1985a) may show that the formation of morphologically distinct spherule clusters occurs long before viable spherules arise. This means that the protein and mRNA species described in the study of Putzer et al. (1984; Fig. 7-7) belong to the class of early, spherulation-specific proteins.

Molecular cloning of spherule-specific mRNAs. In order to study at the molecular level the induction of spherulation proteins, Bernier et al. (1985b) have synthesized cDNA from poly A+ RNA isolated from spherulating microplasmodia. The cDNA was cloned into the vector pBR322 and transformed into E. coli; from 740 clones analyzed, 35 were selected which gave a strong hybridization signal with [^{32}P]-labeled cDNA from spherulating plasmodia. From this set of clones, those representing the most abundant spherule-specific mRNA species were isolated (pLAV2-1 to pLAV2-4), each representing 1.5-4.5% of all clones). In addition, pLAV5-pLAV8, which code for less abundant mRNA species, were selected for further analysis. Northern blot hybridization of these cDNAs to RNA from spherulating plasmodia revealed: (1) The mRNAs detected by these probes accumulated in a time-dependent manner during spherulation; maximum concentration was observed at 18-36 hr of the transition. (2) With the exception of pLAV2-3, no signal was detectable in RNA from growing plasmodia; this strongly suggests that spherulation is accompanied by the transcriptional activation of these genes. (3) All eight mRNA species investigated were induced also by sublethal concentrations of Fe SO_4, indicating that they were not starvation-specific transcripts.

Hybridization of pLAV2-1 to pLAV2-4 to Southern blots of genomic DNA digested with three different restriction enzymes suggested that there was only one copy of each gene present in the Physarum genome. It seems possible that the processes of sporulation and spherulation share common gene functions. This hypothesis may be checked by hybridizations of cloned cDNAs (e.g., SMP11, pLAV2-1 to pLAV2-4) to RNA from both pathways.

Perspective

Analysis of biological stress responses is carried out in many organisms with various kinds of environmental stress signals. The most advanced of these systems is the heat-shock response investigated mainly in the fruit fly, Drosophila (Craig, 1985). Heat-shock-activated genes have been isolated, and factors involved in the control of their expression are being characterized (Wu, 1985).

Spherulation in Physarum is induced not only by one specific stress signal, but by all kinds of unfavorable environmental conditions. The response of the plasmodium is the transition into a dormant structure. Therefore, the spherulation system in principle should be useful to answer the following questions: How is stress-induced encystment regulated? What is the primary metabolic response to stress? What is the mechanism of gene activation? What is the function of stress-induced gene products? The Physarum system may specifically help to discriminate between gene products essential for encystment and proteins required for the germination process.

At the present status of *Physarum* research, these questions cannot be answered. This lagging state of research on *Physarum* compared with other systems results from the following: (1) up to very recently, no DNA cloning work has been carried out (see Chapters 22-24); (2) no developmental mutants have been isolated (also see above, *Sporulation*-Perspective); and (3) transformation of *Physarum* amoebae has not been achieved yet (see Chapter 25). For this reason, most of the work on spherulation published so far is of a descriptive nature, presenting correlations of differentiation with metabolic markers. Once mutant selection in plasmodia and transformation of amoebae have been worked out, *Physarum* might facilitate answering the functional questions posed above, because spherulating plasmodia are easily manipulated under laboratory conditions.

The strategy chosen by some of the investigators (Bernier et al., 1985a,b) includes cloning of abundant spherulation-specific mRNAs and, at a later stage, the cloning of the corresponding genomic sequences. In the future, this strategy should allow the analysis of mechanisms of gene activation during encystment. The isolation of genomic DNA fragments may result in the identification of *cis*-acting regulatory elements common to starvation- or iron-induced genes and provides a basis to investigate changes in DNA modification or chromatin organization. Thus, as in the case of photoinduced sporulation, the signal chain may be analyzed starting at the gene level. DNA cloning and expression studies should also allow the localization of regulated gene products in the spherulating plasmodium and thus may give a clue to an understanding of their function.

SUMMARY

Differentiation of *Physarum* plasmodia is controlled by environmental conditions: ultraviolet blue or red light induces the highly synchronous, irreversible differentiation into fruiting bodies when applied to competent plasmodia at a sufficient dose. Short light pulses (min range) are not sufficient for induction even at high doses. In the dark, macroplasmodia convert into sclerotia consisting of spherule clusters. Spherulation is investigated mainly in suspensions of microplasmodia. It is induced by starvation alone or by various kinds of environmental stress. Unlike sporulation, spherulation in *Physarum* plasmodia is a reversible encystment and not a terminal differentiation process. The plasmodium-spherule transition is inhibited by continuous illumination with blue light.

In both systems, the early metabolic responses to the inducing signals are unknown. Spherulation in microplasmodia is accompanied by a drastic increase in superoxide dismutase activity pointing toward the involvement of oxidative stress defenses during this transition. The presumptive generation of oxygen-free radicals might be related to the increase of cyanide-insensitive respiration observed in starving plasmodia.

Changes in gene expression have been analyzed by the separation of radioactive proteins obtained by in vivo and in vitro labeling. Both systems are characterized by the induction of differentiation-specific, and loss of plasmodial, protein and mRNA species. A prominent plasmodial protein inhibited during both developmental pathways was identified as actin. In both systems, the inhibition of actin mRNA expression was found to be of a transient nature. Two of the proteins induced during sporulation were identified as tubulins (α- and β-type) and may be components of the mitotic spindle during a mitotis occurring at about 12 hr after induction. So far, none of the spherulation-specific gene products have been identified unambiguously; according to two-dimensional gel analysis, some of these gene products are likely to be cell wall proteins. Expression of mRNA sequences during spore and spherule formation was investigated in Northern blot and hybridization experiments with cloned homologous cDNAs and heterologous probes (actin, tubulins). In all cases reported, regulation of the translatable mRNAs correlated with the regulation of mRNA sequences; this indicates control at the level of transcription or mRNA stability.

Studies on sporulation and spherulation in _Physarum_ plasmodia are severely restricted because of the lack of developmental mutants and of a transformation system. They are presently focused on cloning and sequencing of regulated genes and their cDNAs. This approach has become feasible as a result of recent advances in the cloning of DNA fragments unstable in conventional hosts (see Chapter 22). By use of transcription in isolated nuclei, active and inactive chromatin may be isolated and used for a detailed analysis of differential gene organization and chromatin composition (see Chapter 19). With this approach, analysis of the intracellular signal chain is initiated at its end at the gene level. Once an efficient transformation system is established (see Chapter 25) studies can be extended and may lead to the identification of as yet unknown gene products involved in the regulation of development and to an understanding of their function.

ACKNOWLEDGMENT

Work of the authors was supported by the Deutsche Forschungsgemeinschaft.

REFERENCES

Adler, P. N., Davidoff, L. S., and Holt, C. E., 1975, Life cycle variants of *Physarum polycephalum* that lack the amoeba stage, *Science*, 190:65.

Allen, R. G., Newton, R. K., Farmer, K. J., and Nations, C., 1985, Effects of the free radical generator paraquat on differentiation, superoxide dismutase, glutathione and inorganic peroxides in microplasmodia of *Physarum polycephalum*, *Cell Tissue Kinet.*, in press.

Bernier, F., Pallotta, D., and Lemieux, G., 1985b, Molecular cloning of the abundant spherule specific mRNAs of *Physarum polycephalum*, in preparation.

Bernier, F., Seligy, U. L., Pallotta, D., and Lemieux, G., 1985a. Changes in gene expression during spherulation in *Physarum polycephalum*, *Can. J. Biochem. Cell Biol.*, in press.

Berry-Lowe, S. L., and Meagher, R. B., 1985, Transcriptional regulation of a gene encoding the small subunit of ribulose-1,5-biphosphate carboxylase in soybean tissue is linked to the phytochrome response, *Mol. Cell. Biol.*, 5:1910.

Chet, I., and Rusch, H. P., 1969, Induction of spherule formation in *Physarum polycephalum* by polyols, *J. Bacteriol.*, 100:673.

Craig, E. A., 1985, The heat shock response, *CRC Crit. Rev. Biochem.*, 18:239.

Daniel, J. W., and Eustache, J., 1972, The effect of light on the permeability of a myxomycete, *FEBS Lett.*, 26:327.

Daniel, J. W., and Rusch, H. P., 1962, Method for inducing sporulation of pure cultures of the myxomycete *Physarum polycephalum*, *J. Bacteriol.*, 83:234.

Dee, J., 1982, Genetics of *Physarum polycephalum*, *in*: "Cell Biology of *Physarum* and *Didymium*," Vol. 1, H. C. Aldrich and J. W. Daniel, eds., Academic Press, New York.

Gehring, W. J., 1985, The homeo box: a key to understanding of development? *Cell*, 40:3.

Gorman, J. A., and Wilkins, A. S., 1980, Developmental phases in the life cycle of *Physarum* and related myxomycetes, *in*: "Growth and Differentiation in *Physarum polycephalum*," W. F. Dove and H. P. Rusch, eds., Princeton University Press, Princeton.

Guttes, E., Guttes, S., and Rusch, H. P., 1961, Morphological observations on growth and differentiation of *Physarum polycephalum* grown in pure culture, *Dev. Biol.*, 3:588.

Haeder, D.-P., and Schreckenbach, T., 1984, Phototactic orientation in plasmodia of the acellular slime mold, *Physarum polycephalum*, *Plant Cell Physiol.*, 25:55.

Haugli, F. B., Cooke, D., and Sudbery, P., 1980, The genetic approach in the analysis of the biology of *Physarum polycephalum*, *in*: "Growth and Differentiation in *Physarum polycephalum*," W. F. Dove and H. P. Rusch, eds., Princeton University Press, Princeton.

Huettermann, A., 1982, Enzyme and protein synthesis during differentiation of Physarum polycephalum, in: "Cell Biology of Physarum and Didymium", Vol II, H. Aldrich and J. W. Daniel, eds., Academic Press, New York.

Kim, S. K., and Wold, B. J., 1985, Stable reduction of thymidine kinase activity in cells expressing high levels of anti-sense RNA, Cell, 42:129.

Laane, M. M., and Haugli, F. B., 1976, Nuclear behaviour during meiosis in the myxomycete Physarum polycephalum, Norw. J. Bot., 23:7.

Lodish, H. F., Blumberg, D. D., Chisholm, R., Chung, S., Coloma, A., Landfear, S., Barklis, E., Lefebvre, P., Zuker, C., and Mangiarotti, G., 1982, Control of gene expression, in: "The Development of Dictyostelium discoideum," W. F. Loomis, ed., Academic Press, New York.

Mancinelli, A. L., 1980, The photoreceptors of the high irradiance responses of plant photomorphogenesis, Photochem. Photobiol., 32:353.

Mohr, H., 1972, "Lectures on Photomorphogenesis," Springer, Berlin.

Nations, C., Allan, R. G., Farmer, K. J., Toy, P. L., and Sohal, R. S., 1985, Superoxide dismutase activity during the plasmodial life cycle of Physarum polycephalum, Experientia, in press.

Ng, R., and Abelson, J., 1980, Isolation and sequence of the gene for actin in Saccharomyces cerevisiae, Proc. Natl. Acad. Sci., U.S.A., 77:3912.

Pahlic, M., 1985, Multiple forms of actin in Physarum polycephalum, Eur. J. Cell. Biol., 36:169.

Piechaczyk, M., Yang, J.-Q., Blanchard, J.-M., Jeanteur, P., and Marcu, K. B., 1985, Posttranscriptional mechanisms are responsible for accumulation of truncated c-myc RNAs in murine plasma cell tumors, Cell, 42:589.

Pratt, H. L., 1979, Phytochrome: function and properties, Photochem. Photobiol. Rev., 4:59.

Putzer, H., Verfuerth, C., Claviez, M., and Schreckenbach, T., 1984, Photomorphogenesis in Physarum: induction of tubulins and sphorulation-specific proteins and of their mRNAs, Proc. Natl. Acad. Sci., U.S.A., 81:7117.

Putzer, H., Werenskiold, K., Verfuerth, C., and Schreckenbach, T., 1983, Blue light inhibits slime mold differentiation at the mRNA level, EMBO J., 2:261.

Rakoczy, L., 1980, Effect of blue light on metabolic processes, development and movement in true slime molds, in: "The Blue Light Syndrome," H. Senger, ed., Springer, Berlin.

Raub, T. J., and Aldrich, H. C., 1982, Sporangia, spherules and microcysts, in: "Cell Biology of Physarum and Didymium," Vol II, H. Aldrich and J. W. Daniel, eds., Academic Press, New York.

Rothstein, R. J., 1983, One-step gene disruption in yeast, Meth. Enzymol., 101:202.

Sauer, H., 1982, "Developmental Biology of Physarum," Cambridge University Press, Cambridge.

Schaefer, E., and Fukshansky, L., 1984, Action spectroscopy, in: "Techniques in Photomorphogenesis", H. Smith and M. G. Holmes, eds., Academic Press, New York.

Schedl, T., Burland, T. G., Gull, K., and Dove, W. F., 1984, Cell cycle regulation of tubulin RNA level, tubulin protein synthesis, and assembly of microtubules in Physarum, J. Cell Biol., 99:155.

Schmidt, W., 1980, Physiological blue light reception, in: "Structure and Bonding," J. D. Dunitz et al., eds., Springer, Berlin.

Schreckenbach, T., 1984, Phototaxis and photomorphogenesis in Physarum polycephalum plasmodia, in: "Blue Light Effects in Biological Systems," H. Senger, ed., Springer, Berlin.

Schreckenbach, T., and Verfuerth, C., 1982, Blue light influences gene expression and motility in starving microplasmodia of Physarum polycephalum, Eur. J. Cell Biol., 28:12.

Schreckenbach, T., Walckhoff, B., and Verfuerth, C., 1981, Blue light receptor in a white mutant of Physarum polycephalum mediates inhibition of spherulation and regulation of glucose metabolism, Proc. Natl. Acad. Sci., U.S.A., 78:1009.

Senger, H., and Briggs, W. R., 1981, The blue light receptor(s): primary reactions and subsequent metabolic changes, Photochem. Photobiol. Rev., 6:1.

Staiger, D., 1985, Lichtabhängige Expression eines Sporulations-spezifischen Proteins in Physarum polycephalum (Light-dependent expression of a sporulation-specific protein in Physarum polycephalum), Diploma Thesis, University of Tuebingen.

Werenskiold, A. K., 1985, Lichtabhängige Aktin- und Tubulinexpression in Physarum polycephalum (Light-dependent expression of actin and tubulins in Physarum polycephalum), Ph.D. Thesis, University of Munich.

Wormington, W. M., Cho, C. G., and Weaver, R. F., 1975, Sporulation-inducing factor in slime mould Physarum polycephalum, Proc. Natl. Acad. Sci., U.S.A., 256:413.

Wormington, W. M., and Weaver, R. F., 1976, Photoreceptor pigment that induces differentiation in the slime mold Physarum polycephalum, Proc. Natl. Acad. Sci., U.S.A., 73:3896.

Wu, C., 1985, An exonuclease protection assay reveals heat-shock element and TATA box DNA-binding proteins in crude nuclear extracts, Nature, 317:84.

Zechel, K., 1980, Isolation of polymerization-competent cytoplasmic actin by affinity chromatography on immobilized DNase I using formamide as eluant, Eur. J. Biochem., 110:343.

Chapter 8: BIOLOGICAL ASPECTS OF MOTILITY

K.-E. Wohlfarth-Bottermann

Institute for Cytology and Micromorphology
University of Bonn
Bonn, FRG

INTRODUCTION

Locomotion of plasmodia of up to several m^2 in size and with a thickness of several mm is by migration of up to 1 cm/hr, even under conditions of starvation for many days. The basis of this considerable mass transport are the oscillating contractile movements of the plasmodial strands, namely, of their ectoplasmic tubes, which contain the force-generating cytoplasmic actomyosin fibrils. Plasmodial strands with a length of more than 5 cm can be isolated from the plasmodium and measured tensiometrically with respect to their longitudinal contractile activities.

Moreover, their radial contractile activities can be registered in situ by tensiometric, but also by other, noninvasive methods. The experimental possibilities, advantages, and shortcomings have been summarized in the context of a NATO Advanced Study Institute (Wohlfarth-Bottermann, 1983).

ENERGY SUPPLY

The energy-metabolic supply of the rhythmic contractile activity plays a crucial role in the oscillator. Both glycolysis and respiration are obviously involved in the physiological interplay between contractile proteins as effectors, calcium regulatory systems, and the energy-producing metabolism, all mutually interdependent because of feedback regulatory relationships (Korohoda et al., 1983). According to this model, the spatial propagation of the oscillatory activity can be achieved by a transport of ATP, ADP, P_i, and Ca^{++}, as well as mitochondria and contractile proteins in the flowing endoplasm. The function of mitochondria seems to be crucial for the oscillatory cycle; it can be shown (Baranowski, 1985) that there is an alternative, KCN-insensitive pathway of respiration sensitive to salicylhydroxamic acid (SHAM) (Schonbaum et al., 1973). An inhibition of both respiratory pathways under conditions of an undisturbed glycolysis results in a reversible blockage of the rhythmical contractile activity. This would mean that, contrary to previous belief, glycolytic oscillations are not responsible for the primary origin of oscillations, but that the respiratory energy supply is of predominant importance as a "main wheel" of the complex oscillator.

PERIODIC REVERSAL

The periodic reversal of endoplasmic shuttle streaming in plasmodial strands is a hydrostatic pressure flow that follows differences in hydrostatic pressure within different regions of a plasmodium. These oscillatory changing pressure differences are generated by coordinated contractile phenomena of the strands (Wohlfarth-Bottermann, 1983). In spite of this "normal" (i.e., passive) endoplasmic mass translocation, we should keep in mind that the cytoplasm is also able to perform movement phenomena that can hardly be explained by a pressure flow. Thus, the living cytoplasm can, if necessary, do more than be "inactively" transported; it can move itself by generating the needed motive force at <u>any</u> site.

The ability of living cytoplasm to contract at all cellular sites is based on the ubiquitous cytoplasmic actomyosin. Its spatial arrangement in plasmodial strands resembles the location of muscle actomyosin in a muscle tube. The isotonically contracting fibrils in an ectoplasmic tube exert pressure on the interior endoplasmic channel, thus inducing a translocation of endoplasm in the direction of a lower pressure. This is the "normal" mode of endoplasmic mass transport in macroplasmodia. Owing to this well-established mechanism and to the ability to follow the contractile process by the measurement of several microscopic and tensiometric parameters on plasmodial strands, these objects are of special value in creating clear-cut experimental conditions to analyze the physiology of cytoplasmic actomyosin. This is a central problem in current research on the cytoskeleton. It is obvious that macroplasmodia represent a most favorable material because their size allows the application of additional controlling techniques, such as tensiometry (Wohlfarth-Bottermann, 1983).

MOLECULAR STUDIES

Biochemical studies have revealed in vitro, in an admirably clear-cut manner, the capabilities of proteins, i.e., their possible function (Chapters 9-13,15,16). Such functional activities, however, have to be confirmed in vivo, in order for us to know what the proteins are really doing in their natural environment. In other words, biochemistry of pure proteins shows us life as it could be, not necessarily as it is. Two methods showing aspects of the function of cytoplasmic actomyosin in vivo were considered during this Workshop:

Microinjection

The injection of fluorochrome-labeled proteins into the living cell, followed by fluorescence-microscopic analysis of the incorporation into the corresponding structures, e.g., cytoplasmic actomyosin fibrils, permits an analysis of their dynamic behavior (Chapter 14). This new method is still under development and to date, unfortunately, applicable only to very small, special stages (microplasmodia and caffeine droplets), but not to macroplasmodia. The normal plasmodial strands offer numerous other experimental possibilities, are better investigated with respect to their behavior, and show more reliably the regular intrinsic contraction phenomenon. Thus, it is important to adapt the injection of fluorochrome-labeled proteins to macroplasmodia. The dynamic behavior of the fibrils in normal plasmodial strands displays in macroscopic form the behavior and motive force generation of the cytoplasmic actomyosin in other eukaryotic cell types, none of which offer such favorable experimental conditions as the large plasmodial strands of *Physarum*. The application of "fluorescence analog cytochemistry" to normal plasmodial strands will permit coordination with established registration methods, such as infrared registration of the radial and/or tensiometric measurements of longitudinal contractile activities of the strands, the most important coordination parameters to interpret results of this in vivo molecular technique.

Cell-free Models

Reactivation experiments on cytoplasmic actomyosin in cell-free models should inform us about the physiological conditions of cytoplasmic actomyosin contraction in vivo and, if possible, also about the nature of the oscillator that drives the contraction activity and thereby the shuttle-streaming phenomenon. At present, it seems of central interest to clarify the question whether Ca^{++} induces or inhibits contraction of cytoplasmic actomyosin (Yoshimoto and Kamiya, 1984; Wohlfarth-Bottermann, 1984; Wohlfarth-Bottermann et al., 1983). Two approaches were described with "cell-free models" during this meeting: (1) contraction experiments with cytoplasmic actomyosin fibrils in frozen sections (Pies and Wohlfarth-Bottermann,

1984; Pies and Wohlfarth-Bottermann, unpublished); and (2) contraction experiments with glycerol-extracted protoplasmic drops (Achenbach and Wohlfarth-Bottermann, unpublished).

Whereas in cryosections one visualizes directly the isotonic contraction of the fibrils after incubation in a contraction solution containing ATP and different concentrations of Ca^{++}, the contraction of glycerol-extracted drops is monitored by changes in the volume of the drops. The new cell-free model in the form of frozen sections has the advantage that there is no essential extraction of protein components from the environment of the fibrils.

After several days of glycerol extraction, however, we can be sure that regulatory components are at least partially extracted. In spite of these differing methodological approaches, both methods indicated (in contrast to previous experience with cross-striated muscle) that a relatively high Ca^{++} concentration inhibits contraction in nonextracted and short-term extracted models, a finding that is in agreement with the biochemical data of Kohama (Chapter 10). Thus, several observations indicate that Ca^{++} inhibits rather than activates *Physarum* actomyosin ATPase. According to Kessler (Chapter 11), Ca^{++} seems to bind to a light polypeptide chain of myosin, thus possibly regulating actomyosin activity. Nevertheless, the problem of the function of Ca^{++} remains rather complex if one also considers its action on actin-regulating proteins (see Chapters 9,12,13).

In contrast to nonextracted and short-term extracted models, Ca^{++} allows the reactivation of actomyosin ATPase after several days of glycerol extraction. We are at present far from a complete understanding of the regulation of the dynamic processes of contraction and relaxation in cytoplasmic actomyosin by Ca^{++}.

The evaluation of the results gained from cell-free models concerning the effects of Ca^{++} is, moreover, complicated by the 2- to 3-nm filament system (titin, connectin; Chapters 15 and 16), which performs condensation reactions at mM concentrations of divalent cations. Such condensation phenomena may have contributed to the forces registered in previous reactivation studies of cytoplasmic actomyosin. Thus, this problem has to be considered when one evaluates earlier reactivation experiments. A participation of 2- to 3-nm filaments can be excluded in frozen sections (Pies and Wohlfarth-Bottermann, unpublished) by a presaturation of the contractile substrate with 20 mM Mg^{++} and, in the case of the drop models (Achenbach and Wohlfarth-Bottermann, unpublished), by a presaturation with 5 mM Mg^{++} and 0.1 mM Ca^{++} before the addition of actomyosin contraction solutions. These divalent cations result in a condensation of the 2- to 3-nm filament system before the contraction of the actomyosin is assessed.

WHAT WE SHOULD KNOW

The appended five tables attempt to list current central problems that should be addressed in order for us to broaden our knowledge concerning the physiology of cytoplasmic actomyosin in general, the nature of the contractile apparatus in *Physarum*, as well as the unknown oscillator driving the periodic intrinsic contractile phenomena.

It must be admitted that the cytoplasmic motility oscillator in *Physarum* may be of special interest only for the biology of this organism, but it should be emphasized that a deeper insight may further improve the excellent possibilities to investigate the physiology of cytoplasmic actomyosin of eukaryotes with the unique advantages of *Physarum* as a model system.

REFERENCES

Achenbach, F., and Wohlfarth-Bottermann, K. E., Reactivation of cell-free models of endoplasmic drops of *Physarum polycephalum* after glycerol extraction at low ionic strength, *Eur. J. Cell Biol.*, in press.

Baranowski, Z., 1985, Alternative pathway of respiration in *Physarum polycephalum* plasmodia, *Cell Biol. Intern. Reports*, 9:85.

Kessler, D., Eisenlohr, L. C., Lathwell, M. J., Huang, J., Taylor, H. C., Godfrey, S. D., and Spady, M. L., 1980, *Physarum* myosin light chain binds calcium, *Cell Motil.*, 1:63.

Korohoda, W., Shraideh, Z., Baranowski, Z., and Wohlfarth-Bottermann, K. E., 1983, Energy metabolic regulation of oscillatory contraction activity in *Physarum*, *Cell Tissue Res.*, 231:675.

Pies, N. J., and Wohlfarth-Bottermann, K. E., 1984, Reactivation of NBD-phallacidin-labelled actomyosin fibrils in cryosection of *Physarum polycephalum*: a new cell-free model, *Cell Biol. Intern. Reports*, 8:1065.

Pies, N. J., and Wohlfarth-Bottermann, K. E., Reactivation of a cell-free model from *Physarum polycephalum*: studies on cryosections indicate an inhibitory effect of Ca^{++} on cytoplasmic actomyosin contraction, *Eur. J. Cell Biol.*, in press.

Schonbaum, G. R., Bonner Jr., W. D., Storey, B. T., and Bahr, J. T., 1971, Specific inhibition of the cyanide insensitive respiratory pathway in plant mitochondria by hydroxamic acids, *Plant Physiol.*, 47:124.

Wohlfarth-Bottermann, K. E., 1983, Dynamic cellular phenomena in *Physarum* possibly accessible to laser techniques, *in*: The Application of Laser Light Scattering to the Study of Biological Motion, J. C. Earnshaw and M. W. Steer, eds., NATO ASI Series, Series A: Life Sciences, Vol. 59, p. 501, Plenum Press, New York and London.

Wohlfarth-Bottermann, K. E., Cell free models from Physarum--pitfalls and improvements, in: Proceedings of the 10th Yamada Conference 1984, "Cell Motility: Mechanism and Regulation", H. Ishikawa, H. Sato, S. Hatano, eds., University of Tokyo Press, Tokyo, in press.

Wohlfarth-Bottermann, K. E., Shraideh, Z., and Baranowski, Z., 1983, Contractile and structural reactions to impediments of Ca^{++}-homeostasis in Physarum polycephalum, Cell Struct. Funct., 8:255.

Yoshimoto, Y., and Kamiya, N., 1984, ATP- and calcium-controlled contraction in a saponin model of Physarum polycephalum, Cell Struct. Funct., 9:135.

Table 8-1. Biophysical Interaction of Actin and Myosin and its Regulation

Known	We Need to Know	Techniques To Be Addressed
Chemomechanical energy transformation by actin and myosin	Is it a sliding mechanism as in cross-striated muscle?	Observation of single molecules in the fluorescence and/or darkfield microscope
Topographical arrangement of cytoplasmic actomyosin	Its interaction with other cytoskeletal components Its dynamic behavior in vivo	Refined morphological methods to reveal and identify the cytoskeletal elements in situ Injection of fluorescently or otherwise labeled probes into macroplasmodia
There are several actin-regulating proteins, e.g., Fragmin, Profilin, Cap 42, etc.	Are all these also involved in vivo? What is their interaction and their immunocytochemical localization?	Microinjection of fluorescently labeled proteins Biochemistry Immunocytochemistry Genetics (see Chapter 17)
Myosin is regulated by Ca^{++} in vitro	What is the regulatory sequence in vivo?	Cell-free models with varying degrees of extraction

Table 8-1 (continued)

Known	We Need to Know	Techniques To Be Addressed
Ca^{++} is of predominant importance for the contraction mechanism	Free Ca^{++} concentration in vivo and its change during the contraction cycle Which Ca^{++} stores are mainly involved in control levels of intracellular free Ca^{++}? Which mechanism regulates their possible rhythmicity?	Microinjection of Ca-sensitive dyes Cell fractionation Revival of isolated Ca^{++}-pumps
In the endoplasm, actin is predominantly depolymerized	Which factor keeps the actin depolymerized? What is the nature of the signal to start actin polymerization, e.g., in the first minute after the generation of a protoplasmic drop?	
In the test tube, a lot of proteins ("actin-binding proteins") and other substances bind to actin	We need methods to check whether this occurs also in vivo, i.e., to check which of them are of real physiological importance	

Table 8-2. Nature and Function of the Superthin Filaments

Known	We Need to Know	Techniques To Be Addressed
There is a network of *matrix* filaments ("2-3 nm, titin-, connectin-like) in the endoplasm as well as in the ectoplasm	Spatial distribution of *Physarum* titin-like protein by light- and electron microscopical immunocytochemistry in different stages of *Physarum* (indirect immunofluorescence, immuno-gold technique)	Antibodies to the 2- to 3-nm filament protein
	Is there a difference in the lattice of this protein system in the ectoplasm as compared with its form in the endoplasm?	
	Is this protein system thus able to perform dynamic changes in vivo?	
	Does it interact with actin and myosin in vivo?	

Table 8-3. Synchronization of Contractile Activities and Motive Force Generation

Known	We Need to Know	Techniques To Be Addressed
The signal for the blue light reaction is propagated by the endoplasmic stream, i.e., the first appearance of the signal is registered in the actual streaming direction	What is the nature of the signal within the endoplasm? How can this signal modulate the contractile machinery?	
The fastest, registrable (1 min) reaction to blue light is a change in the contraction period immediately after the onset of illumination	What is the nature of the blue light receptor, and where is it located?	Are there mutants without a blue light reaction?
Chemical and physical data concerning the properties of the flowing endoplasm are nearly completely lacking	E.g., a more detailed knowledge about the velocity profile and the behavior of the flowing endoplasm, also in order to understand its transformation into ectoplasm	Laser analysis of the endoplasmic flow in combination with tensiometrical registrations of contractile activities

Table 8-4. Energy Metabolic Basis for Contractile Activity and Nature of the Oscillator

Known	We Need to Know	Techniques To Be Addressed
Energy supply from glycolysis and respiration	More details	Normal biochemistry of the energy metabolism
Locomotory phenomena are based on shuttle streaming. Mass transport is generated by contractile activity, which is driven by an oscillator with a frequency in the min-range	Which are the essential parts of the oscillator? (cytoplasmic actomyosin, Ca^{++}-pumps, glycolysis, respiration?) In which way may they interact?	
The pathway of signal transmission for synchronization of contractile activity is the endoplasmic flow, which leads to a continuing exchange of endoplasm and ectoplasm	Is the nature of this signal mechanical or chemical, or both? In the case of a chemical signal: Is it the oscillator itself, which may be transported within the endoplasm?	

Table 8-5. Techniques and Varia

Known	We Need to Know	Techniques To Be Addressed
The nonsterile mass culture method of large plasmodia (Camp) is antique and time-consuming; for motility problems, the axenic culture method is of limited value because of the small size of the plasmodia	We need a more efficient food source than oat-flakes in order to reduce the time expenditure for large-scale surface cultures	
Cell fractionation is impeded by the clumping activity of the slime in the homogenates and is thus more difficult compared with normal cells	We need to know how to prepare fractions while avoiding the difficulties of the plasmodial slime cover	Slime-free mutants
Sometimes experimental results are difficult to reproduce e.g., because plasmodia change gradually in cultures		Deposition in the American Type Culture Collection and citation of the number in each paper Use of frozen gametes to form plasmodia at periodic intervals (see Chapters 2,18)

Table 8-5 (continued)

Known	We Need to Know	Techniques To Be Addressed
Extraction to obtain cell-free models implies destruction of the contractile apparatus	Stabilization of cytoplasmic actomyosin before and during extraction procedures	
Injection of fluorescently and otherwise labeled compounds contribute to our knowledge of their function in vivo, e.g., fluorescently labeled actin	We need to know how to better overcome the strong tendency to sequestrate injected substances into lysosomes, in order to be able to apply this technique to plasmodial strands	Improvement of microinjection methods (see Chapters 20 and 21)

Chapter 9: ACTIN, MYOSIN, AND THE ASSOCIATED-PROTEINS FROM THE <u>PHYSARUM</u> PLASMODIUM

Sadashi Hatano

Institute of Molecular Biology
Nagoya University
Nagoya, JAPAN

INTRODUCTION

The plasmodium of *Physarum polycephalum* is widely used for the biochemical study of actin, myosin, and related proteins, for two good reasons. First, the cytology of protoplasmic streaming and sol-gel transformations in plasmodia has been intensively studied (Chapter 8; Kamiya, 1959); second, relatively large amounts of plasmodia can be easily obtained. Ten plastic buckets of 10 L each are used for the cultivation of surface plasmodia in order to collect 100-200 g of material every 2 days.

Loewy (1952) first showed the presence of an actomyosin-like protein in an extract of the *Physarum* plasmodium, and it was partially purified by others (T'so et al., 1956; Nakajima, 1960).

Hatano isolated and purified actin (Hatano and Oosawa, 1966) and myosin (Hatano and Tazawa, 1968; Hatano and Ohnuma, 1970) from *Physarum* plasmodia. Adelman and Taylor (1969) also succeeded in isolating actin from the same material. These pioneering works stimulated investigators in the field of cell motility to pursue the study of actin and myosin in other non-muscle cells. It has been shown that actin and myosin are ubiquitous in eukaryotic cells and play key roles for a variety of cell functions, not only in cell motility, but also in cell adhesion and cell morphogenesis.

ACTIN-ASSOCIATED PROTEINS

In the course of study on actin from *Physarum* plasmodia, Hatano found that crude extracts from plasmodia contain factors that inhibit actin polymerization. This was shown in the following ways: (1) *Physarum* G-actin in the crude extract does not polymerize to F-actin, whereas the G-actin, once purified, polymerizes to F-actin, the structure of which is similar to that of F-actin from striated muscle. (2) The polymerization of purified G-actin is inhibited by the addition of small amounts of *Physarum* crude extract.

It took a long time, however, until several protein factors were isolated and characterized (see also Chapters 12 and 13). Hasegawa et al. (1980) first isolated and purified a Ca-dependent protein factor of 42K called fragmin. Hinssen (1981) also isolated a Ca^{2+}-dependent, actin-modulating protein, identical with fragmin, from *Physarum* plasmodia. Protein factors of 95K (gelsolin [Yin and Stossel, 1979]; villin [Bretscher and Weber, 1980]), which are functionally very similar to fragmin, have been isolated from mammalian cells.

FRAGMIN

The activity of fragmin is regulated by Ca^{2+} concentrations in the μM range, which is the physiological concentration of Ca^{2+} in plasmodia. Fragmin cuts F-actin into short fragments at Ca^{2+} concentrations higher than 10^{-6} M, when fragmin is added to F-actin in substoichiometric molar ratios (for example, 1 fragmin to 10 G-actins). This process is rapid, suggesting that the fragmin effect on F-actin is not a passive but an active one (Chapter 12). Fragmin enhances the polymerization of G-actin, and short F-actins form after polymerization. When more fragmin is added, shorter F-actin filaments form (Hasegawa et al., 1980).

Sugino and Hatano (1982) showed that the initial rate of the nucleation for the actin polymerization is enhanced in proportion to the concentration of fragmin. It is assumed that fragmin binds to G-actin to form a 1-to-1 complex. This complex becomes the nucleus for the actin polymerization. Many nuclei form, so that the polymerization is much enhanced.

As is well known, F-actin has polarity. When F-actin is treated with muscle heavy meromyosin, F-actin is decorated to form arrowhead-like structures. It has been shown that the elongation of F-actin occurs from both terminal ends, but the rate of elongation is much faster at the barbed than at the pointed end. However, elongation of the nucleus (the G-actin-fragmin complex) occurs only from the pointed end, since fragmin caps the barbed end. Association of the short F-actins that form is also inhibited by the capping of the barbed ends by fragmin.

$$\text{G-actin + fragmin} \xrightarrow{Ca^{2+}} \text{G-actin·fragmin}$$

$$\text{G-actin·fragmin + n G-actin} \rightleftharpoons \text{short F-actins (capped with fragmin)}$$

$$\text{short F-actins (capped with fragmin)} \not\rightarrow \text{F-actin}$$

CAP 42A AND B

Maruta and Isenberg (1984) isolated from plasmodia two additional capping proteins of 42K called Cap 42a and Cap 42b. Unlike fragmin, however, neither Cap 42a nor Cap 42b has F-actin cutting activity. Interestingly, Cap 42b shows its capping activity only when it is fully phosphorylated (Chapter 13).

PROFILIN

Ozaki et al. (1983) isolated from plasmodia a second type of protein factor that regulates actin polymerization in a Ca^{2+}-independent manner. The physicochemical and functional properties of this protein were found to be similar to those of profilins, which have been isolated from sources such as mammalian tissue culture cells (Markey, et al., 1978), Acanthamoeba (Reichstein and Korn, 1979), and sea urchin eggs (Mabuchi and Hosoya, 1982), so this protein factor was termed *Physarum* profilin. *Physarum* profilin regulates actin polymerization, but in a way different from fragmin. The polymerization of purified *Physarum* G-actin is inhibited by *Physarum* profilin. Both the initial rate and the extent of the polymerization are reduced, depending on the concentration of profilin.

Ozaki and Hatano (1984) found that profilin reversibly binds to G-actin and forms an unpolymerizable complex of one G-actin and one profilin. The dissociation constant (K_D) for the complex is estimated to be 2 μM from the apparent increase of the critical concentration and the decrease of the initial rate of actin polymerization in the presence of profilin.

$$\text{G-actin + profilin} \overset{K_D = 2\ \mu M}{\rightleftharpoons} \text{G-actin·profilin (unpolymerizable complex)}$$

When G-actin and profilin are mixed in a test tube and the polymerization is induced on addition of salts, equilibria among F-actin, G-actin (at the critical concentration), profilin, and G-actin profilin complex are established.

$$\text{F-actin} \rightleftarrows \underset{\text{(critical concentration)}}{\text{G-actin}} + \text{profilin} \overset{K_D}{\rightleftarrows} \text{G-actin}\cdot\text{profilin}$$

The content of F-actin in solution can be calculated from both the critical concentration of G-actin (0.2 μM) and the dissociation constant (2 μM) for the G-actin profilin complex. When actin and profilin are mixed at 50 μM each, which is the physiological concentration of each protein, it is estimated that 91% of the actin polymerizes to F-actin. However, if the critical concentration is estimated to be increased from 0.2 to 1.5 μM by capping of the barbed end by fragmin, for example (Wegner and Isenberg, 1983), the equilibria are shifted toward formation of the unpolymerizable complex, so that the content of F-actin decreases from 91% to 55%. This possibility was first suggested by Tobacman et al. (1983) for *Acanthamoeba* profilin. Thus, one can expect to induce large changes of the F-actin content in the mixed solution of actin and profilin under the physiological conditions of the plasmodium. It is assumed that profilin serves to maintain a reserve of G-actin in the plasmodium.

GELATION FACTOR

The third factor -- a gelation factor (high molecular weight [HMW], actin-binding protein) of F-actin solution -- was isolated from *Physarum* plasmodia by Sutoh et al. (1984). The *Physarum* gelation factor is a dot-shaped, flexible molecule similar to spectrin and to related proteins isolated from erythrocytes and other cell types. The molecules cross-link F-actin to form a network structure of F-actin (F-actin gel).

$$\text{F-actin} + \text{HMW actin-binding protein} \longrightarrow \text{F-actin gel}$$

Both the in vitro functions and the possible physiological roles of these three types of actin-associated proteins are briefly summarized in Table 9-1.

CONTRACTION-RELAXATION CYCLE OF ACTOMYOSIN

It is considered at present that the motive force for the protoplasmic streaming of the plasmodium is generated by the contraction of the ectoplasmic gel layer. The inner sol (endoplasm) is squeezed into the plasmodial strands by the contraction of the gel layer of the peripheral part of the plasmodium, in which there are many actomyosin fibers. The actomyosin fiber has been shown to be a bundle of 10^3-10^9 of F-actin fibers to which myosin molecules and/or myosin oligomers are attached (Chapter 8). Such actomyosin fibers are birefringent under the polarizing microscope. This birefringence displays a dynamic behavior, appearing and disappear-

Table 9-1. Actin-associated Proteins from *Physarum* Plasmodium

G-actin $\underset{\gamma,\beta}{\rightleftarrows}$ F-actin $\underset{\alpha}{\rightleftarrows}$ Network structure or bundle of F-actin

Type	Protein	Ca-Dependency	In Vitro Functions	Possible Physiological Roles
γ	Profilin	-	Formation of unpolymerizable complex of G-actin	Storage of G-actin Regulation of gel-sol transformation
β	Fragmin, Cap 42a, Cap 42b	+	Promotion of nucleation Capping of barbed end of F-actin Fragmentation of F-actin (by fragmin)	Nucleation site for actin polymerization Unidirectional elongation of F-actin Solation of F-actin gel Relaxation of actomyosin
α	HMW actin-binding protein, myosin	-	Formation of network structure of F-actin and/or bundle of F-actin Formation of actomyosin fiber*	Gelation and/or bundling of F-actin Contraction of actomyosin

* By superprecipitation of F-actin and myosin. Another type of F-actin-binding protein has been isolated by Ogihara and Tonomura (1982).

ing according to the contraction and relaxation cycle of the gel layer (Nakajima and Allen, 1965; Kamiya, 1973; Sato et al., 1981).

Recently, Ishigami (personal communication) confirmed that the contraction and relaxation cycle of the gel layer is closely associated with the periodic appearance of the birefringent fibers. The birefringent fiber increases in number when the gel layer contracts and decreases during relaxation.

Matsumura and Hatano (1978) showed that superprecipitation of actomyosin synthesized from purified Physarum F-actin and myosin is a reversible process. The extent of the superprecipitation is regulated by the concentration of ATP in the medium. When the ATP concentration is high (0.1-1 mM), F-actin and small myosin filaments are separately present (the clearing phase of the superprecipitation). When the ATP concentration is decreased to 1 μM, F-actin forms the bundles of F-actin decorated with myosin molecules and/or myosin oligomers.

$$\text{clearing phase (relaxation)} \underset{}{\overset{\text{ATP}}{\rightleftharpoons}} \text{superprecipitation (contraction)}$$

$$\text{F-actin (separate F-actin)} + \text{myosin (small, thick filament)} \rightleftharpoons \text{actomyosin fiber (birefringent fiber, bundle of F-actin)}$$

Ca^{2+} REGULATION OF ACTOMYOSIN

Ca^{2+} does not show any effects on the superprecipitation of actomyosin synthesized from purified F-actin and myosin. However, several reports have indicated that Ca^{2+} concentrations on the μM range regulate the contraction of the actomyosin in plasmodia. Hatano and Oosawa (1970) prepared caffeine drops (small fragments of the plasmodium; see Chapter 14) by treatment with a 10 mM caffeine solution. They showed that a Ca^{2+} concentration of more than 10^{-6} M in the medium induces the contraction and relaxation cycle of granular cytoplasm in the caffeine drop. The movement ceases, and the granular cytoplasm expands in the caffeine drop when the Ca^{2+} concentration is less than 10^{-7}M.

Yoshimoto and Kamiya (1981) speculated that the Ca^{2+} concentration in the plasmodial strand periodically changes according to the cycle of tension development of the strand. The Ca^{2+} concentration reaches a maximum when the tension goes to a minumum, and vice versa. They used Ca-ionophore-treated strands and measured the Ca^{2+} concentration released from the strand into the medium. If the Ca^{2+} concentration measured in this way really reflects the concentration of free Ca^{2+} in the strand, these observations suggest that Ca^{2+} inhibits the activity of the actomyosin system in the plasmodium.

The idea that Ca^{2+} inhibits rather than stimulates actomyosin activity is supported by recent discoveries that superprecipitation of a natural actomyosin (myosin B) is inhibited by Ca^{2+} in the μM range (Kohama et al., 1980). The phosphorylation of myosin (Nachmias, 1981; Ogihara et al., 1983) and/or the binding of Ca^{2+} to myosin (Chapters 10 and 11) inhibit the actin-activated ATPase of *Physarum* myosin.

Sugino and Matsumura (1983) showed that when the Ca^{2+} concentration in the medium is more than 10^{-6} M, fragmin reduces tension development by the actomyosin threads made from purified *Physarum* F-actin and myosin.

Thus, the Ca^{2+} effect on the actomyosin system of plasmodia appears to be complicated. Ca^{2+} in the μM range activates actomyosin in the caffeine drop, whereas Ca^{2+} at the same concentration inhibits the actin-activated ATPase of myosin through the phsophorylation of myosin, the binding of Ca^{2+} to myosin, and/or the fragmentation of F-actin by fragmin.

The author (Hatano et al., 1985) has suggested that the Ca^{2+} sensitivity of *Physarum* actomyosin is biphasic; that is, when the concentration of Ca^{2+} increases from 10^{-7} M to 10^{-6} M, the actomyosin system is activated by an unknown factor(s) such as troponin in the striated muscle. After contraction, the actomyosin would be relaxed by the inhibition of actin-activated myosin ATPase and/or by the fragmentation of F-actin in the presence of more than 10^{-6} M of Ca^{2+}. This idea has been supported by a recent experiment by Yoshimoto and Kamiya (1984). The tension generation of a saponin-treated plasmodial strand depends on the Ca^{2+} concentration in the medium. The optimal Ca^{2+} concentration is 10^{-7} M, and the tension decreases when the Ca^{2+} concentration is either above or below this value.

WHAT SHOULD BE INVESTIGATED

Ca^{2+} Concentration in Situ

It is essential to know the concentration of Ca^{2+} in the plasmodium and its change during the contraction and relaxation cycle. One possible way to investigate the free Ca^{2+} concentration is to microinject Ca^{2+}-sensitive fluorescent dyes such as quin 2 or the Ca^{2+}-sensitive fluorescent protein, aequorin, into plasmodia. This line of study is being carried out in the author's laboratory.

State of Actin in Situ

The state of actin in the endoplasm is also unknown. Ishigami and Hatano (unpublished results) showed that about 40% of actin is extractable from the plasmodium by treatment with 0.2% Triton X-100. The birefringent fibers are well-preserved in the Triton-treated plasmodium (Triton cytoskeleton), and they contract on addition of ATP. Thus, most of the actin exists as F-actin in the Triton-treated plasmodium. F-actin forms bundles of F-actin or birefringent fibers in the gel layer of the plasmodium.

It has been shown by electron microscopy that F-actin is not observed in the plasmodial endoplasm or in fresh droplets isolated from the endoplasm, but F-actins soon appear in the droplets in accordance with the transformation from the sol state of the endoplasm to the gel state. These results suggest that the G-F transformation is involved in the sol-gel transformation (Isenberg and Wohlfarth-Bottermann, 1976). It is not known, however, in what states actin exists in the endoplasm. Does actin in the endoplasm exist in unpolymerizable states, such as a complex of G-actin and profilin? In what states and in what percentages is actin extracted from the endoplasm and the ectoplasm by Triton treatment?

The microinjection of fluorescently labeled actin into the plasmodium will be a very useful technique to observe the dynamic behavior of actin in the contraction-relaxation cycle and the sol-gel transformation (Chapters 20 and 21).

REFERENCES

Adelman, M. R., and Taylor, E. W., 1969, Isolation of an actomyosin-like protein complex from slime mould plasmodium and the separation of the complex into actin- and myosin-like fractions. Biochemistry, 8:4976.

Bretscher, A., and Weber, K., 1980, Villin is a major protein of the microvillus cytoskeleton which binds both G and F actin in a calcium-dependent manner, Cell, 20:839.

Hasegawa, T., Takahashi, S., and Hatano, S., 1980, Fragmin: a calcium ion-sensitive regulatory factor on the formation of actin filaments. Biochemistry, 19:2677.

Hatano, S., and Ohnuma, J., 1970, Purification and characterization of myosin A from the myxomycete plasmodium, Biochim. Biophys. Acta, 205:110.

Hatano, S., and Oosawa, F., 1966, Isolation and characterization of plasmodium actin, Biochim. Biophys. Acta, 127:488.

Hatano, S., and Oosawa, F., 1970, Specific effect of Ca^{++} on movement of plasmodial fragments obtained by caffeine treatment, Exp. Cell Res., 61:199.

Hatano, S., Ozaki, K., Sugino, H., and Masuda, H., 1985, Mechanisms of regulation of actin polymerization by profilin and fragmin and their possible physiological roles in plasmodium of Physarum polycephalum, in: "Cell Motility, Mechanisms and Regulation", H. Ishikawa, S. Hatano, and H. Sato, eds., Univ. Tokyo Press, Tokyo, in press.

Hatano, S., and Tazawa, M., 1968, Isolation, purification and characterization of myosin B from myxomycete plasmodium. Biochim. Biophys. Acta, 154:507.

Hinssen, H., 1981, An actin modulating protein from Physarum polycephalum. I. Isolation and purification, Eur. J. Cell Biol., 23:225.

Isenberg, G., and Wohlfarth-Bottermann, K.E., 1976, Transformation of cytoplasmic actin. Importance for the organization of the contractile gel reticulum and the contraction-relaxation cycle of cytoplasmic actomyosin, Cell Tissue Res., 173:495.

Kamiya, N., 1959, Protoplasmic streaming, Protoplasmatologia 8, 3a:1.

Kamiya, N., 1973, Contractile characteristics of the myxomycete plasmodium, in: "Proceedings of IV International Biophysics Congress" (Moscow, 1972), Symposial paper 1, pp. 447-465.

Kohama, K., Kobayashi, K., and Mitani, S., 1980, Effect of Ca ion and ADP on superprecipitation of myosin B from slime mold, Physarum polycephalum, Proc. Jpn Acad., 56B:591.

Loewy, A. G., 1952, An actomyosin-like substance form the plasmodium of a myxomycete, J. Cell Comp. Physiol., 40:127.

Mabuchi, I., and Hosoya, H., 1982, Actin-modulating proteins in the sea urchin egg. II. Sea urchin egg profilin, Biomed. Res., 3:465.

Markey, F., Lindberg, U., and Eriksson, L., 1978, Human platelets contain profilin, a potential regulator of actin polymerizability, FEBS Lett., 88:75.

Maruta, H., and Isenberg, G., 1984, Ca^{2+}-dependent actin-binding phosphoprotein in Physarum polycephalum. Subunit b is DNase I-binding and F-actin capping protein, J. Biol. Chem., 259:5208.

Matsumura, F., and Hatano, S., 1978, Reversible superprecipitation and bundle formation of plasmodium actomyosin, Biochim. Biophys. Acta, 533:511.

Nachmias, V. T., 1981, Physarum myosin light chain one: A potential regulatory factor in cytoplasmic streaming, Protoplasma, 109:13.

Nakajima, H., 1960, Some properties of a contractile protein in the slime mold, Protoplasma, 52:413.

Nakajima, H., and Allen, R. D., 1965, The changing pattern of birefringence in plasmodia of the slime mold, Physarum polycephalum, J. Cell Biol., 25:361.

Ogihara, S., Ikebe, M., Takahashi, K., and Tonomura, Y., 1983, Requirement of phosphorylation of myosin heavy chain for thick filament formation, actin activation of Mg^{2+}-ATPase activity, and Ca^{2+}-inhibitory superprecipitation, J. Biochem., 93:205.

Ogihara, S., and Tonomura, Y., 1982, A novel 36,000-dalton actin-binding protein purified from microfilaments in Physarum plasmodia which aggregates actin filaments and blocks actin-myosin interaction, J. Cell Biol., 93:604.

Ozaki, K., and Hatano, S., 1984, Mechanism of regulation of actin polymerization by Physarum profilin, J. Cell Biol., 98: 1919.

Ozaki, K., Sugino, H., Hasegawa, T., Takahashi, S., and Hatano, S., 1983, Isolation and characterization of Physarum profilin, J. Biochem., 93:295.

Reichstein, E., and Korn, E. D., 1979, Acanthamoeba profilin. A protein of low molecular weight from Acanthamoeba castellanii that inhibits actin nucleation, J. Biol. Chem., 254:6174.

Sato, H., Hatano, S., and Sato, Y., 1981, Contractility and protoplasmic streaming preserved in artificially induced plasmodial fragments, the "caffeine drops", Protoplasma, 109:187.

Sugino, H., and Hatano, S., 1982, Effect of fragmin on actin polymerization: Evidence for enhancement of nucleation and capping of the barbed end, Cell Motil., 2:457.

Sugino, H., and Matsumura, F., 1983, Fragmin induces tension reduction of actomyosin threads in the presence of micromolar levels of Ca^{2+}, J. Cell Biol., 96:199.

Sutoh, K., Iwane, M., Matsuzaki, F., Kikuchi, M., and Ikai, A., 1984, Isolation and characterization of a high molecular weight actin-binding protein from Physarum polycephalum plasmodia, J. Cell Biol., 98:1611.

Tobacman, L. S., Brenner, S. L., and Korn, E. D., 1983, Effect of Acanthamoeba profilin on the pre-steady state kinetics of actin polymerization and on the concentration of F-actin at steady state, J. Biol. Chem., 258:8806.

Ts'o, P. O. P., Eggman, L., and Vinograd, J., 1956, The isolation of myxomyosin, and ATP-sensitive protein from the plasmodium of a myxomycete, J. Gen. Physiol., 39:801.

Wegner, A., and Isenberg, G., 1983, Twelve-fold difference between the critical monomer concentrations of the two ends of actin filaments in physiological salt conditions, Proc. Natl. Acad. Sci., U.S.A., 80:4922.

Yin, H. L., and Stossel, T. P., 1979, Control of cytoplasmic actin gel-sol transformation by gelsolin, a calcium-dependent regulatory protein, Nature, 281:583.

Yoshimoto, Y., and Kamiya, N., 1981, Simultaneous oscillations of Ca^{2+} efflux and tension generation in the permealized plasmodial strand of Physarum. Cell Motil., 1:433.

Yoshimoto, Y., and Kamiya, N., 1984, ATP- and calcium-controlled contraction in a saponin model of Physarum polycephalum, Cell Struct. Funct., 9:135.

Chapter 10: INHIBITORY CA^{2+}-REGULATION OF THE PHYSARUM ACTOMYOSIN SYSTEM

Kazuhiro Kohama[1] and Setsuro Ebashi[2]

[1]Department of Pharmacology
University of Tokyo
Tokyo, JAPAN

[2]National Institute for Physiological Sciences
Okazaki, JAPAN

INTRODUCTION: A BRIEF HISTORICAL SKETCH OF THE CA^{2+} CONCEPT

Ca^{2+} is now accepted as the most fundamental regulator of intracellular processes in general. This crucial role was first recognized in the research on muscle contraction. The first event that could somehow be related to the present Ca^{2+} concept was the famous finding of Ringer concerning the indispensable nature of Ca^{2+} for cardiac contractility (Ringer, 1983). Biological sciences at that time, however, had not yet reached a level to evaluate this important finding properly; they interpreted Ca^{2+} as a mere factor to maintain the physiological state of cells in competition with K^{+}.

The first person who clearly recognized the crucial role of Ca^{2+} in the contemporary sense was Heilbrunn (1940), who showed Ca^{2+}-induced shortening of injured frog muscle fibers. His observation was substantiated by Kamada and Kinosita (1943) and also by himself (Heilbrunn and Wiercinski, 1947); they demonstrated that Ca^{2+}, injected into the myoplasm of a frog fiber through a micropipette, induced local contracture, or shortening of the fiber.

Prior to Heilbrunn's work, some investigators had already noticed the contraction-inducing action of Ca^{2+} on myoplasm (Chambers and Hale, 1932; Keil and Sichel, 1936). Strangely enough, none of them claimed that Ca^{2+} should somehow be involved in physiological processes, in spite of their having had evidence equivalent to that of Heilbrunn.

The history from the age of Heilbrunn to the discovery of the troponin-tropomyosin system (Ebashi, 1963; Ebashi and Kodama, 1965) has been described in detail (Ebashi and Endo, 1968). This review article was written under the belief that the troponin system should be a sole regulatory mechanism of muscle contraction, but this idea was soon dispelled by the discovery of myosin-linked regulation (Kendrick-Jones et al., 1970). Smooth muscle has been shown to be controlled by an entirely different system from the above two kinds of muscle (cf. Ebashi et al., 1976; Hartshorne and Siemankowski, 1981). In addition, a number of regulatory factors have been proposed (Marston, 1982; Sobue et al., 1982; Ebisawa and Nonomura, 1985).

In the meantime, the discovery of Ca^{2+}-regulation in a metabolic process, i.e., phosphorylase b kinase (Ozawa et al., 1967), together with that of troponin led Kakiuchi to the discovery of 'modulator protein' (Kakiuchi 1969, 1970); this factor was also discovered independently by Cheung through a different approach (Cheung, 1970) and termed calmodulin.

It is now known that Ca^{2+}-regulation is characterized by a great diversity in the mode of its action. Even the apparently established belief of muscle researchers that Ca^{2+} is primarily an activator of the contractile system is now disproven by the studies on the actomyosin system of *Physarum*, as described below.

If we extend our scope to cell motility that is independent of the actomyosin system, the variety of modes of action of Ca^{2+} will become more pronounced. It is surprising that living organisms have created numerous complicated mechanisms to deal with this simple substance. Perhaps the development of Ca^{2+}-regulatory systems is an important aspect of the evolution of life.

CA^{2+} AS A REGULATOR OF PLASMODIAL CYTOPLASMIC STREAMING

The idea that Ca^{2+} is involved in the regulation of the plasmodial cytoplasmic streaming has been supported by a number of physiological observations (e.g., Hatano, 1970; Ridgway and Durham, 1976; Ueda et al., 1978; Yoshimoto et al., 1981; Yoshimoto and Kamiya, 1984).

Early observations claimed that Ca^{2+} should stimulate cytoplasmic streaming by analogy with Ca^{2+}-activation of muscle contraction (Hatano, 1970; Ridgway and Durham, 1976; Ueda et al., 1978). Recent reports concerning Ca^{2+}-dependent contraction of chemically skinned plasmodial strands, however, have suggested that the increase in cytoplasmic Ca^{2+} concentration from 10^{-7} M to 10^{-5} M inhibits contractile activity (Yoshimoto et al., 1981; Yoshimoto and Kamiya, 1984). Similar observations suggesting an inhibitory

Ca^{2+}-control of cytoplasmic streaming of internodal cells of Characeae were reported (Hayama et al., 1979; Hayama and Tazawa, 1980; Williamson and Ashley, 1982; Tominaga et al., 1983; Shimmen et al., 1984).

INHIBITION BY CA^{2+} OF ACTIN-MYOSIN-ATP INTERACTION OF PLASMODIUM

Early studies on the Ca^{2+}-control of the actin-myosin-ATP interaction in plasmodium were made with crude preparations of native actomyosin. The observation that the Mg-ATPase activity of this preparation was increased in the presence of Ca^{2+} led the authors to the conclusion that Ca^{2+} activated the interaction (Nachmias and Asch, 1974; Kato and Tonomura, 1975).

We have also observed a marked elevation of the Mg-ATPase activity of crude native actomyosin preparations by Ca^{2+} (Fig. 10-1). However, when the preparation was purified by repeated washing, the effect of Ca^{2+} on the activity was eventually reversed, i.e., the Mg-ATPase activity of purified native actomyosin was decreased with increase in Ca^{2+} (Fig. 10-1). As shown in Table 10-1, a crude extract of plasmodia is abundant in Ca^{2+}-activated soluble ATPase(s), which contains an ATP pyrophosphohydrolase (Table 10-1; see Kawamura and Nagano, 1975); the apparent Ca^{2+}-activation of the crude actomyosin preparation was owing to the contamination by the soluble ATPase(s). Accordingly, we have reached the conclu-

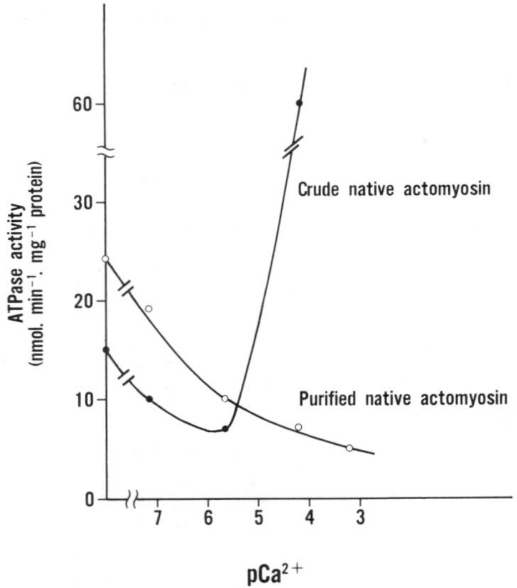

Fig. 10-1. Mg-ATPase activity of crude (●) and purified (○) native actomyosin. For assay conditions, see the legend to Table 1.

Table 10-1. Soluble and Actomyosin ATPase Activities of Physarum Plasmodia

Conditions		ATPase activity (nmol/min/ mg protein)	Total ATPase activity (nmol/min/ 100g plasmodia)
Supernatant	1×10^{-4} M EGTA	11.2	3.1×10^4
	5×10^{-5} M Ca^{2+}	57.2	16.0×10^4
Purified native actomyosin	1×10^{-4} M EGTA	15.7	0.14×10^4
	5×10^{-5} M Ca^{2+}	4.7	0.04×10^4

Plasmodia of Physarum polycephalum were allowed to grow on rolled oats. Fresh migrating sheets were harvested, homogenized in high salt (0.5 M NaCl, 10 mM EGTA, 0.1 mM DTT, and protease inhibitors) at pH 8.4, and centrifuged at 100,000 X *g* for 1 hr. The supernatant was made to pH 6.5, mixed with 5 vol of water, and centrifuged at 10,000 X *g* for 20 min (Hatano and Tazawa, 1968). The supernatant was assayed for ATPase activity. The precipitate was dissolved in the high salt and used as crude native actomyosin. The crude native actomyosin was mixed with five volumes of cold water to precipitate native actomyosin (see Fig. 10-1). Purified native actomyosin was obtained by repeating cycles of dissolution and precipitation of the crude actomyosin in high and low salt, respectively (Kohama et al., 1980). ATPase activities were determined by the pH-stat method (White, 1982) in 0.5 mM Mg-ATP, 1.0 mM Mg^{2+}, 30 mM KCl, and 0.1 mM EGTA-Ca buffer (0.1 mM EGTA plus various concentrations of $CaCl_2$) at pH 7.50 and 25°C.

sion that Ca^{2+} controls the actin-myosin-ATP interaction by inhibiting its Mg-ATPase activity (Fig. 10-1; Kohama et al., 1980).

MYOSIN-LINKED NATURE OF CA^{2+}-INHIBITION OF ACTIN-MYOSIN-ATP INTERACTION

Studies for the next two years on the Ca^{2+}-inhibition of the actin-myosin-ATP interaction of Physarum were carried out with Kendrick-Jones at the Laboratory of Molecular Biology, Cambridge (Kohama and Kendrick-Jones, 1982; Kohama et al., 1983; cf. Ogihara et al., 1983). The main conclusions reached were as follows:

1. The decrease of actin-activated ATPase activity of plasmodial myosin (p-myosin; see Fig. 10-2e for myosin purity) with an

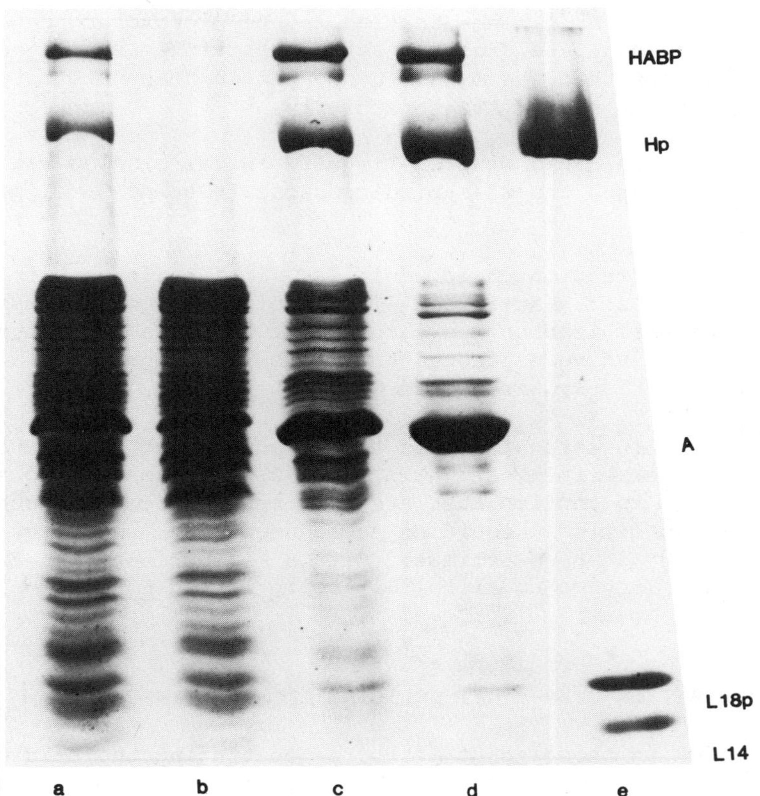

Fig. 10-2. SDS PAGE of proteins from steps in p-myosin purification procedures. Purified native actomyosin (cf. Table 10-1) was dissolved in ATP solution (20 mM ATP, 5 mM $NaHCO_3$, 1 mM EGTA and 2 mM DTT at pH 6.9) and mixed with concentrated Mg acetate to a final concentration of 0.1 M. After being stirred for 1 hr, the mixture was centrifuged at 100,000 X g for 30 min to remove actin and related proteins. The supernatant was mixed with 2 vol of water to yield precipitation of myosin (Kohama, 1981). Purified p-myosin was obtained by repeating dissolution in ATP solution, treatment with Mg acetate and recovery of myosin as a precipitate. (a), high salt extract (cf. Table 10-1); (b) supernatant (cf. Table 10-1); (c) crude native actomyosin; (d) purified native actomyosin; (e) purified p-myosin. HABP, high molecular weight actin-binding protein (Sutoh et al., 1984); Hp, p-myosin heavy chain; A, actin; L18p, 18K light chain of p-myosin; L14, 14K light chain. SDS PAGE was performed with an 8% upper separation gel to analyze Hp and 16% lower separation gel to analyze A, L18p and L14.

increase in the Ca^{2+} concentration in the micromolar range could be observed when pure skeletal actin substitutes for *Physarum* actin. This indicates that the myosin component is responsible for the Ca^{2+}-inhibition of the ATPase activity.

2. Myosin could bind Ca^{2+}; half-maximal binding was in the micromolar range, the Ca^{2+}-binding capacity being 2 mol per mol myosin.

3. The fraction enriched in 14K light chain conferred Ca^{2+}-inhibition on actin-activated ATPase activities of desensitized p-myosin (desensitization was carried out by alpha-chymotrypsin treatment); this finding suggested that this light chain would be involved in this myosin-linked Ca^{2+}-inhibition.

Further investigation has been conducted at the University of Tokyo. The experiment shown in Fig. 10-3 (Kohama and Kohama, 1984) was intended to confirm that previous results, carried out under arbitrary conditions, could be reproduced under physiological conditions, which had been estimated with a ^{31}P-nuclear magnetic resonance study (Kohama et al., 1984) using living *Physarum* plasmodia (Fig. 10-3, inset).

CA^{2+}-INHIBITION OF P-MYOSIN SLIDING ALONG ACTIN-CABLES

Following the elegant method demonstrating the sliding of myosin-coated beads along actin-cables of *Nitella* or *Chara* internodal cells (Sheetz and Spudich, 1983; Shimmen and Yano, 1984), p-myosin was allowed to move along actin-cables at the rate of 1.4 ± 0.08 μm/sec (mean ± standard error, n=11) in the absence of Ca^{2+} (Fig. 10-4). This velocity was reduced to 0.40 ± 0.08 μm/sec (n=11) in the presence of Ca^{2+}.

Since the beads coated with myosin of rabbit skeletal muscle moved at a velocity independent of Ca^{2+} concentration (Shimmen and Yano, 1985), p-myosin is certainly responsible for the Ca^{2+} control.

CA^{2+}-INHIBITION OF ACTIN-ACTIVATED ATPASE ACTIVITY OF AMOEBAL MYOSIN OF *PHYSARUM*

We have isolated myosin from *Physarum* amoebae (a-myosin) by applying procedures used for purifying p-myosin. The actin-activated ATPase activity of a-myosin is inhibited by Ca^{2+} essentially in the same way as that of p-myosin (Fig. 10-5). Thus, a-myosin shares its Ca^{2+} regulation characteristics with p-myosin.

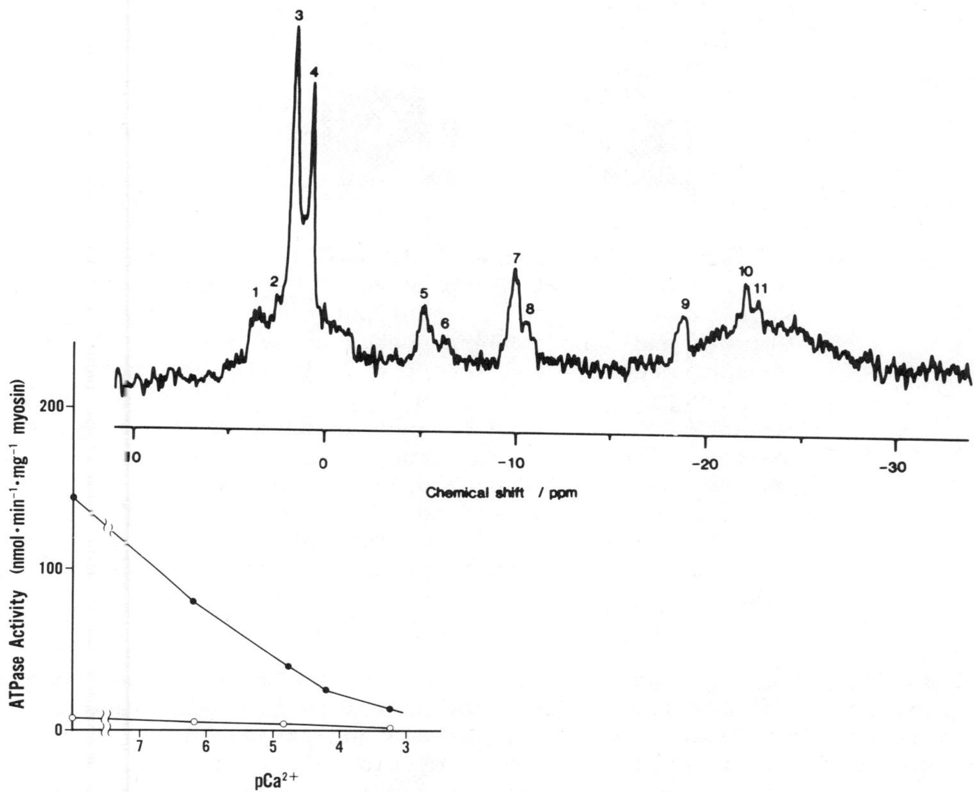

Fig. 10-3. ATPase activities of purified p-myosin in the presence (●) and absence (○) of skeletal muscle actin as a function of Ca^{2+} concentration under intracellular conditions, i.e., 30 mM KCl, 1.0 mM Mg^{2+}, 0.5 mM Mg-ATP and 0.1 mM EGTA-Ca buffer (Kohama and Kohama, 1984). These concentrations of KCl and Ca^{2+} were adopted from Anderson (1964) and Ridgway and Durham (1976), respectively. The concentrations of Mg^{2+} and Mg-ATP and the pH were estimated from the result of ^{31}P nuclear magnetic resonance study on living plasmodia (Inset, Kohama et al., 1984).

STAGE-SPECIFIC MYOSINS

a-Myosin, like p-myosin, consists of a 250K heavy chain and 18K and 14K light chains according to SDS PAGE (Fig. 10-6) and shows a consensus myosin shape, i.e., a two-headed and long-tailed structure (not shown; see Ogihara et al., 1983; Kohama et al.,

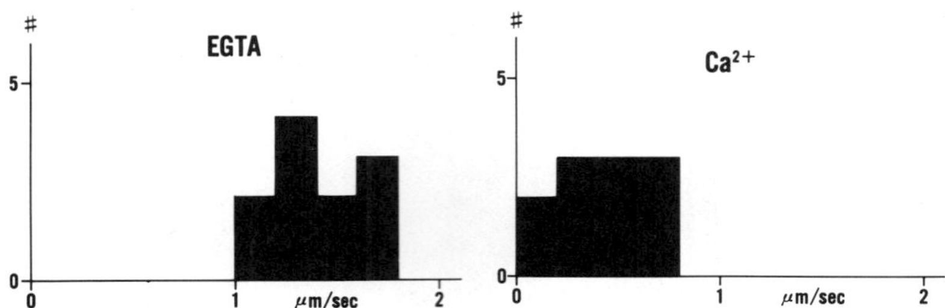

Fig. 10-4. Ca^{2+}-inhibition of the sliding of the p-myosin along actin-cables. Actin-cables in a Chara internodal cell were exposed by the internal perfusion developed by Shimmen and Yano (1984). Latex beads coated with p-myosin were injected into the cell with EGTA solution (3 mM $MgCl_2$, 1mM ATP, 1 mM DTT, 30 mM PIPES 7.0, 170 mM sorbitol, and 5 mM EGTA adjusted to pH 7.0 by 66 mM KOH) or with a Ca^{2+} solution (3 mM $MgCl_2$, 1 mM ATP, 1 mM DTT, 30 mM PIPES pH 7.0, 170 mM sorbitol, 5 mM EGTA and 5 mM $CaCl_2$ adjusted to pH 7.0 by 76 mM KOH). Velocities (μm/sec) of the movement of the beads were observed under Nomarski optics (Kohama and Shimmen, 1985).

1983 for p-myosin shape). However, the peptide maps of heavy chain (Fig. 10-7) and 18K light chain (not shown) of a-myosin are quite different from those of p-myosin (Kohama and Takano-Ohmuro, 1984). A polyclonal antibody against the a-myosin heavy chain specifically reacted with a-myosin heavy chain, and a monoclonal antibody against the p-myosin heavy chain reacted only with the p-myosin chain (Fig. 10-6). Two-dimensional gel electrophoresis, combining IEF and SDS PAGE, also shows a difference between amoebal and plasmodial 18K light chains (Fig. 10-7). These results indicate that both the heavy chain and the 18K light chain are distinct between a-myosin and p-myosin.

On the other hand, the amoebal 14K light chain is not distinguishable from the plasmodial 14K light chain in two-dimensional IEF-SDS PAGE (Fig. 10-8) or in its Ca^{2+}-binding activity (Fig. 10-9). This is in accord with the reports of Kessler et al., 1980 (see Chapter 11). The amoebal 14K chain is not distinct from the plasmodial counterpart in its peptide map (not shown), or reactivity with antibodies against plasmodial 14,000 Mr light chain (Fig. 10-6).

Thus, a-myosin appears to consist of heavy chains and 18K light chains different from those of p-myosin, with 14K light chains identical with those of plasmodia. The fact that the 14K light chain is common between p-myosin and a-myosin gives further support

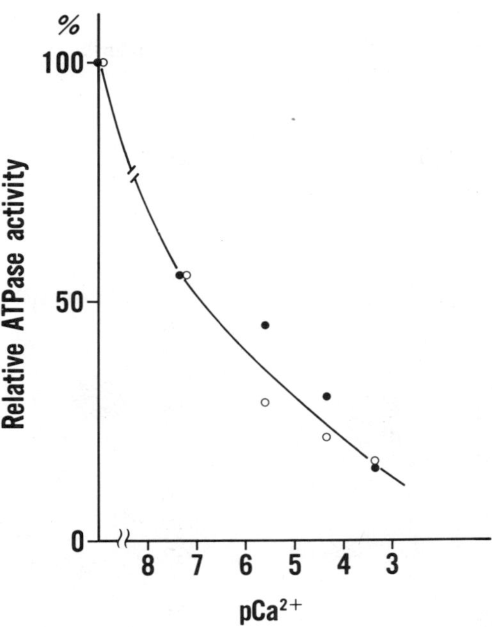

Fig. 10-5. Inhibitory effect of Ca^{2+} on actin-activated ATPase activity of amoebal myosin. Myosin was prepared from heterothallic myxoamoebae (TM4; Uyeda and Furuya, 1985) by essentially the same method as for preparing plasmodial myosin (cf. Table 10-1, Fig. 10-2). The amoebal myosin was subjected to the examination of skeletal muscle actin-activated ATPase activity under the conditions described in the legend to Table 1. The ATPase activities at 100% were 150.0 (●) and 132.0 (○) nmol $min^{-1}mg^{-1}$ myosin.

to the suggestion that this light chain is responsible for Ca^{2+} regulation.

CONCLUSION

Since the establishment around 1960 of the role of Ca^{2+} in muscle contraction, it had been the conviction of muscle biologists that the primary role of Ca^{2+} in the actomyosin-dependent contractile processes is an active one. However, work with Physarum polycephalum has revealed that Ca^{2+} can be an inhibitor of its actin-myosin-ATP interaction.

This inhibitory Ca^{2+} control seems to be exerted through myosin, in which the 14K light chain may play a crucial role. How-

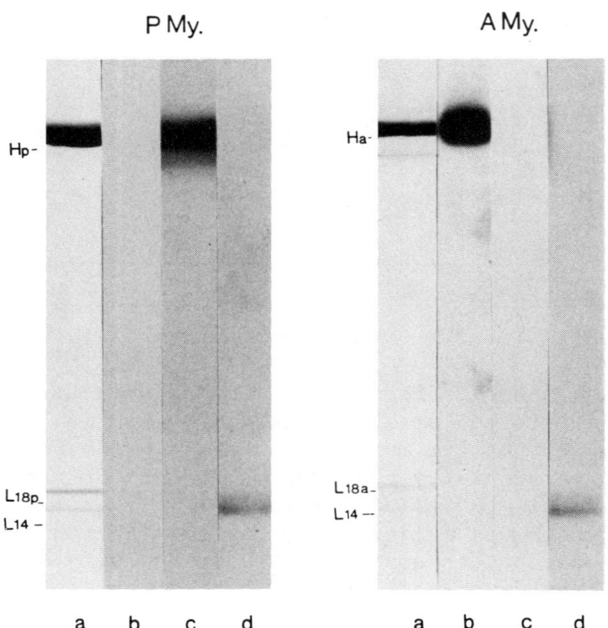

Fig. 10-6. Immunoblots of a-myosin (AMy) and p-myosin (PMy). Heavy chains of a-myosin (Ha) and p-myosin (Hp) and 14K (L14) light chain of p-myosin were purified by the preparative SDS PAGE, and used to immunize guinea pigs (Ha, L14) or mice (Hp). Antibody against Hp is monoclonal, but others are polyclonal. (a) protein stain; (b) immunostains with anti-Ha; (c) with anti-Hp; (d) with anti-L14.

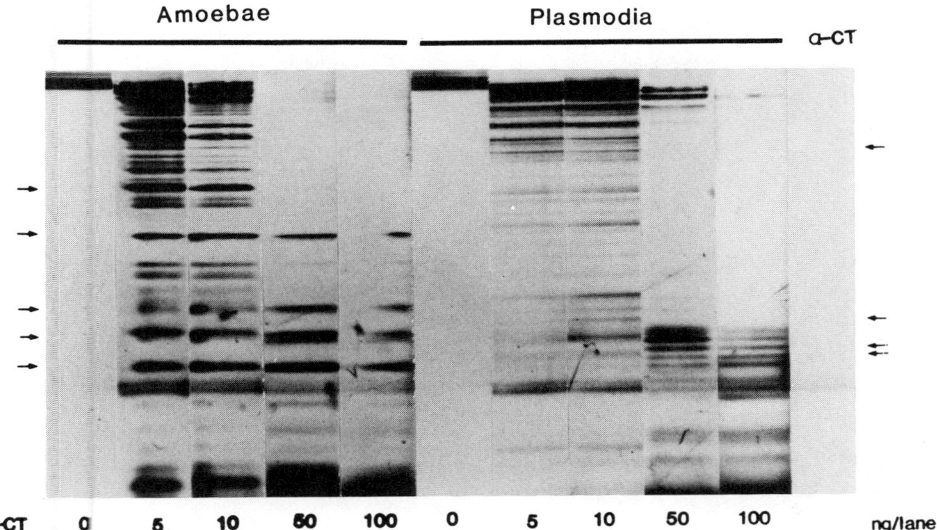

Fig. 10-7. Alpha-chymotryptic peptide map of amoebal and plasmodial myosin heavy chains by the method of Cleveland et al. (1977). Differences in the peptides were shown by the arrows (Kohama and Takano-Ohmuro, 1984).

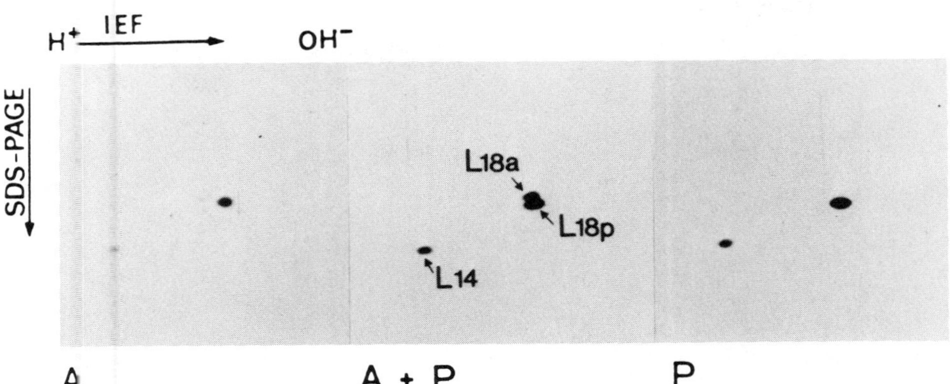

Fig. 10-8. Two-dimensional IEF-SDS PAGE of myosin light chains of a-myosin (A), p-myosin (P), and their mixture (A+P). L18a, 18K light chain of a-myosin; L18p, 18K light chain of p-myosin; L14, 14K light chain (Kohama and Takano-Ohmuro, 1985).

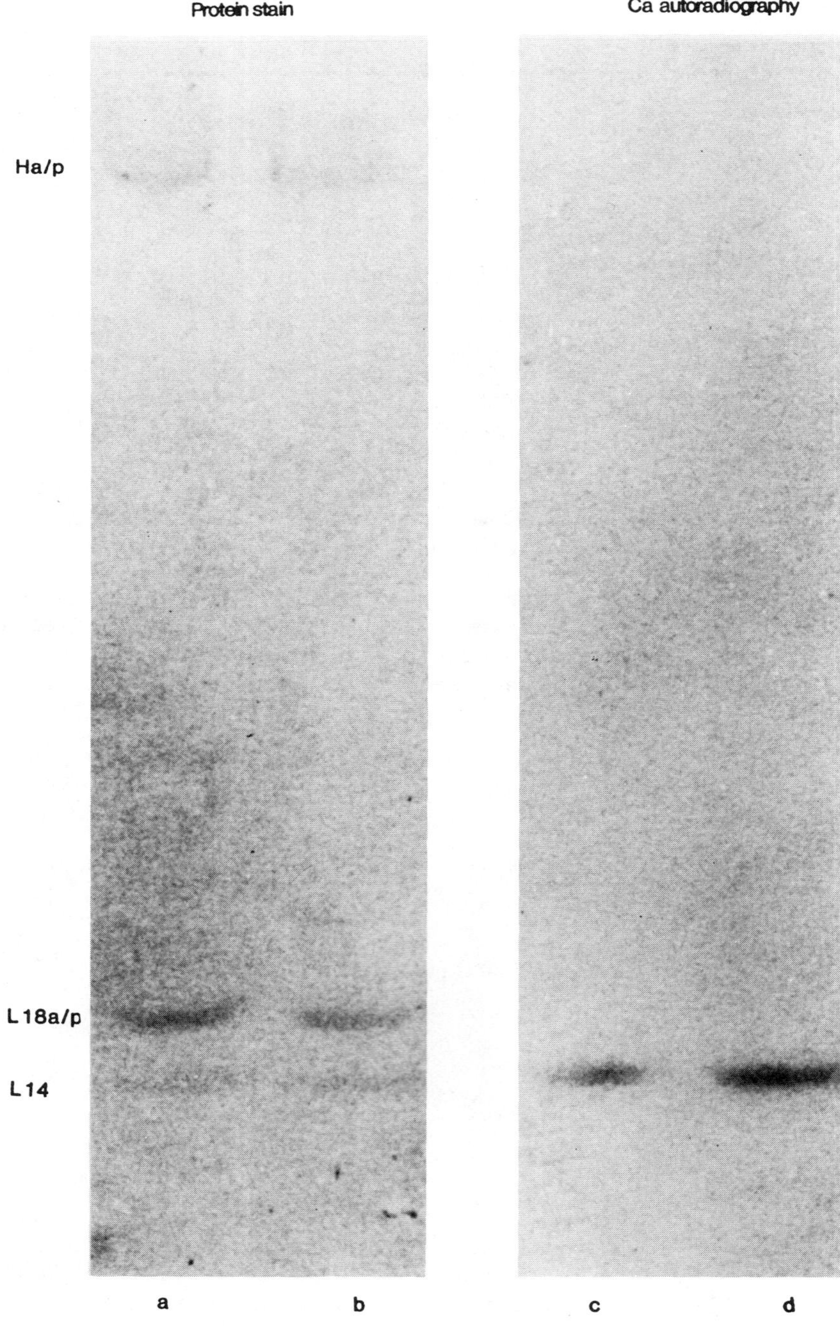

Fig. 10-9. Ca^{2+}-binding to 14K light chain. a-Myosin (a,c) and p-myosin (b,d) were subjected to SDS PAGE followed by Western blots (a,b). The blots were incubated in the solution of 60 mM KCl, 5 mM $MgCl_2$, 10 mM imidazole-HCl pH 6.8 and $^{45}CaCl_2$ and were then subjected to autoradiography (c,d) (Maruyama et al., 1983).

ever, there is a strong indication that the actin side also participates in Ca^{2+}-regulation[1].

The actomyosin system of the amoebal form of Physarum is controlled by Ca^{2+} in the same manner as that of the plasmodial form; this may reflect the fact that the 14K light chain is shared by both amoebal and plasmodial myosins. Other myosin gene products, i.e., heavy chain and 18K light chain, are quite distinct between these two myosins.

ACKNOWLEDGMENTS

This work was partly supported by grants from the Yamada Science Foundation, the Aino Hospital Foundation, and the Iatrochemical Foundation, and also by a Grant-in-Aid for Scientific Research from the Ministry of Education, Science and Culture of Japan.

REFERENCES

Anderson, T. P., 1964, Regional differences in ion concentration in migrating plasmodia, in: "Primitive Motile System in Cell Biology," R. E. Allen and N. Kamiya, eds., p. 128, Academic Press, New York.

Chambers, R., and Hale, H. P., 1932, The formation of ice in protoplasm, Proc. Roy. Soc., 110 B:336.

Cheung, W. Y., 1970, Cyclic 3',5' nucleotide phosphodiesterase: demonstration of an activator, Biochem. Biophys. Res. Commun., 38:533.

Cleveland, D. W., Fischer, S. G., Kirschner, M. W., and Laemmli, U. K., 1977, Peptide mapping by limited proteolysis in sodium dodecyl sulfate and analysis by gel electrophoresis, J. Biol. Chem., 252:1102.

Ebashi, S., 1963, Third component participating in the superprecipitation of 'natural actomyosin', Nature, 200:1010.

Ebashi, S., and Endo, M., 1968, Calcium ion and muscle contraction, Progr. Biophys. Mol. Biol., 18:123.

Ebashi, S., and Kodama, A., 1965, A new protein factor promoting aggregation of tropomyosin, J. Biochem., 58:107.

Ebashi, S., Nonomura, Y., Toyo-oka, T., and Katayama, E., 1976, Regulation of muscle contraction by the calcium-troponin-tropomyosin system, in: "Calcium in Biological Systems," C. J. Duncan, ed., p. 349, Cambridge University Press, London.

---

[1] For the role of actin and actin-related proteins in the Ca^{2+}-regulation of Physarum actomyosin, see also Kohama (1981), Sugino and Matsumura (1983), Kohama and Kohama (1984), and Chapter 9.

Ebisawa, K., and Nonomura, Y., 1985, Enhancement of actin-activated myosin ATPase by an 84K Mr actin-binding protein in vertebrate smooth muscle, J. Biochem., 98:1127.

Hartshorne, D. J., and Siemankowski, R. F., 1981, Regulation of smooth muscle actomyosin, Ann. Rev. Physiol., 43:519.

Hatano, S., 1970, Specific effect of Ca^{2+} on movement of plasmodial fragment obtained by caffeine treatment, Exp. Cell Res., 61:199.

Hatano, S., and Tazawa, M., 1968, Some properties of a contractile protein in a myxomycete plasmodium, Biochim. Biophys. Acta, 154:507.

Hayama, T., Shimmen, T., and Tazawa, M., 1979, Participation of Ca^{2+} in cessation of cytoplasmic streaming induced by membrane excitation in Characeae internodal cells, Protoplasma, 99:305.

Hayama, T., and Tazawa, M., 1980, Ca^{2+} reversibly inhibits active rotation of chloroplasts in isolated cytoplasmic droplets in Chara, Protoplasma, 102:1.

Heilbrunn, L. V., 1940, The action of calcium on muscle protoplasm, Physiol. Zool., 13:88.

Heilbrunn, L. V., and Wiercinski, F. J., 1947, The action of various cations on muscle protoplasm, J. Cell. Comp. Physiol., 29:15.

Kakiuchi, S., Yamazaki, R., and Nakajima, H., 1969, Studies on brain phosphodiesterase (2), Bull. Japan. Neurochem. Soc., 8:17.

Kakiuchi, S., Yamazaki, R., and Nakajima, H., 1970, Properties of a heatstable phosphodiesterase activating factor isolated from brain extract, Proc. Jpn Acad., 46B:587.

Kamada, T., and Kinosita, H., 1943, Disturbances initiated from naked surface of muscle protoplasm, Japan. J. Zool., 10:469.

Kato, T., and Tonomura, Y., 1975, Ca^{2+}-sensitivity of actomyosin ATPase purified from Physarum polycephalum, J. Biochem., 77:1127.

Kawamura, M., and Nagano, K., 1975, Calcium ion-dependent ATP pyrophosphohydrolase in Physarum polycephalum, Biochim. Biophys. Acta, 397:207.

Keil, E. M., and Sichel, F. J. M., 1936, The injection of aqueous solution, including acetylcholine, into the isolated muscle fiber, Biol. Bull., 71:402.

Kendrick-Jones, J., Lehman, W., and Szent-Györgyi, A. G., 1970, Regulation in molluscan muscles, J. Mol. Biol., 54:313.

Kessler, D., Eisenlohr, L. C., Lathwell, M. J., Huang, J., Taylor, H. C., Godfrey, S. D., and Spady, M. L., 1980, Physarum myosin light chain binds calcium, Cell Motil., 1:63.

Kohama, K., 1981, Ca-dependent inhibitory factor for the myosin-actin-ATP interaction of Physarum polycephalum, J. Biochem., 90:1829.

Kohama, K., Craig, R., Kohama, T., and Kendrick-Jones, J., 1983, Characterization of Ca^{2+}-sensitive Physarum myosin, Europ. J. Cell Biol., suppl., 1:25.

Kohama, K., and Kendrick-Jones, J., 1982, Negative Ca^{2+}-sensitivity of actin-activated Mg-ATPase activity of myosin from Physarum polycephalum, J. Muscle Res. Cell Motil., 3:491.

Kohama, K., Kobayashi, K., and Mitani, S., 1980, Effects of Ca ion and ADP on superprecipitation of myosin B from slime mold, Physarum polycephalum, Proc. Jpn. Acad., 56B:591.

Kohama, K., and Kohama, T., 1984, Myosin confers inhibitory Ca^{2+}-sensitivity on actin-myosin-ATP interaction of Physarum polycephalum under physiological conditions, Proc. Jpn. Acad., 60B:435.

Kohama, K., and Shimmen, T., 1985, Inhibitory Ca^{2+}-control of movement of beads coated with Physarum myosin along actin-cables in Chara internodal cells, Protoplasma, 129:88.

Kohama, K., and Takano-Ohmuro, H., 1984, Stage specific myosins from amoeba and plasmodium of slime mold, Physarum polycephalum, Proc. Jpn. Acad., 60B:431.

Kohama, K., and Takano-Ohmuro, H., 1985, Stage specific myosins from amoeba and plasmodium of Physarum polycephalum, Develop. Growth Diff., 27:510.

Kohama, K., Tanokura, S., and Yamada, K., 1984, ^{31}P nuclear magnetic resonance studies of intact plasmodia of Physarum polycephalum, FEBS Lett., 76:161.

Marston, S. B., 1982, The regulation of smooth muscle contractile proteins, Prog. Biophys. Mol. Biol., 41:1.

Maruyama, K., Mikawa, T., and Ebashi, S., 1983, Detection of calcium binding proteins by ^{45}Ca autoradiography on nitrocellulose membrane after sodium dodecyl sulfate gel electrophoresis, J. Biochem., 95:511.

Nachmias, V. T., and Asch, A., 1974, Actin mediated calcium dependency of actomyosin in a myxomycete, Biochem. Biophys. Res. Comm., 60:654.

Ogihara, S., Ikebe, M., Takahashi, K., and Tonomura, Y., 1983, Requirement phosphorylation of Physarum myosin heavy chain for thick filament formation, actin activation of Mg^{2+}-ATPase activity, and Ca^{2+}-inhibition, J. Biochem., 93:205.

Ozawa, E., Hosoi, K., and Ebashi, S., 1967, Reversible stimulation of muscle phosphorylase b kinase by low concentration of calcium ions, J. Biochem., 61:531.

Ridgway, E. B., and Durham, A. C. H., 1976, Oscillations of calcium ion concentrations in Physarum polycephalum, J. Cell Biol., 69:223.

Ringer, S., 1883, A further contribution regarding the influence of the blood on the contraction of the heart, J. Physiol., 4:29.

Sheetz, J. P., and Spudich, J. A., 1983, Movement of myosin-coated fluorescent beads on actin cables in vitro, Nature, 303:31.

Shimmen, T., Tominaga, Y., and Tazawa, M., 1984, Involvement of Ca^{2+} and flowing endoplasm in recovery of cytoplasmic streaming after K^{+}-induced cessation, Protoplasma, 121:178.

Shimmen, T., and Yano, Y., 1985, Ca^{2+} regulation of myosin sliding along Chara actin bundles mediated by native tropomyosin, Proc. Jpn. Acad., 61B:86.

Shimmen, T., and Yano, Y., 1984, Active sliding movement of latex beads coated with skeletal muscle myosin on Chara actin bundles, Protoplasma, 121:132.

Sobue, K., Morimoto, K., Inui, M., Kanda, K., and Kakiuchi, S., 1982, Control of actin-myosin interaction of gizzard smooth muscle by calmodulin- and caldesmon-limited flip-flop mechanisms, Biomed. Res., 3:188.

Sugino, H., and Matsumura, F., 1983, Fragmin induces tension reduction of actomyosin threads in the presence of micromolar levels of Ca^{2+}, J. Cell Biol., 96:199.

Sutoh, K., Iwane, M., Matsuzaki, F., Kikuchi, M., and Ikai, A., 1984, Isolation and characterization of a high molecular weight actin-binding protein from Physarum polycephalum plasmodia, J. Cell Biol., 98:1611.

Tominaga, Y., Shimmen, T., and Tazawa, M., 1983, Control of cytoplasmic streaming by extracellular Ca^{2+} in permeabilized Nitella cell, Protoplasma, 116:75.

Ueda, T., Gotz von Olenhusen, K., and Wohlfarth-Bottermann, K.-E., 1978, Reaction of the contractile apparatus in Physarum to injected calcium, ATP, ADP, and 5'-AMP, Cytobiologie, 18:76.

Uyeda, T. Q. P., and Furuya, M., 1985, Cytoskeletal changes visualized by fluorescence microscopy during amoeba-to-flagellate and flagellate-to-amoeba transformation in Physarum polycephalum, Protoplasma, 126:221.

White, H. D., 1982, Special instrumentation and techniques for kinetic studies of contractile systems, Methods Enzymol., 85B:698.

Williamson, R. E., and Ashley, C. C., 1982, Free Ca^{2+} and cytoplasmic streaming in the alga Chara, Nature, 296:647.

Yoshimoto, Y., and Kamiya, N., 1984, ATP and calcium-controlled contraction in a saponin model of Physarum polycephalum, Cell Struct. Funct., 9:135.

Yoshimoto, Y., Matsumura, F., and Kamiya, N., 1981, Simultaneous oscillations of Ca^{2+} efflux and tension generation in the premeabilized plasmodial strand of Physarum, Cell Motil., 1:432.

Chapter 11: PHYSARUM MYOSIN BINDS CA^{2+}: RESULTS FROM ELECTROPHORESIS AND EQUILIBRIUM DIALYSIS EXPERIMENTS

Dietrich Kessler[1] and Beverly K. Dolberg[2]

[1]Department of Biology
Colgate University
Hamilton, NY USA

[2]Department of Biology
Haverford College
Haverford, PA USA

INTRODUCTION

We wish to understand the molecular mechanism of calcium-regulated cytoplasmic movements in the plasmodium of Physarum polycephalum. Early studies demonstrated that the contractile activity of the cytoplasm, presumably an expression of the actomyosin ATPase activity, is influenced by Ca^{2+} (Hatano, 1970; Matthews, 1977). Recently it has become evident that the state of polymerization and aggregation of the actin filaments may also be Ca^{2+}-regulated (see Chapter 9).

We have been studying the properties of Physarum myosin. Like myosins from most other sources, Physarum myosin appears to be a hexamer containing six polypeptide subunits divided into three classes: a class of heavy chains and two classes of light chains called LC-1 and LC-2 in Physarum (Kessler, 1982). Ca^{2+} binds with high affinity to one of the polypeptide subunits of the myosin, the LC-2 class of light chains (Kessler et al., 1980). We are currently studying the Ca^{2+}-binding properties of the myosin in physiological conditions, i.e., micromolar Ca^{2+} and millimolar Mg^{2+} concentrations.

Our goal is to discover whether Ca^{2+} binding to myosin regulates actomyosin contraction. Recently, Kohama and Kendrick-Jones (1982) and Kohama and Ebashi (Chapter 10) have proposed that the actomyosin ATPase activity in this organism is negatively regulated by Ca^{2+}, i.e., inhibited by high Ca^{2+} concentration (10^{-5} M) and enhanced by low Ca^{2+} (10^{-7}M). Here we will examine the evidence for Ca^{2+} binding to Physarum myosin in light of this hypothesis of inhibitory Ca^{2+} regulation of Physarum actomyosin ATPase activity.

PHYSARUM MYOSIN LC-2 BINDS CA^{2+}: SDS PAGE

Calmodulin, troponin C, and several other Ca^{2+}-binding proteins display more rapid mobility during polyacrylamide gel electrophoresis in sodium dodecyl sulfate (SDS PAGE) if Ca^{2+} is present in the sample (Van Eldik et al., 1980).

We observed this calcium-dependent electrophoretic mobility shift occurring with one class of Physarum myosin light chains, LC-2 (Kessler et al., 1980), using samples of Physarum myosin-enriched actomyosin purified by the procedure of Nachmias (1982). We employed the discontinuous SDS PAGE system of Laemmli (1970), using 15% or 17% acrylamide in running gels cast in 1-mm-thick slabs (Studier, 1973). Samples of Physarum actomyosin were boiled 1-3 min in Laemmli sample buffer containing 1 mM EGTA [ethylene glycol bis (β-aminoethyl ether) N,N'-tetraacetic acid]. Before electrophoresis, stock solutions of $CaCl_2$, Cd $(NO_3)_2$, or $LaCl_3$ were added to some samples to a final concentration of 5 mM. No attempt was made to change the cation concentration of the electrophoretic buffer in the reservoirs or the gel itself. Rabbit skeletal muscle myosin was run in the same conditions, and the relative molecular weights (M_r) of the light chains in the samples were calculated on the assumption of molecular weights of 25,000 for rabbit myosin alkali-1 light chain and 16,000 for alkali-2 light chain (Weeds and Lowey, 1971).

We calculated a M_r of 17,700 for Physarum myosin LC-1 in the presence or absence of Ca^{2+}, very similar to the value of 18,000 reported by Kohama and Ebashi (Chapter 10). With our SDS PAGE system, Physarum myosin LC-2 showed a Ca^{2+}-dependent electrophoretic mobility shift: 16,900 in EGTA and 16,100 in Ca^{2+}. Cd^{2+} and La^{+3} gave the same effect as Ca^{2+}. However, the addition of Mg^{2+} up to 200 mM in the sample did not influence the electrophoretic mobility of LC-2.

Burgess et al. (1980) have devised an electrophoretic system in which either 0.1 mM EDTA or $CaCl_2$ is present not only in the sample, but also in the gel and reservoir buffers. The relative mobilities of the polypeptide bands in the sample are compared with protein standards in each gel. In these conditions, the Ca^{2+} dependent mobility shift of Ca^{2+}-binding proteins is enhanced. For example, bovine brain calmodulin (16,700 MW by sequence analysis) has an M_r of 15,400 after electrophoresis in the 0.1 mM $CaCl_2$ gel and 21,000 in 0.1 mM EDTA. Burgess (personal communication) ran samples of our Physarum myosin-enriched actomyosin in this gel system and found Physarum myosin LC-2 has an M_r of 15,400 in Ca^{2+} and 16,500 in EDTA.

This difference in Ca^{2+}-dependent electrophoretic mobility shift between bovine brain calmodulin and Physarum myosin LC-2, and the fact that J. R. Dedman (personal communication) found no immunological crossreaction between Physarum myosin and antiserum directed against rat testis calmodulin, both support the conclusion that Physarum myosin LC-2 is not identical with calmodulin.

Our electrophoretic experiments strongly suggest that Physarum myosin LC-2, but not LC-1, is a Ca^{2+}-binding protein. Kohama and Ebashi (Chapter 10) also report a Ca^{2+}-dependent mobility shift during electrophoresis for Physarum myosin LC-2. Their reported value of 14,000 M_r for LC-2 in the presence of Ca^{2+} may reflect the choice of molecular weight standards used by different investigators to calculate the relative molecular weight of the polypeptides.

PHYSARUM MYOSIN BINDS CA^{2+}: EQUILIBRIUM DIALYSIS

Using the equilibrium dialysis method of Potter et al. (1983), we compared the Ca^{2+}-binding properties of purified rabbit skeletal muscle myosin with Physarum myosin-enriched actomyosin (Kessler and Dolberg, 1985). We verified the results of Holroyde et al. (1979) that rabbit myosin binds 1.5 to 2 moles of Ca^{2+} per mole of myosin with a binding constant of about 2.6 X 10^7 M^{-1} in the absence of Mg^{2+} (Fig. 11-1). In the presence of 0.03 mM Mg^{2+}, the Ca^{2+} binding is greatly reduced. At physiological Mg^{2+} concentration (1 mM), no Ca^{2+} binding can be detected, since the Ca^{2+}-binding sites on rabbit myosin are fully occupied by Mg^{2+}. For this reason they are referred to as nonspecific Ca^{2+}-binding sites or Ca^{2+}/Mg^{2+} binding sites and are believed not to function in the regulation of skeletal muscle actomyosin ATPase activity by Ca^{2+}.

With the same equilibrium dialysis technique, we found Physarum myosin-enriched actomyosin binds about 3.8 moles of Ca^{2+} per mole of myosin in the absence of Mg^{2+}. Evidence for binding sites with two different affinities was obtained with approximate binding constants of 1.2 X 10^7 M^{-1} and 1.04 X 10^6 M^{-1}. We estimated the myosin concentration by densitometric analysis of stained polypeptide bands after SDS PAGE. The actin contamination of the sample was 20-30% of the total protein. In the presence of 2 mM Mg^{2+}, a binding constant for Ca^{2+} of 5 X 10^6 M^{-1} was estimated with Physarum myosin-enriched actomyosin, but only 0.36 mole Ca^{2+} was bound per mole myosin (Fig. 11-2). The data points in Fig. 11-2 are more scattered than those in Fig. 11-1 due to the difficulty in obtaining high Physarum myosin concentrations, resulting in less accurate estimates of the binding constants and number of binding sites.

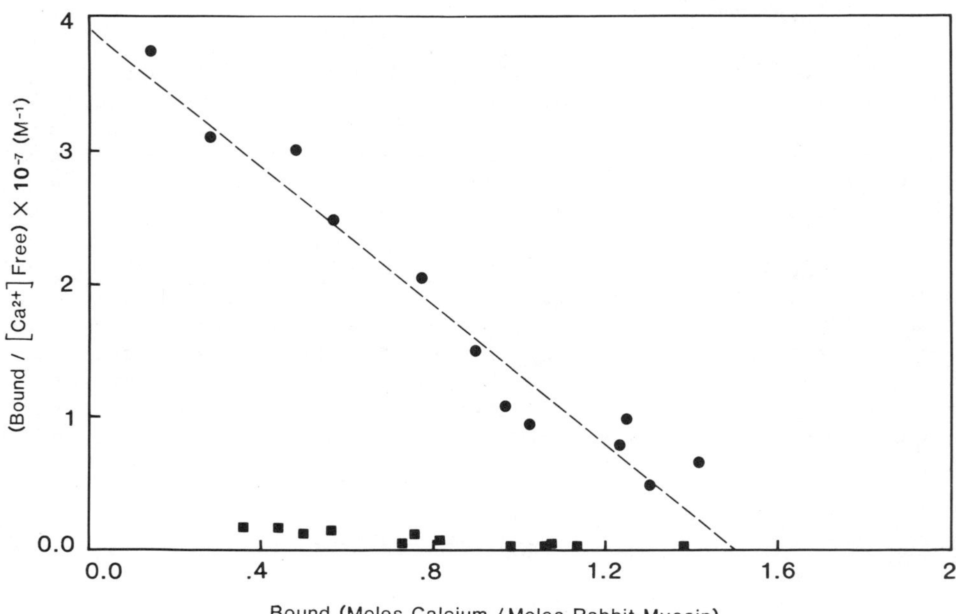

Fig. 11-1. Scatchard plot of Ca^{2+} binding to rabbit skeletal muscle myosin during equilibrium dialysis in the presence and absence of 0.03 mM Mg^{2+}. OmM Mg^{2+} (●): K_{assoc} = 2.57 X 10^7 M^{-1}, n = 1.5. 0.03 mM Mg^{2+} (■): K_{assoc} = 1.88 X 10^6 M^{-1}, n = 1.14. 1.00 mM Mg^{2+}: no discernible binding with this technique.

Nitrocellulose transfers of SDS PAGE gels of this preparation may be incubated in a solution containing ^{45}Ca and autoradiographed by use of the technique of Maruyama et al. (1984). Only the polypeptide band containing *Physarum* myosin LC-2 binds Ca^{2+} by this method. On the assumption that all classes of polypeptides responsible for Ca^{2+} binding in the native *Physarum* actomyosin sample are still capable of Ca^{2+} binding after SDS PAGE and transfer to nitrocellulose, our results show that LC-2 is the site of Ca^{2+}

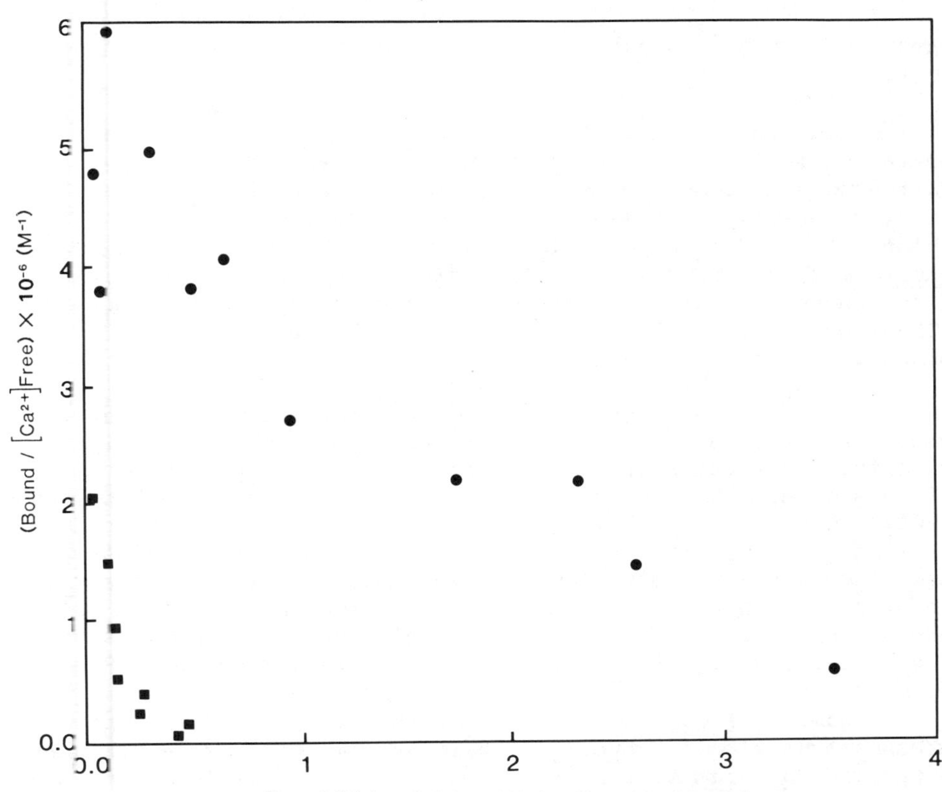

Fig. 11-2. Scatchard plot of Ca^{2+} binding to Physarum myosin-enriched actomyosin during equilibrium dialysis in the presence (■) and absence (●) of 2 mM Mg^{2+}. The slope of the line for binding in the absence of Mg^{2+} may indicate two sets of binding sites with slightly different affinities, K_1 = ca 2 X 10^6 M^{-1}, n = ca 0.9 and K_2 = ca 9 X 10^5 M^{-1}, n_2 = ca 4. The addition of Mg^{2+} gives a K_{assoc} of ca 5 X 10^6 M^{-1} and n = ca 4.

binding in the equilibrium dialysis experiment. The presence of a Ca^{2+}-binding contaminant in our preparation appears to be ruled out by the absence of any polypeptide band other than myosin LC-2 binding ^{45}Ca after autoradiography.

NATURE OF THE CA^{2+}-SENSITIVE SYSTEM REGULATING ACTOMYOSIN ATPase ACTIVITY IN *PHYSARUM*

Kohama has recently found biochemical evidence that Ca^{2+} can act as an inhibitor of the *Physarum* actomyosin ATPase activity. Furthermore, a good deal of the inhibitory activity involves Ca^{2+} binding directly to the myosin, probably to LC-2 (Kohama and Ebashi, Chapter 10). This hypothesis is supported by our evidence that *Physarum* myosin LC-2 is a Ca^{2+}-binding polypeptide. It is the first indication with a purified contractile system that actomyosin ATPase activity can be inhibited rather than activated by Ca^{2+} binding to a myosin light chain.

The theory requires a new molecular model to explain how movements of the light chains of the myosin upon Ca^{2+} binding interfere with the interaction of myosin with actin. In *Physarum*, both classes of myosin light chains may function in the regulatory process. In support of this idea, the two classes of light chains from *Physarum* myosin appear to share properties that ordinarily are associated exclusively with the regulatory class of light chains in myosins from other systems: *Physarum* myosin LC-2 binds Ca^{2+}, whereas LC-1 is phosphorylated and can occupy the binding site of the regulatory light chain in scallop myosin (Nachmias, 1981a,b).

A number of questions regarding the role of Ca^{2+} in actomyosin ATPase regulation in *Physarum* remain unanswered. As better characterized *Physarum* myosin and actomyosin preparations become available, the stoichiometry of Ca^{2+} binding to the myosin can be more accurately determined. When the presence of Ca^{2+}-binding contaminants is unequivocally excluded, and the problem of damage or loss of the light chains during purification is avoided, then the model for regulation of *Physarum* actomyosin ATPase activity by direct binding of Ca^{2+} to myosin LC-2 can be fully accepted. The role of phosphorylation in modifying the actomyosin ATPase activity, as well as the possibility of an additional Ca^{2+} regulatory protein in the plasmodium (Kohama, 1981), can be more carefully explored.

ACKNOWLEDGMENTS

This work was partly supported by grants from the National Science Foundation (PCM-8016620) and the Research Corporation (C-1157) to D. Kessler.

REFERENCES

Burgess, W. H., Jemiolo, D. K., and Kretsinger, R. H., 1980, Interaction of calcium and calmodulin in the presence of sodium dodecyl sulfate, Biochim. Biophys. Acta, 623: 257.

Hatano, S., 1970, Specific effect of Ca^{2+} on movement of plasmodial fragment obtained by caffeine treatment, Exp. Cell Res., 61: 199.

Holroyde, M. J., Potter, J. D., and Solaro, R. J., 1979, The calcium binding properties of phosphorylated and unphosphorylated cardiac and skeletal myosins, J. Biol. Chem., 254: 6478.

Kessler, D., 1982, Plasmodial structure and motility, in: "Cell Biology of Physarum and Didymium", Vol. I, H. C. Aldrich and J. W. Daniel, eds., Academic Press, New York.

Kessler, D., and Dolberg, B. K., 1985, Physarum myosin binds calcium: equilibrium dialysis results, J. Cell Biol., 101: 162a.

Kessler, D., Eisenlohr, L. C., Lathwell, M. J., Huang, J., Taylor, H. C., Godfrey, S. D., and Spady, M. L., 1980. Physarum myosin light chain binds calcium, Cell Motil., 1:63.

Kohama, K., 1981, Ca-dependent inhibitory factor for the myosin-actin-ATP interaction of Physarum myosin, J. Biochem., 90: 1829.

Kohama, K., and Kendrick-Jones, J., 1982, Negative Ca^{2+}-sensitivity of actin-activated Mg-ATPase activity of myosin from Physarum polycephalum, J. Muscle Res. Cell Motil., 3: 491.

Laemmli, U. K., 1970, Cleavage of structural proteins during the assembly of the head of bacteriophage T_4, Nature, 227: 680.

Maruyama, K., Mikawa, T., and Ebashi, S., 1984, Detection of calcium binding proteins by ^{45}Ca autoradiography on nitrocellulose membrane after sodium dodecyl sulfate gel electrophoresis, J. Biochem., 95: 511.

Matthews, L. M., Jr., 1977, Calcium ion regulation in caffeine derived microplasmodia of Physarum polycephalum, J. Cell Biol., 72: 502.

Nachmias, V. T., 1981a, Hybrids of Physarum myosin light chains and desensitized scallop myofibrils, J. Cell. Biol., 90: 408.

Nachmias, V. T., 1981b, Physarum myosin light chain one: A potential regulatory factor in cytoplasmic streaming, Protoplasma, 109:13.

Nachmias, V. T., 1982, Purification of myosin from Physarum polycephalum, in: "Cell Biology of Physarum and Didymium", Vol. II, H. C. Aldrich, and J. W. Daniel, eds., Academic Press, New York.

Potter, J. D., Strang-Brown, P., Walker, P. L., and Iida, S., 1983, Ca^{2+} binding to calmodulin, Methods Enzymol., 102: 135.

Studier, F. W., 1973, Analysis of bacteriophage T7 early RNAs and proteins on slab gels, J. Mol. Biol., 79: 237.

Van Eldik, L. J., Piperno, G., and Watterson, D. M., 1980, Comparative biochemistry of calmodulins and calmodulin-like proteins, Ann. N.Y. Acad. Sci., 356: 36.

Weeds, A. G., and Lowey, S., 1971, Substructure of myosin molecule. II. The light chains of myosin, J. Mol. Biol., 61: 701.

Chapter 12: KINETICS OF MODULATOR-ACTIN INTERACTIONS: A COMPARISON OF *PHYSARUM* FRAGMIN WITH ACTIN MODULATORS FROM DIFFERENT MUSCLE TYPES

Horst Hinssen[1*], Frank E. Engels[1], and Jochen D'Haese[2]

[1]Institute of Molecular Biology
Austrian Academy of Sciences
Salzburg, AUSTRIA

[2]Institute of Zoology II
University of Düsseldorf, FRG

*Present address:
California Institute of Technology
Pasadena, CA USA

INTRODUCTION

Fragmin is one of several actin-binding proteins found in the plasmodia of *Physarum polycephalum* and has been shown to be a powerful regulatory protein of the polymeric state of actin (Hasegawa et al., 1980; Hinssen 1981a,b; Sugino and Hatano 1982; see also Chapters 9 and 13). It is thought to be part of an actin-associated system in plasmodia, which enables rapid transitions of filamentous into non-filamentous or low-polymeric actin and vice versa. Such transitions are required for the dynamic endoplasm-ectoplasm conversions. Indeed, fragmin has been localized immunocytochemically to the fibrillar system of the plasmodial ectoplasm, thus co-localizing with actin and myosin (Osborn et al., 1983).

Fragmin belongs to a functionally distinct type of actin-binding protein that has also been found in other non-muscle cells, such as gelsolin in macrophages (Yin et al., 1980), blood platelets (Bryan and Kurth, 1984), villin in intestinal brush border cells (Glenney et al., 1981), severin in *Dictyostelium* (Yamamoto et al., 1982), and a similar 45K protein in sea urchin eggs (Wang and Spudich, 1984). Because of their relatively complex pattern of interaction with actin, the term "actin modulator" has been suggested for this type of protein (Hinssen, 1981a). Actin modulators interact with either G- or F-actin in a Ca-dependent manner; they can each promote the formation of short filaments by nucleating actin polymerization and reduce effectively the lengths of pre-existing filaments.

Recent investigations have shown that these proteins exist not only in highly motile or morphogenetically very active cells, but also in muscle cells with their comparatively stable morphology, where gross changes of the polymeric state of actin at high rate do not occur (Hinssen et al., 1984; Hinssen et al., 1985). At least a portion of the actin modulator material in smooth and skeletal muscle cells is incorporated into the myofibrillar system (Hinssen et al., 1985). So far the relationship between actin modulators from such different cell types as Physarum and muscle is mainly functional, on the basis of their effects on actin polymerization. They differ, however, in many other respects, e.g., molecular weight, peptide maps, and immunological properties.

In this communication, we have investigated the interaction of fragmin and actin in more detail, especially with regard to the effects on F-actin, and at the same time we have directly compared fragmin with actin modulators we have purified from various muscle types.

FRAGMENTATION OF ACTIN FILAMENTS

Fragmin and the following actin modulators from various muscles were purified to homogeneity: PSAM (pig stomach actin modulator) from mammalian smooth muscle; ChGAM (chicken gizzard actin modulator) from avian smooth muscle; RSAM (rabbit skeletal actin modulator) from mammalian cross-striated muscle; and EWAM (earthworm actin modulator) from the obliquely striated muscle of the annelid Lumbricus terrestris. Fragmin has a molecular weight of 42K (Hasegawa et al., 1980; Hinssen 1981a), whereas the vertebrate muscle modulators have apparent molecular weights of 85, 87, and 89K for ChGAM, PSAM, and RSAM, respectively. EWAM shows two components of 44 and 46K in SDS electrophoresis. All modulators form strong complexes with G-actin for which a stoichiometry of 1:1 was found in the case of fragmin (Hasegawa et al., 1980), whereas two actins per modulator molecule were bound by all other modulators, including EWAM, as was determined by gel chromatography. One of the actins from the 2:1 complexes is easily released at $[Ca^{++}] \leq 10^{-7}$ M, leaving a stable 1:1 complex (data not shown).

A most specific property of actin modulator, which (in the case of Physarum) clearly distinguishes fragmin from the related capping proteins Cap 42a and Cap 42b (Maruta et al., 1984; see also Chapter 13), is the ability to rapidly shorten actin filaments, supposedly by a direct-severing action. This reaction occurs so fast that its time course could be followed closely only by the specifically modified viscometric assay shown in Fig. 12-1A, which allowed measurements at 2-sec intervals.

The decrease of actin viscosity after addition of the modulators (Fig. 12-1A) reflects the concomitant decrease in actin

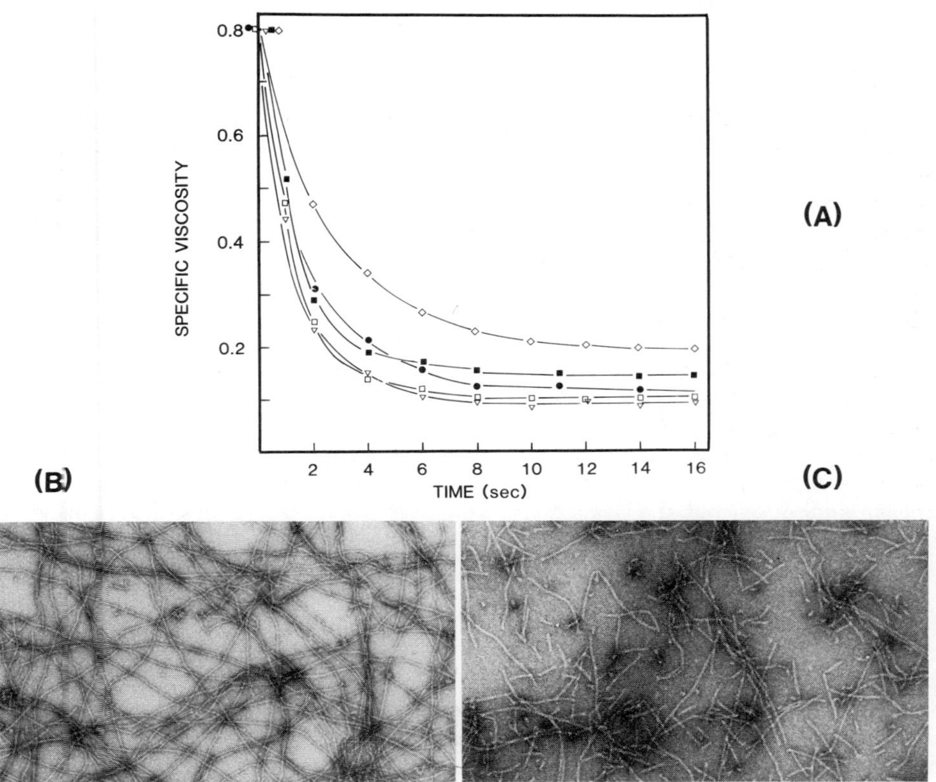

Fig. 12-1. Time course of modulator-actin interactions. (A) Decrease of actin viscosity after addition of Physarum fragmin (●) or muscle actin modulators (PSAM▽; ChGAM□; RSAM◇; EWAM■). Modulators were rapidly mixed with skeletal muscle F-actin (1 mg/ml in 2 mM $MgCl_2$, 100 mM KCl, 10 mM imidazole, pH 7.4, and 0.2 mM $CaCl_2$) at a molar ratio of 1:100 (modulator:actin). The reactions were stopped after different time intervals by addition of EGTA to adjust [Ca^{++}] to 10^{-7} M. The 0-sec time-points were obtained by addition of EGTA prior to the modulators. (B,C) Electron micrographs of negatively stained actin samples reacted with fragmin at 0 sec (B, control) and 6 sec (C). 50,000 X.

filament length. This is clearly shown in the electron micrographs from negatively stained actin samples (Fig. 12-1B,C). The reaction rate is very similar for all proteins investigated, and steady state is usually obtained within 6-8 sec, with RSAM slightly slower than the other modulators. After fragmentation of actin filaments, the modulators remain at one end of an actin fragment, as was shown for PSAM by immuno-gold labeling with anti-PSAM on negatively stained actin fragments (not displayed) and also for fragmin by indirect evidence (Sugino and Hatano, 1982).

The rapid reaction of all modulators with F-actin shown here strongly favors a direct severing of filaments as the mechanism of interaction. This requires a binding of the modulator to actin molecules at random positions within the filament and without preference for capping of the terminal actins. The alternative possible mechanism for a shortening of actin filaments--a combined filament capping and nucleation of polymerization--requires only a binding of the modulator to terminal actin molecules and uses the intrinsic polymerization-depolymerization equilibrium mechanisms. However, this type of reaction can be excluded for the modulator-actin interaction because it is incompatible with the observed reaction rate. It is also unlikely that a spontaneous fragmentation of actin filaments and a subsequent capping by the modulator causes the observed filament shortening. Though spontaneous fragmentation occurs (Wegner and Savko, 1982), it is heavily dependent on the salt conditions and observed only in the presence of divalent cations and the absence of potassium ion. We have found no such dependence in the modulator-actin interactions.

The precise molecular mechanism of severing is not clear yet. Both an allosteric effect or a competitive type of reaction is conceivable. In the first case the binding of the modulator to the actin would decrease its affinity for one of the adjacent actins in the filament, whereas in the second case both the modulator and actin would compete for the same binding site at one actin molecule and, because of its stronger affinity, the modulator would replace the actin. In each case, the internal stability of the filament is reduced, which leads to a separation into two fragments, to one of which the modulator remains bound.

CALCIUM-DEPENDENCE AND REVERSIBILITY

All five actin modulators investigated are completely Ca-dependent with regard to their interaction with actin. Gel filtration and sedimentation experiments reveal no binding between modulators and actin at $[Ca^{++}] \lesssim 10^{-7}$ M. Fig. 12-2A shows the activity of Physarum fragmin and several muscle modulators at various Ca concentrations, as determined from their effects on the viscosity of F-actin.

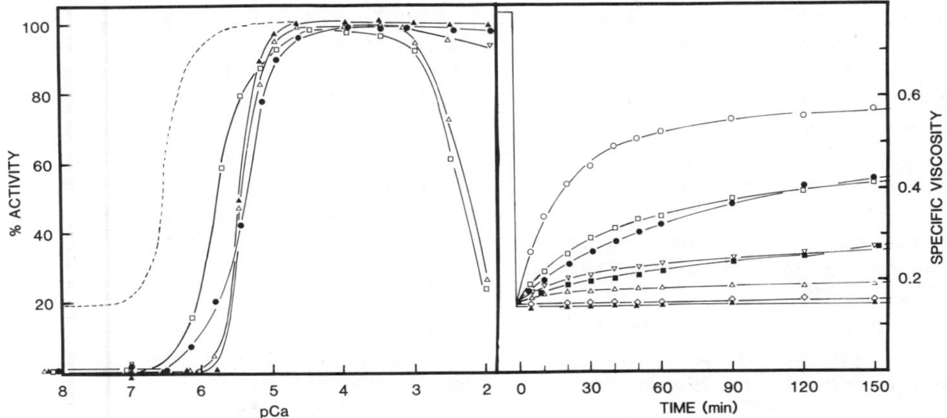

Fig. 12-2. Ca-dependence and reversion of modulator-actin interactions. (A) Physarum fragmin (●) or muscle modulators (RSAM▲; EWAM□; PSAM, CHGAM△; CHGAM▽, where different from PSAM) were added at a molar ratio of 1:100 to F-actin (1 mg/ml in 2 mM $MgCl_2$, 100 mM KCl, 10 mM imidazole pH 7.2. To obtain Ca^{++}-concentrations between 10^{-8} and 10^{-5} M, 3 mM of different Ca/EGTA buffers were added; for higher Ca^{++}-concentrations $CaCl_2$ was added). After steady state was attained, the viscosities of the samples were determined. The relative activity was calculated from the viscosity data, with the maximal decrease obtained for each modulator as 100% value. The dotted line shows the Ca-activation of vertebrate smooth muscle actomyosin ATPase. The crude actomyosin preparation from pig stomach contained myosin light-chain kinase as a Ca^{++}-sensitive compound, which activates the actomyosin ATPase by phosphorylation of a myosin light chain. The reaction was measured under the same conditions as the actin-modulator reaction. (B) Fragmin (closed symbols) or PSAM (open symbols) were added to skeletal muscle F-actin (conditions as in A except that 0.05 mM $CaCl_2$ was present in all samples). This resulted in a rapid decrease in actin viscosity. After steady state was attained, EGTA was added at various concentrations to readjust [Ca^{++}] to the following values: ◇ 3 X 10^{-7}M; ▲△ 5 X 10^{-8}; ▽ 1 X 10^{-8} M; □■ 6 X 10^{-9} M; ○● 2 X 10^{-9}M. Viscosity changes were recorded as a function of time after adjusting [Ca^{++}].

All the modulators are activated within a relatively narrow range of Ca-concentrations: in the case of the vertebrate muscle modulators clearly between 10^{-6} and 10^{-5} M, but for fragmin and EWAM slightly lower. For comparison and as a reference to another Ca-activated process within the contractile system, we have measured the activation of the actomyosin-ATPase from vertebrate smooth muscle by the myosin light chain kinase (MLCK)-dependent phosphorylation of myosin under the same conditions as the modulator-actin interaction. The results show that, at least in the case of vertebrate muscle, the activations of the contractile mechanism and of the actin modulator are not simultaneously occurring processes in the cell. Only PSAM and EWAM reveal a significant inactivation at Ca-concentrations about 10^{-4}M.

A reversion of the effects of actin modulators on actin, namely, a release of the modulator from the fragmented actin filaments, would result in instant reannealing of the fragments to long filaments and a concomitant increase in viscosity. As shown from the curves in Fig. 12-2B, a reversion is in fact obtained after decrease of the Ca-concentration. However, in sharp contrast to the on-reaction, no immediate increase in viscosity is observed when the Ca-concentration is adjusted to just below the threshold level for activation of the modulators. Only at a considerably lower Ca-concentration is there a slow viscosity rise. With a further decrease in Ca-concentration, the rate of viscosity increase increases as well but, even at the lowest value measured, the reversion still takes about 30 min, and the original viscosity is not fully restored. The reversion is slower for fragmin than for PSAM (Fig. 12-2B). The two other vertebrate muscle modulators are similar to PSAM, whereas EWAM resembles fragmin in that respect (not shown). It is noteworthy that the addition of tropomyosin from skeletal muscle considerably accelerates the reversion of this modulator-actin interaction. From recent experiments with PSAM and tropomyosin (unpublished results), it has become evident that a complete recovery of viscosity and reannealing of filaments can be obtained within 15 min, even at Ca-concentrations very close to the threshold level of modulator activation. It is conceivable that actin-binding proteins--not necessarily tropomyosin--exist in vivo, which may induce a full reversion at physiologically significant rates.

CONCLUSIONS

We have found that the disintegration of actin filaments by fragmin and of the functionally related muscle actin modulators is a rapid reaction that generates short actin fragments or oligomers within a few seconds. Their average size is determined by the amount of modulator added. Activation of the modulators for interaction with actin requires Ca^{++} in physiological concentrations.

The low reversibility of the reaction may be due to a cooperative formation of a strong complex between modulator, Ca^{++}, and actin, from which Ca^{++} cannot easily be removed by EGTA. However, reversion can be accelerated considerably by other actin-binding proteins such as tropomyosin.

The proteins from different sources show remarkably small variations in their interaction with actin, in spite of the fact that they were isolated from cells that are evolutionarily very distant and where actin has quite different functional aspects. The differences observed do not suggest a functional distinction between non-muscle and muscle modulators, because differences are also observed among the muscle modulators themselves. Also, the molecular weight difference between the ca 40K and ca 90K modulators is not reflected in their function. However, considering the fact that 40K modulators have been found also in Dictyostelium (Yamamoto et al., 1982) and sea urchin eggs (Wang and Spudich, 1984), and that 90K modulators such as gelsolin and villin have been reported only from vertebrate cells, it may be speculated that the former type is preferentially or exclusively expressed in nonvertebrates, whereas the 90K type may be restricted to vertebrates.

REFERENCES

Bryan, J., and Kurth, M. C., 1984, Actin-gelsolin interactions: Evidence for two actin-binding sites, J. Biol. Chem., 259:7480.

Glenney, J. R., Kaulfus, Ph., and Weber, K., 1981, F-actin assembly modulated by villin: Ca^{++}-dependent nucleation and capping of the barbed end, Cell, 24:471.

Hasegawa, T., Takahashi, S., Hayashi, H., and Hatano, S., 1980, Fragmin: A calcium ion sensitive regulatory factor on the formation of actin filaments, Biochemistry, 19:2677.

Hinssen, H., 1981a, An actin-modulating protein from Physarum polycephalum. I. Isolation and purification, Eur. J. Biol., 23:225.

Hinssen, H., 1981b, An actin-modulating protein from Physarum polycephalum. II. Ca^{++}-dependence and other properties, Eur. J. Cell Biol., 23:234.

Hinssen, H., Small, J. V., and Sobieszek, A., 1984, A Ca^{++}-dependent actin modulator from vertebrate smooth muscle, FEBS Lett., 166:90.

Hinssen, H., Engels, F. E., and Small, J. V., 1985, The role of actin modulators in smooth and skeletal muscle, J. Muscle Res. Cell Motil., 6:117.

Maruta, H., Knoerzer, W., Hinssen, H., and Isenberg, G., 1984, Regulation of actin polymerization by non-polymerizable actin-like proteins, Nature, 312:424.

Osborn, M., Weber, K., Naib-Majani, W., Hinssen, H., Stockem, W., and Wohlfarth-Bottermann, K.-E., 1983, Immunochemistry of the acellular slime mould Physarum polycephalum. III. Distribution of myosin and the actin-modulating protein (fragmin) in sandwiched plasmodia, Eur. J. Cell Biol., 29:179.

Sugino, H., and Hatano, S., 1982, Effect of fragmin on actin polymerization: evidence for enhancement of nucleation and capping of the barbed end, Cell Motil., 2:457.

Yamamoto, K., Pardee, J. D., Reidler, J., Stryer, L., and Spudich, J. A., 1982, Mechanism of interaction of Dictyostelium severin with actin filaments, J. Cell Biol., 95:711.

Yin, H. L., Zaner, K. S., and Stossel, T. P., 1980, Ca^{++} control of actin gelation: interaction of gelsolin with actin filaments and regulation of actin gelation, J. Biol. Chem., 255:9494.

Wang, L., and Spudich, J. A., 1984, A 45,000-mol-wt protein from unfertilized sea urchin eggs severs actin filaments in a calcium-dependent manner and increases the steady-state concentration of non-filamentous action, J. Cell Biol., 99:844.

Wegner, A., and Savko, P., 1982, Fragmentation of actin filaments, Biochemistry, 21:1909.

Chapter 13: A NONPOLYMERIZABLE ACTIN DERIVATIVE REGULATES ACTIN POLYMERIZATION BY CAPPING THE FAST-GROWING END OF ACTIN FILAMENTS

Hiroshi Maruta

Department of Biology
Yale University
New Haven, CT, USA

INTRODUCTION

Physarum contains at least four different actin-binding proteins of 42 K (Hasegawa et al., 1980; Hinssen, 1981; Maruta et al., 1983; Maruta and Isenberg, 1983, 1984; Maruta et al., 1984). They are called actin, Cap 42(a), Cap 42(b), and fragmin, on the basis of both their biological activities and molecular masses. Actin is the major polymerizable protein that forms a filament by self-assembly. Three other proteins do not form any filament but instead regulate the elongation of actin filament by capping the fast-growing end of actin filaments. Only fragmin has an additional F-actin severing activity that leads to the fragmentation of actin filaments in a calcium-dependent manner. Cap 42(a) requires calcium for its capping activity. Cap 42(b) is a phosphoprotein that requires calcium for its capping activity only when it is fully phosphorylated by a specific kinase from Physarum. This kinase is highly specific for Cap 42(b) and does not phosphorylate any other proteins. Only actin serves as a substrate for cAMP-dependent protein kinase, whereas only fragmin serves as a substrate for pp60 Src, a tyrosine-specific kinase, which is solely responsible for the transformation of cells by Rous sarcoma virus. Therefore, the four proteins of 42 K are functionally distinguishable.

Interestingly, however, the following data suggest that the F-actin capping protein Cap 42(b) is structurally related to the major polymerizable actin, whereas the F-actin capping protein Cap 42(a) is structurally related to the F-actin severing protein fragmin: (1) both actin and Cap 42(b) bind DNase I; (2) peptide maps as well as amino acid sequences of these two DNase I-binding proteins are almost identical; (3) fragmin and Cap 42(a), which do not bind DNase I, are almost identical in both peptide maps and amino acid sequences but differ from actin; (4) the three F-actin-capping proteins share at least one common antigenic determinant with actin; and (5) ATP is bound to all the four proteins, whereas UTP is bound only to Cap 42(a) and Cap 42(b).

Since Physarum contains at least five distinct actin genes (Schedl and Dove, 1982) it is possible that Cap 42(b) and actin are derived from two separate actin genes. However, it is equally possible that Cap 42(b) is a product of a post-translational modification of actin. Thus, we have proposed a new concept that Cap 42(b), a minor nonpolymerizable actin variant or derivative, can regulate the self-assembly of the major polymerizable actin (Maruta et al., 1984).

How can an actin-related protein bind (or cap) only one specific end of actin filament to block elongation? Actin polymerization occurs in a head-to-tail fashion (Wegner, 1976), so that at least two distinct binding sites on each actin molecule are involved in the monomer-monomer interaction during the polymerization. The one binding site resides in the "kappa" region near the center of the molecule that contains the lysine 191, and the other binding site resides in the "omega" region near the C-terminus that contains the cysteine 374 (Sutoh, 1984; Elzinga and Phelan, 1984). Thus, the specific interaction of the "kappa" site in one actin molecule with the "omega" site in the next molecule appears to be the basic mechanism underlying the head-to-tail polymerization of actin. This means that if there is an "omega-minus" type of mutant actin that has an impaired "omega" site but an intact "kappa" site, this molecule certainly would fail to polymerize by itself but would still be able to bind via its intact "kappa" site to the "omega" site of a normal actin sitting at one end of actin filament. This would block the elongation of the filament at this end because this univalent molecule cannot accept any more actin to propagate the filament.

Which binding site of the actin must be impaired to cap the fast-growing end of an actin filament? Noncovalent binding of ATP to each actin molecule, but not its hydrolysis, is essential for the polymerization (Cooke and Murdoch, 1973). The ATP binding site of actin resides in or near the "omega" site (Hegyi et al., 1985). Furthermore, it has been suggested that ATP (or ATP-G-actin) caps selectively the fast-growing end of actin filaments, indicating that the "omega" site is exposed only at the fast-growing end and buried at the slow-growing end of actin filaments (Carlier et al., 1984). Therefore, if the "omega" site is selectively impaired in an actin molecule, this modified actin, called "omega-minus" actin, is able to bind only the fast-growing end of an actin filament to block the further elongation (see Fig. 13-1).

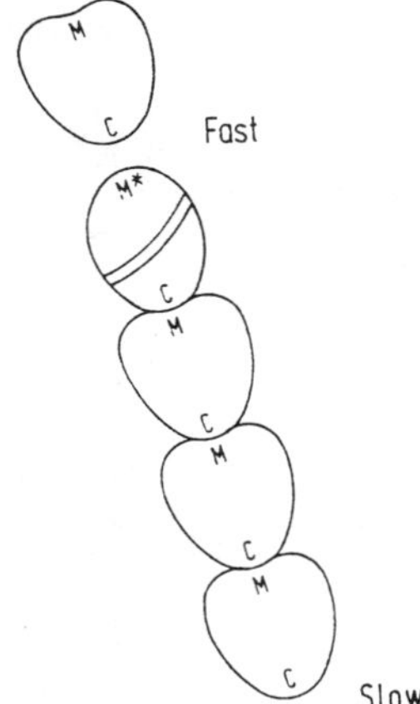

Fig. 13-1. Capping the fast-growing end of actin filament by an "omega-minus" actin, C: "kappa" site; M: "omega" site; M*: altered omega site.

In this paper, experimental evidence is provided to suggest that a covalent binding of ATP to a polymerizable actin can also lead to the creation of a nonpolymerizable actin, which is able to cap only the fast-growing end of an actin filament and block the elongation in a calcium-dependent manner.

EXPERIMENTAL PROCEDURES

Covalent Binding of ATP to G-Actin

Rabbit skeletal muscle actin (1.0 mg/ml) in G buffer containing 10 mM-Tris-HCl, pH 7.5, 1 mM dithiothreitol, 0.2 mM calcium, and 1 mM ATP was irradiated by an ultraviolet light at 0°C for 60 min under the conditions described previously for the direct photoaffinity labeling of the ATPase catalytic sites in myosin heavy chain, actin, and the three F-actin capping proteins of 42 K (Maruta and Korn, 1981; Maruta et al., 1984), as well as the exchangeable GTP-binding site in β tubulin (Hesse et al., 1985). After 60-min irradiation, about 10% of the actin was covalently derivatized by ATP. As a control, G actin was irradiated by the ultraviolet light in the absence of ATP. As shown in Fig. 13-2, most of the G actin lost the polymerizability after 60-min irradiation with or without ATP.

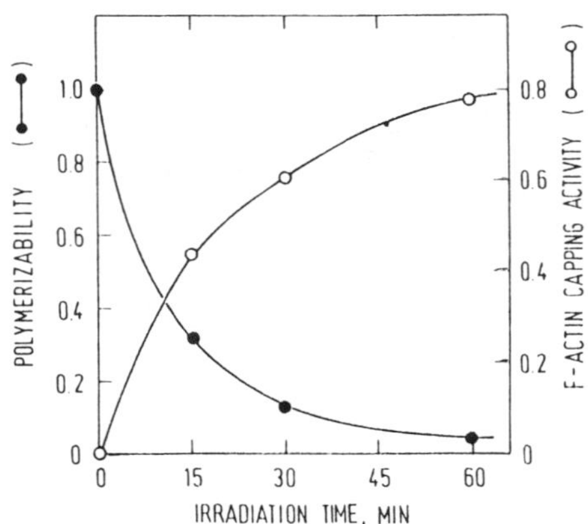

Fig. 13-2. Polymerizability and F-actin capping activity of actin irradiated by ultraviolet light in the presence of ATP.

F-Actin Capping Activity of the "Modified" Actin

Interestingly, however, only the "modified" actin (ATP=actin) preparation, which had been ultraviolet-irradiated in the presence of ATP, showed a significant level of F-actin capping activity, which leads to a reduction of a low-shear viscosity of an F-actin solution (see Fig. 13-2), as measured according to the procedure described previously (Maruta and Isenberg, 1983). The control actin preparation that had been irradiated in the absence of ATP did not show any significant F-actin capping activity. These results indicate that G actin covalently bound to ATP is able to cap an end of actin filament. To determine which end of the actin filament is capped, short actin filaments decorated with S-1 fragment of myosin were used as nuclei for actin polymerization in the presence or absence of the "modified" actin, as previously described (Maruta and Isenberg, 1983). In the presence of the "modified" actin, growth of the actin filament occurred only at the slow-growing "pointed" end, and no filament growth took place at the fast-growing "barbed" end. This indicates that the "modified" actin caps preferentially the fast-growing end, as do the three F-actin capping proteins of 42 K from Physarum (Maruta and Isenberg, 1983; Maruta and Isenberg, 1984). Like Cap 42 (b) in the nonphosphorylated form, the "modified" actin did not require calcium for its capping activity (see Fig. 13-3). However, this "modified" actin did not serve as a substrate for either Cap 42(b) kinase or pp60 Src, indicating that it is distinct from both Cap 42(b) and fragmin.

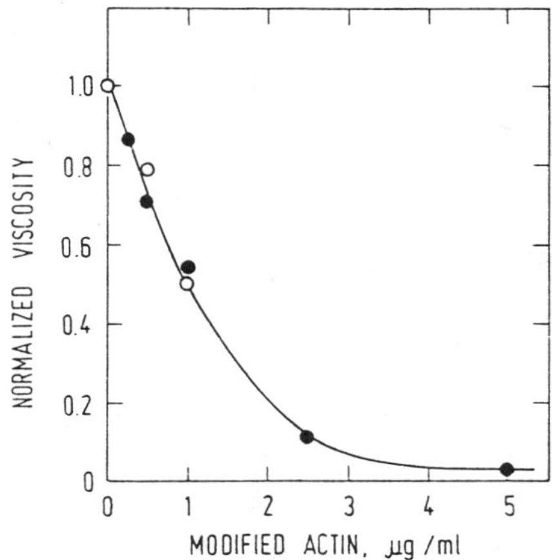

Fig. 13-3. Effect of calcium on F-actin capping activity of "modified" actin. closed circles: + 0.2 mM calcium; open circles: + 1.0 mM EGTA.

Ca^{2+}-dependent Association of the "Modified" Actin with Normal Actin

Gel-filtration of the "modified" actin preparation has revealed that the F-actin capping activity resides in an actin dimer. The dimer could be dissociated into monomers on DEAE-cellulose in the presence of 10 M formamide according to the procedure described previously (Maruta and Isenberg, 1984). As shown in Fig. 13-4, the isolated "modified" actin monomer required calcium for its capping activity; this suggested a possibility that the "modified" actin monomer had been associated with a regular G actin in a calcium-dependent manner to form a heterodimer that no longer requires calcium for its capping activity. To test this possibility, the isolated "modified" actin monomer was incubated with the regular G actin at 25°C for 15 min in the presence of either 0.2 mM calcium or 1 mM EGTA, and then the mixture was subjected to gel-filtration on Sephadex G-150. As shown in Fig. 13-5, in the presence of calcium (circle), most of the F-actin capping activity was recovered in the fractions where actin dimer was expected to appear. This indicates that most of the "modified" actin monomer formed a heterodimer with the regular G-actin, and this dimer did not require calcium for its capping activity. By contrast, in the absence of calcium (triangles), most of the F-actin capping activity was recovered in the fraction where actin monomer was expected to appear. This indicates that the "modified" actin monomer was not able to form the dimer and still required calcium for its capping activity. These results confirmed that calcium is essential for the formation of a heterodimer between the "modified" actin monomer and the reg-

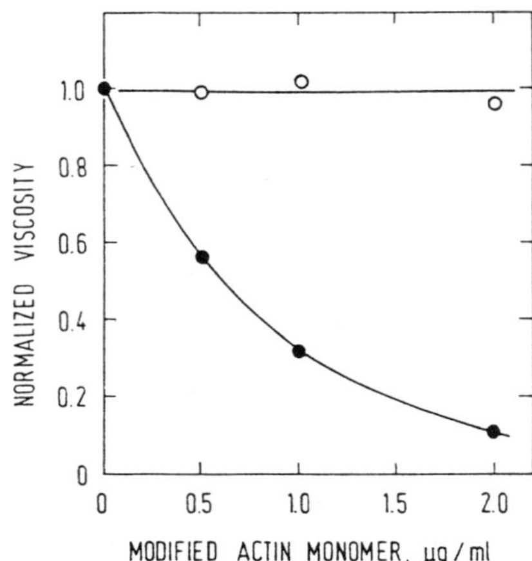

Fig. 13-4. Calcium dependency of F-actin capping activity of the "modified" actin monomer isolated after 10 M formamide treatment. closed circles: + 0.2 mM calcium; open circles: + 1.0 mM EGTA.

ular G actin, but not for the binding of the dimer to the end of actin filaments (see Fig. 13-6).

DISCUSSION

Using the direct photoaffinity-labeling technique, I have demonstrated that the "modified" actin, actin covalently bound to ATP, almost completely lost its polymerizability but simultaneously acquired an F-actin capping activity that leads to a calcium-dependent blockage of actin polymerization only at the fast-growing end of actin filament. This behavior of the "modified" actin appears to coincide well with what we expect if an actin monomer has an impaired "omega" site but an intact "kappa" site. The in vitro conversion of the polymerizable actin to a non-polymerizable but capping "omega-minus" derivative by a covalent binding to ATP suggest a possibility that Cap 42(b), which is very closely related to the polymerizable actin, has a significant alteration in the primary structure within or near the "omega" site where ATP and other XTPs are presumed to be bound. Indeed, UTP is covalently bound to Cap 42(b) but not to actin, although both proteins covalently bind ATP (Maruta et al., 1984). Nevertheless, the final proof should await the complete amino acid sequencing of Cap 42(b). Similarly, if one could manage to alter specifically the primary structure within the "kappa" site of actin, one could create a "kappa-minus" actin that would presumably cap only the slow-growing end of an actin filament. In this context, it may be of interest

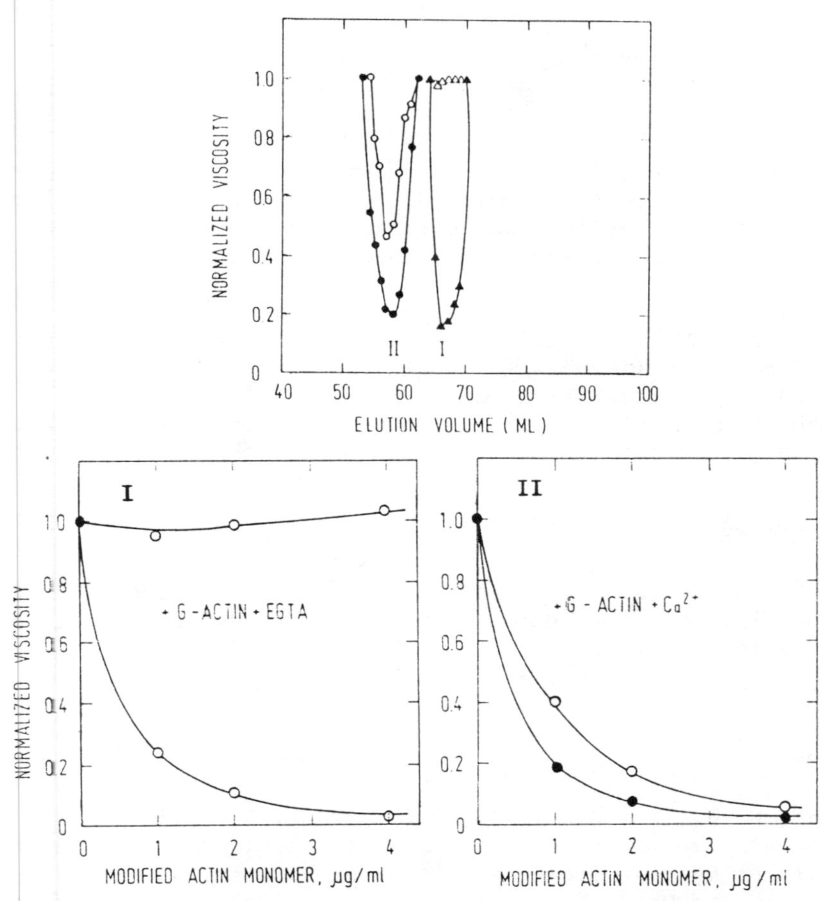

Fig. 13-5. Gel-filtration of a mixture of "modified" actin monomers and G-actin (1:1) after incubation in the presence of either 0.2 mM calcium(II) or 1 mM EGTA(I). F-actin capping activity of both monomer (I) and dimer (II) was assayed in the presence of either 0.2 mM calcium (closed symbols) or 1 mM EGTA (open symbols).

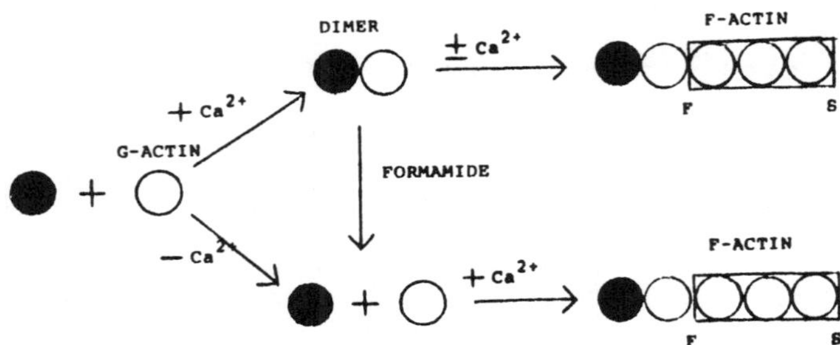

Fig. 13-6. The calcium-dependent formation of a capping heterodimer between the "modified" actin monomer (closed circles) and G-actin (open circles). F: fast-growing end of actin filament; S: slow-growing end of actin filament.

to note that HUT 14, a transformed cell line derived from normal diploid human DK fibroblasts, contains a β actin mutant with a single amino acid substitution at residue 245 (Gly to Asp) not far from the "kappa" site, which is poorly polymerizable and interferes with the polymerization of normal β actin (Kakunaga et al., 1984).

ACKNOWLEDGMENT

The author is very grateful to Dr. Gerhard Isenberg for his consistant support and encouragement throughout this work, which was done in his laboratory at the Max-Planck-Institute for Psychiatry.

REFERENCES

Carlier, M.-F., Pantaloni, D., and Korn, E. D., 1984, Evidence for an ATP cap at the ends of actin filaments and its regulation of the F-actin steady state, J. Biol. Chem., 259:9983.

Cooke, R., and Murdoch, L., 1973, Interaction of actin with analogs of adenosine triphosphate, Biochemistry, 12:3927.

Elzinga, M., and Phelan, J. J., 1984, F-actin is intermolecularly cross-linked by N,N'-p-phenylenedimaleimide through lysine-191 and cysteine-374, Proc. Natl. Acad. Sci., U.S.A., 81:6599.

Hasegawa, T., Takahashi, S., Hayashi, H., and Hatano, S., 1980, Fragmin: a calcium ion sensitive regulatory factor on the formation of actin filaments, Biochemistry, 19:2677.

Hegyi, G., Szilagyi, L., and Elzinga, M., 1985, Photoaffinity labeling of the actin nucleotide-binding site, J. Muscle Res. Cell Motil., in press.

Hesse, J., Maruta, H., and Isenberg, G., 1985, Monoclonal antibodies localize the exchangeable GTP-binding site in β-and not α-tubulins, FEBS Lett., 179:91.

Hinssen, H., 1981, An actin-modulating protein from Physarum polycephalum: isolation, purification, calcium-dependence and other properties, Eur. J. Cell Biol., 23:225.

Kakunaga, T., Leavitt, J., Hirakawa, T., and Taniguchi, S., A point mutation and other changes in cytoplasmic actins associated with the expression of transformed phenotypes, 1984, in: "Cancer Cells", A. J. Levine, G. F. Vande Woude, W. C. Topp, and J. D. Watson, eds., Vol. 1, The Transformed Phenotype, p. 67, Cold Spring Harbor Laboratory, Cold Spring Harbor.

Maruta, H., and Isenberg, G., 1983, Ca^{2+}-dependent actin-binding phosphoprotein in Physarum polycephalum, II. Ca^{2+}-dependent F-actin-capping activity of subunit a and its regulation by phosphorylation of subunit b. J. Biol. Chem., 258:10151.

Maruta, H., and Isenberg, G., 1984, Ca^{2+}-dependent actin-binding phosphoprotein in Physarum polycephalum, Subunit b is a DNase I-binding and F-actin capping protein, J. Biol. Chem., 259:5208.

Maruta, H., Isenberg, G., Schreckenbach, T., Hallmann, R., Risse, G., Shibayama, T., and Hesse, J., 1983, Ca^{2+}-dependent actin-binding phosphoprotein in Physarum polycephalum. I. Ca^{2+}/actin-dependent inhibition of its phosphorylation, J. Biol. Chem., 258:10144.

Maruta, H., Knoerzer, W., Hinssen, H., and Isenberg, G., 1984, Regulation of actin polymerization by nonpolymerizable actin-like proteins, Nature, 312:424.

Maruta, H., and Korn, E. D., 1981, Direct photoaffinity labeling by nucleotides of the apparent catalytic site on the heavy chains of smooth muscle and Acanthamoeba myosins, J. Biol. Chem., 256:499.

Schedl, T., and Dove, W. F., 1982, Mendelian analysis of the organization of actin sequences in Physarum polycephalum, J. Mol. Biol., 160:41.

Sutoh, K., 1984, Actin-actin and actin-deoxyribonuclease I contact sites in the actin sequence, Biochemistry, 23:1942.

Wegner, A., 1976, Head to tail polymerization of actin, J. Mol Biol., 108:139.

Chapter 14: DYNAMICS AND FUNCTION OF MICROFILAMENTS IN PHYSARUM POLYCEPHALUM AS REVEALED BY FLUORESCENT ANALOG CYTOCHEMISTRY (FAC) AND ELECTRON MICROSCOPY

Wilhelm Stockem and Jörg Kukulies

Institute of Cytology
University of Bonn
Bonn, FRG

ABSTRACT

Tetramethylrhodaminyl (TRITC)-phalloidin and isolated muscle or *Physarum* G-actin labeled with various fluorochromes were microinjected into living stages of *Physarum polycephalum* (cell fragments and microplasmodia). Subsequent analysis of the intracellular redistribution of the molecular probes by fluorescence microscopy, video-enhancement, and digital image processing revealed that polymerization-depolymerization and contraction-relaxation cycles of the microfilament system are functionally related to changes in cell shape, protoplasmic streaming activity, and ultrastructural morphology of the specimens. In relaxed cell fragments, TRITC-phalloidin and rhodamine-isothiocyanate (RITC)-actin first diffuse randomly and then are locally incorporated into a thin cortical layer at the internal face of the plasma membrane. During Ca^{2+}-induced contraction, the fluorescent layer starts to detach from the plasma membrane, thus causing separation of the central granuloplasm from the peripheral hyaloplasm. Thin sections of both relaxed and contracted specimens demonstrate that the fluorescent layer in living cell fragments coincides exactly with a sheath of more or less oriented microfilaments. In contrast, RITC-bovine serum albumin injected as a control is excluded from those regions that show intense fluorescence with RITC-actin and TRITC-phalloidin and the presence of an actin network by electron microscopy.

INTRODUCTION

The investigation of living cells with the fluorescence microscope has been impeded in the past by several disadvantages limiting the applicability of this technique, i.e., toxicity of the fluorescent dyes, radiation damage of the specimens, and rapid bleaching of the fluorescent signal. A novel technical approach to overcome these difficulties and to take advantage of the potentially high resolution properties of fluorescence microscopy has recently been proposed as FAC, i.e., fluorescent analog cytochemistry (Wang et al., 1982).

The FAC technique represents a combination of vital high-resolution microscopy, microinjection of fluorochromed probes, and video image intensification as well as digital image processing combined with automatic computer analysis. Thereby, dynamic events such as cell motility, cell-to-cell interaction, endocytosis, cell surface modulation, secretion, or assembly and disassembly of cell organelles have become accessible to direct investigation (for reviews see Taylor and Wang, 1980; Kreis and Birchmeier, 1982; Kukulies and Stockem, 1985a).

RESULTS AND CONCLUSIONS

Production and General Morphology of Cell Fragments

Small cell fragments of *Physarum polycephalum* were derived from isolated veins and protoplasmic drops by treatment with a Tris-buffered 10-15 mM caffeine solution (Hatano, 1970; Sato et al., 1981). Caffeine is presumed to interfere with the interaction of microfilaments and membranes. Consequently, upon addition of the drug, regions where the microfilament system is in close contact with the plasma membrane of the genuine drop (Fig. 14-1A, striated lines) produce numerous surface blebs of hyaline or granular appearance and thereby a 5- to 10-fold increase in the surface area (Fig. 14-1b, arrowheads).

The blebs then start to constrict from the drop and persist in the sample as so-called caffeine droplets or cell fragments. The cell fragments exhibit a range of sizes (50-300 μm) and degrees of hyalo-granuloplasmic separation (Fig. 14-1C) depending on the state of Ca^{2+}-mediated contraction or relaxation (for a detailed description, see Kukulies et al., 1983).

Spatial Organization of the Microfilament System in Cell Fragments

Successful incorporation of the fluorescent analogs RITC-G-actin, TRITC-phalloidin, and RITC-bovine serum albumin (BSA) into cell fragments of *Physarum polycephalum* provided evidence for the involvement of a cortical microfilament system in both motive force generation for cytoplasmic streaming and morphodynamic changes of cytoplasmic organization (Kukulies et al., 1983, 1984; Kukulies and Stockem, 1985b). Microinjection of the fluorescent analogs

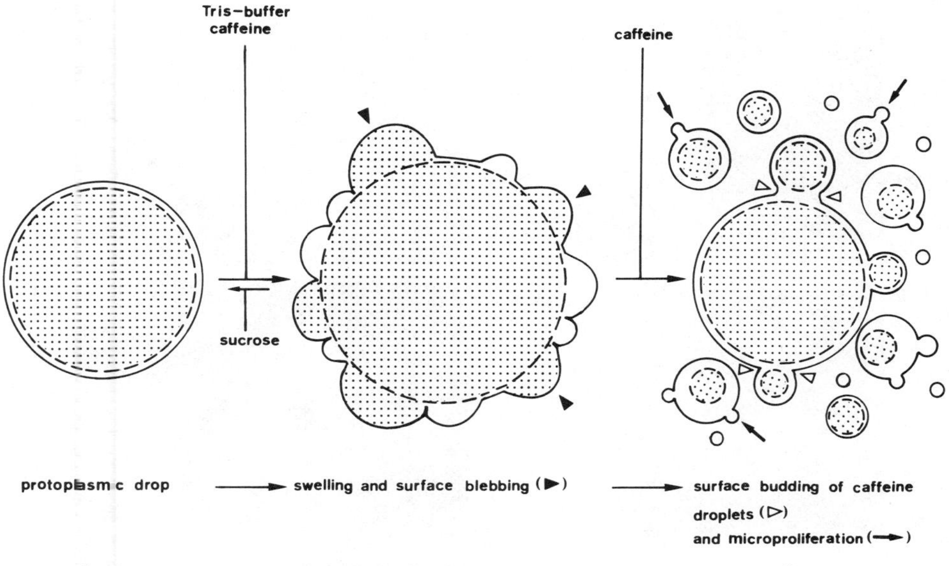

Fig. 14-1. Schematic drawing summarizing the generation of Physarum cell fragments (caffeine droplets) from Tris-buffer/caffeine-treated, isolated protoplasmic drops (A), by surface blebbing (B), and surface budding (C). Dotted areas = granuloplasm; white areas = hyaloplasm; broken lines = filament cortex; continuous lines = plasma membrane. (Reprinted with permission from Kukulies et al., 1983).

does not change the normal motile behavior of cell fragments. During contraction, injected specimens (Fig. 14-2a, c, e) exhibit the same hyalo-granuloplasmic separation and streaming activities as seen in noninjected controls. Examination at defined external Ca^{2+} concentrations shows that under relaxing conditions (1 mM EGTA) the RITC-actin condenses on the internal face of the plasma membrane (Fig. 14-2b, FC_1). Increase of external Ca^{2+} to 1 mM causes contraction, i.e., the fluorescent layer separates from the plasma membrane to delineate the central granuloplasm from the peripheral hyaloplasm (Fig. 14-2d, FC_1). As contraction proceeds, a second fluorescent layer appears underneath the plasma membrane (Fig. 14-2d; FC_2). Microinjection of TRITC-phalloidin into Physarum cell fragments produces a staining pattern exactly identical with

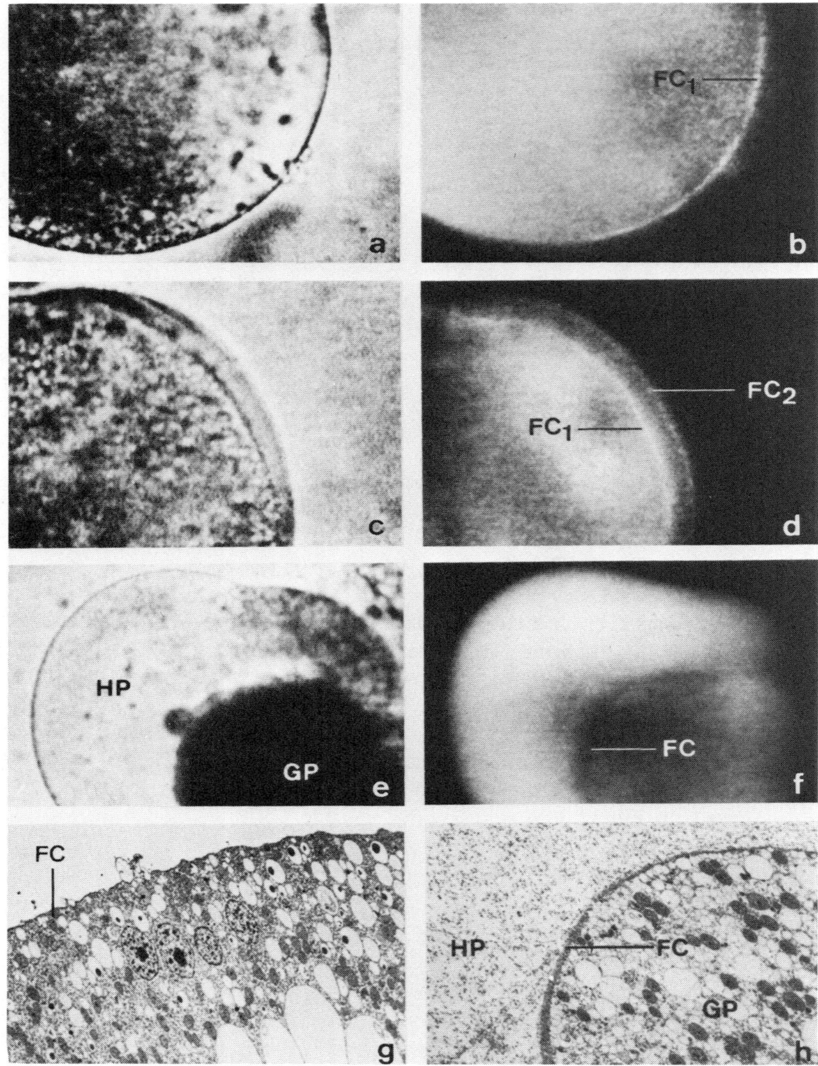

Fig. 14-2. Light (a,c) and fluorescence (b,d) micrographs of a living Physarum cell fragment 10 min after microinjection of RITC-G-actin before (a,b) and 3 min after addition of Ca^{2+} (c,d). Light (e) and fluorescence (f) micrographs of a living Physarum cell fragment 20 min after microinjection of RITC-BSA. Electron micrographs (g,h) of $OsO_4/HgCl_2$-fixed Physarum cell fragments in the relaxed (g) and contracted state (h). FC_1 = primary filament cortex; FC_2 = secondary filament cortex; FC = unlabeled filament cortex; GP = granuloplasm; HP = hyaloplasm. a-d: 475x; e,f: 260x; g,h: 4000x. (Reprinted with permission from Kukulies et al., 1984).

that of injected RITC-G-actin. There is strong evidence that the RITC-G-actin is actually incorporated into single microfilaments together with nonfluorescent endogenous G-actin (Wang and Taylor, 1980), whereas the TRITC-phalloidin is specifically bound to the surface of already preexisting microfilaments (Faulstich et al., 1983).

In contrast to RITC-G-actin and TRITC-phalloidin, the RITC-BSA exhibits a completely different distribution pattern. Relaxed droplets show an even distribution of BSA-fluorescence for more than 1 hr, with no signs of localized incorporation into distinct structures. In contracted droplets, a bright fluorescence prevails in the hyaloplasm, and no RITC-BSA is found at the hyalo-granuloplasmic border (Fig. 14-2f, FC).

Ultrastructural investigation of the microfilament system in relaxed (Fig. 14-2g) and contracted cell fragments (Fig. 2h) reveals that the localization of F-actin is identical with the sites that contain a distinct fluorescence of FITC-actin or TRITC-phalloidin and that simultaneously lack fluorescence of RITC-BSA. Relaxed fragments, when fixed, show a rather uniform distribution of cell organelles and a filament cortex in close contact with the plasma membrane (Fig. 14-2g, FC), whereas contracted specimens when fixed show a prominent cortex at the hyalo-granuloplasmic border (Fig. 14-2h, FC).

Function of the Microfilament System in Cell Fragments

The results obtained with RITC-actin and TRITC-phalloidin clearly demonstrate that the microfilament system of *Physarum* cell fragments serves two different functions (summarized in Fig. 14-3):

(1) Besides cell surface phenomena such as changes in shape by local contraction of membrane-associated microfilaments (B', C'), the cortical layer can also detach from the plasmalemma and separate the central granuloplasm from the peripheral hyaloplasm (A, B, B'). Comparable morphogenetic alterations have also been described for *Amoeba proteus* after application or microinjection of different drugs, including unlabeled phalloidin (Hoffmann et al., 1984).

(2) A general contraction of the filament cortex fails to induce a typical protoplasmic streaming activity except when a local destruction of the layer does occur so that the generated hydraulic pressure is transformed into fountain-like streaming of cell constituents from the granuloplasm into the hyaloplasm (C, C'). By means of successive polymerization and depolymerization of cytoplasmic actin at opposite sites of the cell fragments, a shuttle-like streaming pattern can arise that resembles the corresponding phenomena in *Physarum* macroplasmodia (Kamiya, 1959; Wohlfarth-Bottermann, 1975).

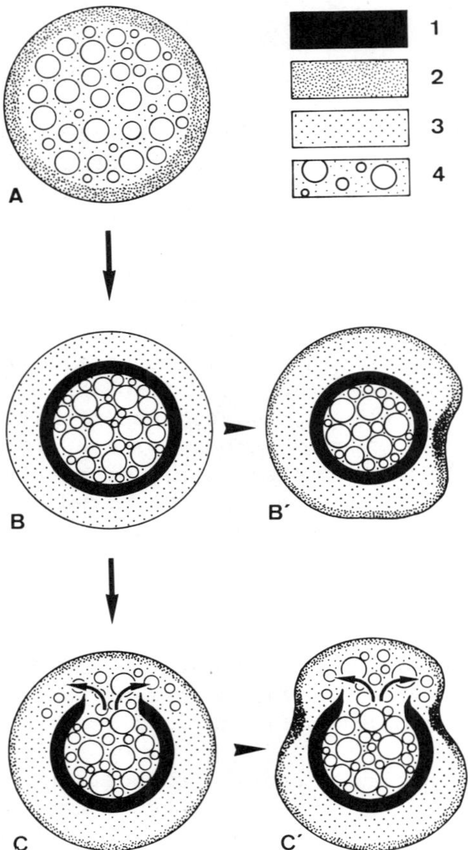

Fig. 14-3. Schematic drawing summarizing the changes in spatial organization and dynamic activity of the microfilament system in Physarum cell fragments. A = relaxed fragment; B,B' = contracted fragments; C,C' = contracted fragments with protoplasmic streaming; 1 = contracted microfilament system; 2 = relaxed microfilament system; 3 = hyaloplasm; 4 = granuloplasm. (Reprinted with permission from Kukulies and Stockem, 1985b).

Concluding Remarks

After visualization of a cortical microfilament system in living cell fragments by microinjection of different fluorescent analogs, the same technique was used to label a corresponding system in more functional stages of the acellular slime mold (Kukulies et al., 1985). The results of this investigation demonstrate that under favorable situations the FAC technique is also applicable to the analysis of contractile and morphodynamic phenomena in axenically cultured Physarum microplasmodia. These experiments have given evidence for the involvement of a membrane-attached filament cortex in motive force generation, both for protoplasmic streaming and for the formation of a plasma membrane invagination system. The cortical filament layer in Physarum seems to be comparable to the microfilament system in the leading edge or ruffling zone of tissue culture cells, whereas the cytoplasmic fibrils are more similar in function to stress fibers.

ACKNOWLEDGMENTS

This work was supported by a grant of the Deutsche Forschungsgemeinschaft (Sto 126/4-2). The authors wish to thank Professor Dr. H. Faulstich (Heidelberg) for the gift of TRITC-phalloidin.

REFERENCES

Faulstich, H., Trischmann, H., and Mayer, D., 1983, Preparation of tetramethylrhodaminyl-phalloidin and uptake of the toxin into short-term cultured hepatocytes by endocytosis, Exp. Cell Res., 144:73.

Hatano, S., 1970, Specific effect of Ca^{2+} on movement of plasmodial fragment obtained by caffeine treatment, Exp. Cell Res., 61:199.

Hoffmann, H. U., Stockem, W., and Gruber, B., 1984, Dynamics of the cytoskeleton in Amoeba proteus. II. Influence of different agents on the spatial organization of microinjected fluorescein-labeled actin, Protoplasma, 119:79.

Kamiya, N., 1959, Protoplasmic streaming, in: "Protoplasmatologia VIII/3/a", L. V. Heilbrunn and F. Weber, eds., p. 1, Springer, Wien.

Kreis, T. E., and Birchmeier, W., 1982, Microinjection of fluorescently labeled proteins into living cells with emphasis on cytoskeletal proteins, Int. Rev. Cytol., 75:209.

Kukulies, J., Brix, K., and Stockem, W., 1985, Fluorescent analog cytochemistry of the actin system and cell surface morphology in Physarum microplasmodia, Eur. J. Cell Biol., in press.

Kukulies, J., and Stockem, W., 1985a, Fluorescent analog cytochemistry of living cells, Zeiss Inform., 98: in press.

Kukulies, J., and Stockem, W., 1985b, Function of the microfilament system in living cell fragments of Physarum polycephalum as revealed by microinjection of fluorescent analogs, Cell Tissue Res., 242:323.

Kukulies, J., Stockem, W., and Achenbach, F., 1984, Distribution and dynamics of fluorochromed actin in living stages of Physarum polycephalum, Eur. J. Cell. Biol., 35:235.

Kukulies, J., Stockem, W., and Wohlfarth-Bottermann, K. E., 1983, Caffeine-induced surface blebbing and budding in the acellular slime mold Physarum polycephalum, Z. Naturforsch., 38c:589.

Sato, H., Hatano, S., and Sato, Y., 1981, Contractility and protoplasmic streaming preserved in artificially induced plasmodial fragments, the "caffeine drops", Protoplasma, 109:187.

Taylor, D. L., and Wang, Y. L. 1980, Fluorescently labelled molecules as probes of the structure and function of living cells, Nature, 284:405.

Wang, Y. L., and Taylor, D. L., 1980, Preparation and characterization of a new molecular cytochemical probe, J. Histochem. Cytochem., 28:1198.

Wohlfarth-Bottermann, K.-E., 1975, Weitreichende fibrilläre Protoplasmadifferenzierungen und ihre Bedeutung für die Protoplasmaströmung. X. Die Anordnung der Actymyosinfibrillen in experimentall unbeeinflußten Protoplasmaadern von Physarum in situ, Protistologica, 11:19.

Chapter 15: A TITIN-LIKE PROTEIN IS PRESENT IN PHYSARUM POLYCEPHALUM -- PRESENT KNOWLEDGE

Dieter Gassner

Institute of Cytology
University of Bonn
Bonn, FRG

INTRODUCTION

Titin (also called connectin) represents a pair of closely related, highly elusive megadalton polypeptides, which together are among the most abundant myofibrillar proteins in skeletal and heart muscles from a wide range of vertebrates and invertebrates (Wang, 1985). It is suggested that titin constitutes a major component of an elastic, endosarcomeric, superthin filament lattice that coexists with myosin and actin filaments and serves as an organizing template or dynamic scaffold (Wang, 1984).

Ozaki and Maruyama (1980) described a completely sodium dodecyl sulfate (SDS)-insoluble protein residue obtained from plasmodia of the slime mold Physarum polycephalum by an elaborate extraction procedure similar to their initial preparation of connectin from muscle (Maruyama et al., 1977). Because of its insolubility, the characterization of the plasmodial protein residue remained at a preliminary state. The filamentous structure of the residue in the electron microscope and the amino acid composition suggested a similarity to muscle titin.

In this study the purification of an uncommon, giant, SDS-soluble protein from Physarum polycephalum is reported, which, on the basis of its features, is suggested to be titin-like.

The giant molecular size of the protein, the difficulty to solubilize the protein in SDS, and its strong tendency to form gel-like, SDS-insoluble aggregates even in denaturants make it understandable that the titin-like protein has escaped attention in previous Physarum protein biochemistry.

SDS-SOLUBLE FORM OF PHYSARUM TITIN-LIKE PROTEIN OBTAINED

In the course of the examination of putative protein candidates as structural components of elastic cell ghosts, which contain 2- to 3-nm filaments and which remain after sequential treatment of different plasmodial stages with Triton X-100, KI, and SDS plus urea (Gassner et al., 1983), we frequently observed faint, very high molecular weight polypeptides in SDS gels. This suggested a similarity to muscle titin (connectin). This study was initiated by the observation that in SDS gel electrophoresis of DNase/RNase-pretreated homogenates of endoplasmic drops, generated by puncturing plasmodial strands, extremely large polypeptides appeared frequently but not always as a distinct doublet, but often also as faint bands and diffuse smears.

For the purification of the extremely large polypeptides, a modified SDS gel filtration procedure or salt fractionation of an SDS extract followed by chromatography was used, as described for the isolation of titin from rabbit or chicken breast myofibrils (Wang, 1982). As starting material, 3- to 4-sec-old endoplasmic drops were used (Fig. 15-1a). Such drops were chosen, first, to avoid most of the SDS-insoluble carbohydrates of the slime layer covering the plasmodial surface and second, to obtain pure, sol-like endoplasm that contains neither actomyosin fibrils (Fig. 15-1b) nor a high amount of polymerized actin at this age (Wohlfarth-Bottermann, 1983).

A purification protocol was developed by the use of direct agarose gel filtration of a DNase/RNase-pretreated SDS extract of these endoplasmic drops or, additionally by a salt precipitation step essentially based on a rather specific and quantitative precipitation of the titin-like protein-SDS complex by addition of NaCl to SDS-solubilized Physarum proteins (Gassner et al., 1985).

Gel Filtration of an SDS Extract

Liquid N_2-frozen drops (2 g fresh weight) were thawed in a solution containing 0.1% Triton X-100, 10 mM EGTA, 5 mM $MgCl_2$, 50 mM Tris-HCl, pH 8.0, in the presence of 500 μg RNase/ml and 500 μg DNase/ml. After homogenization, the extract was incubated for 45 min at room temperature with slight stirring to destroy high molecular weight DNA and RNA which interfere with the protein elution during chromatography. Ca^{++} was added to a concentration of 5 mM, followed by stirring for 10 min. The suspension was centrifuged for 10 min at 10,000 X *g*, and the pellet was discarded. One volume of double-concentrated hot (90°C) sample buffer (10% SDS, 40 mM dithiothreitol [DTT], 10 mM EDTA, 100 mM Tris-HCl, pH 8.0) was added to the yellow and slightly cloudy supernatant. The sample

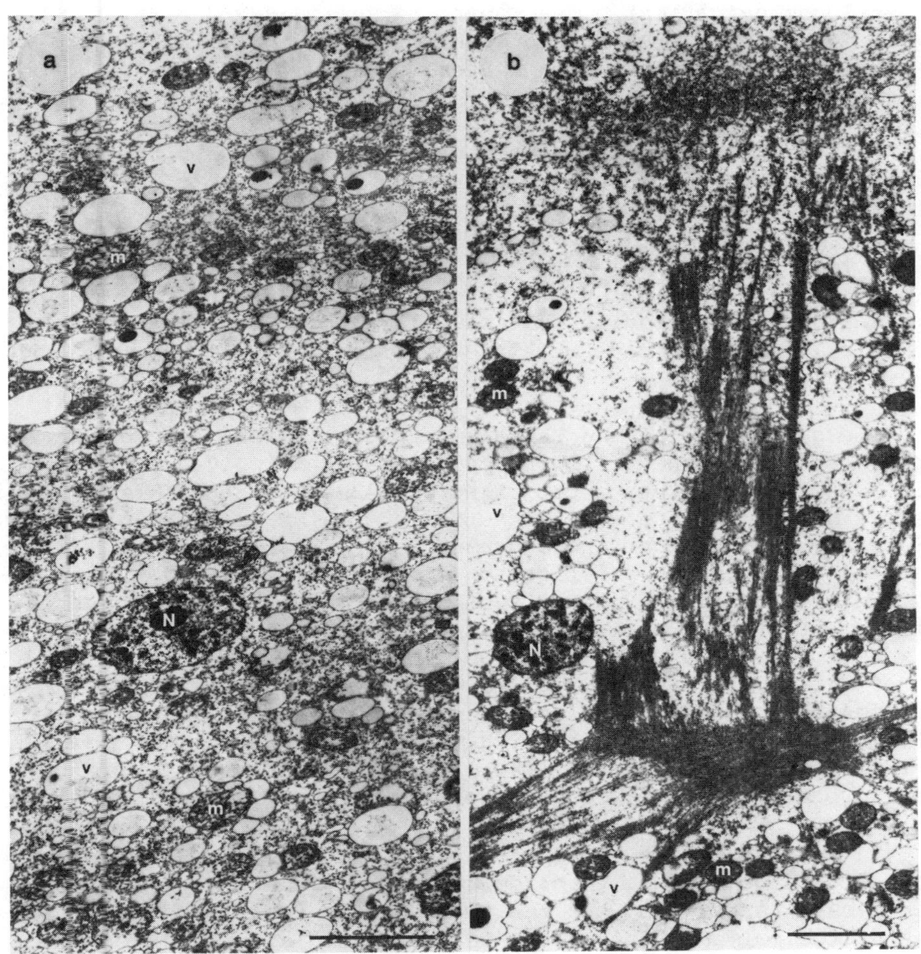

Fig. 15-1. Electron micrographs of thin sections of endoplasmic drops of <u>Physarum</u> <u>polycephalum</u> in the 0-min (a) and 7.5-min (b) stages after osmium/$HgCl_2$-fixation, critical-point drying, and methacrylate embedding. Note the amorphous cytomatrix in (a) and the distinct polygonal actomyosin fibril pattern in (b). Nucleus (N), mitochondria (m), and vacuoles (v). Bars, 3 µm. (Reprinted with permission from Wissenschaftliche Verlagsgesellschaft, Stuttgart)

was then heated in a boiling water bath and stirred for 2 min, cooled, and then centrifuged at 140,000 X g for 120 min (Maruyama, personal communication). The clear and yellow supernatant (ca 8 mL) was applied to a Pharmacia Sepharose CL-2B column (2.6 X 83 cm) equilibrated with elution buffer (0.1% SDS, 5 mM EDTA, 0.5 mM DTT, 100 mM Tris-glycine, pH 8.8).

The absorbance of the eluate was monitored at 280 nm. Fractions were collected at room temperature, and aliquots were checked for protein composition by microslab gradient SDS-PAGE.

Salt Fractionation

Salt fractionation of SDS-solubilized proteins was performed essentially as described for titin purification from muscle (Wang, 1982).

The giant size of the *Physarum* protein in SDS, and thus the large distance to the following polypeptides in polyacrylamide gels (Fig. 15-2, inset, upper right), has made it possible to purify the protein by agarose gel filtration of the DNase/RNase-pretreated SDS extract in the continuous presence of the denaturant SDS. An essential step in the purification procedure was degradation of DNA and RNA that interfered with the protein elution profile in the area of the heavy chain proteins. A typical elution profile of an SDS extract and gel patterns of selected fractions are presented in Fig. 15-2. The high molecular weight protein appears in the ascending shoulder of the second peak (Fig. 15-2, fractions 44-79), followed by degraded high molecular weight protein, which, as titin from muscle (Wang, 1982), appears as a smear below the giant protein after gel electrophoresis (fractions 80-90).

In the descending shoulder of the second peak, myosin heavy chain and an unknown polypeptide of considerable amount and larger than myosin heavy chain are the two most prominent bands. The mass of the polypeptides elutes in the ascending shoulder of the third peak with a major band at 43K quantitatively representing actin. *Physarum* yellow pigment appears mainly in the trailing edge of the third peak. The almost identical purification and fractionation protocols by which titin is purified from myofibrils (Wang, 1982) and the giant protein from *Physarum* support the conclusion that the *Physarum* protein is titin-like.

Some Comments

Peak 1 in the elution profile (Fig. 15-2), always present in front of the giant protein, is nothing but higher aggregates of the giant *Physarum* protein. This was confirmed after SDS-PAGE by comparing the Coomassie blue-stainable protein in the wells of the stacking gels with the protein in the position of the high molecular

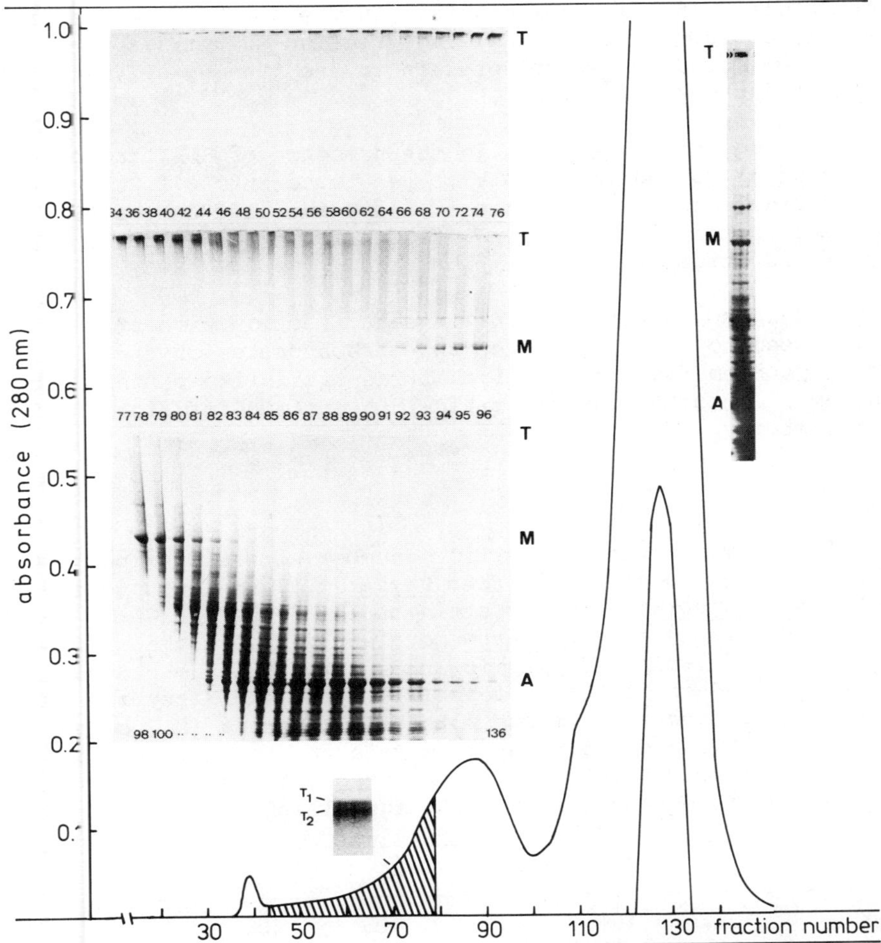

Fig. 15-2. Purification of Physarum titin-like protein. Agarose gel filtration of a DNase/RNase-pretreated SDS extract of 0-min endoplasmic drops. Insets: Upper right: Gel pattern (2.6-12%) of SDS extract subsequently loaded on the Sepharose CL-2B column. Note closely spaced doublet of titin-like polypeptides near the top of the gel. Left: Gel patterns (2.6-15%) of fractions indicated by numbers in the elution profile. Positions of titin-like protein (T), myosin heavy chain (M), and actin (A) are indicated. Shaded area in the elution profile represents the titin-like protein fractions. Titin 1 and 2 (T_1, T_2) polypeptide pattern of pooled fractions in a SDS gel: small inset above the shaded area in the elution profile. (Reprinted with permission from Wissenschaftliche Verlagsgesellschaft, Stuttgart)

weight protein in the running gels. In the stacking gel, staining intensity decreases with increasing fraction number; in the position of the high molecular weight protein in the running gel, staining intensity increases.

This suggests that, even in the presence of SDS, the giant protein forms aggregates unable to penetrate into a 2.6% stacking gel. Furthermore, a comparison of the amino acid composition of the giant protein fractions and of peak 1 protein revealed that they are identical (Table 15-1).

High-speed centrifugation and salt fractionation of the SDS extract seem to remove glycogen and carbohydrates of the slime (a special problem for purification of the titin-like protein from *Physarum*), contaminating the titin-like protein fractions in direct chromatography.

Table 15-1. Amino acid composition (number of residues per 1000 residues) of purified *Physarum* titin-like protein (A), *Physarum* peak 1 protein of the Sepharose CL-2B chromatography (B), *Physarum* covalently-crosslinked matrices of 0-min endoplasmic drops (C), *Physarum* plasmodial SDS-insoluble protein (D) (Ozaki and Maruyama, 1980), connectin from sheep heart myofibrils (E), connectin (F), and titin (G) from chicken breast myofibrils (see also Wang, 1985). (Reprinted with permission from Wissenschaftliche Verlagsgesellschaft, Stuttgart)

	A	B	C	D	E	F	G
Asp	102	101	101	100	105	93	95
Thr	58	55	53	64	49	66	75
Ser	69	64	70	67	66	67	69
Glu	128	130	128	116	133	118	116
Pro	60	63	61	62	65	67	74
Gly	72	70	80	70	87	76	71
Ala	90	87	96	84	80	75	62
Val	58	55	52	60	66	78	85
Meth	20	20	14	18	—	16	10
Ile	45	48	37	49	45	56	59
Leu	86	86	83	88	79	76	67
Tyr	27	29	25	29	31	30	30
Phe	38	37	34	42	34	29	27
Lys	71	75	71	74	75	79	82
His	38	37	34	24	19	18	15
Arg	60	65	60	49	56	50	55

Concentration of the purified giant protein in SDS buffer by ultrafiltration results in SDS-insoluble, gel-like aggregates that do not move into the stacking gel at all. Aggregates are formed in SDS by repeated freezing and thawing; for example, by storage of the sample in a freezer. Because of its tendency to form insoluble aggregates, the Physarum protein is often frustrating to handle.

AMINO ACID COMPOSITION

As discussed above, peak 1 of the gel filtration represents aggregates of the titin-like protein, as demonstrated by the amino acid composition compared with the purified titin-like protein (Table 15-1, columns A, B). The amino acid compositions of covalently cross-linked matrices prepared from 3- to 4-sec-old endoplasmic drops by SDS/urea/DDT extraction and electrophoretic purification are also identical (Table 15-1, column C). Therefore, aggregated titin-like protein seems to be the only subunit of the insoluble matrices (compare Table 15-1, columns A,B,C).

Column D of Table 15-1 represents the amino acid analysis of the SDS-insoluble residue of whole plasmodia (Ozaki and Maruyama, 1980), which is very similar to the SDS-soluble giant protein purified from endoplasmic drops by chromatography. For comparison, amino acid compositions of muscle titin (connectin) of different sources are presented (Table 15-1, columns E,F,G). Note the similarity in amino acid composition of muscle titin and Physarum titin-like protein. The titin-like protein contains more Ala, Leu, and His, and less Thr and Val, than muscle titin (compare Table 15-1, columns G and A). Asp, Glu, Ala, and Leu are the major amino acid components of the giant Physarum protein. Note the high content of proline (6%), an α-helix-breaking amino acid.

MORPHOLOGICAL STUDIES PERFORMED

Is the giant titin-like protein a putative candidate as a structural component for a cytoskeletal filament lattice in Physarum? Indications in this direction come from reconstitution studies by removal of SDS from titin-like protein fractions.

Huge, gel-like aggregates were formed, which after negative staining seem to be composed of linear, superthin, 2- to 3-nm filaments (Fig. 15-3). Self-assembly into filaments apparently occurs during SDS removal. Similar features, as in the reconstituted, pure Physarum titin-like protein gel, can be found in situ in replicas of fixed, critical-point-dried and dry-cleaved specimens and in thoroughly extracted specimens of different Physarum stages (0-min drop, 7.5-min drop, whole plasmodia). This is illustrated for 0-min endoplasmic drops (used as the starting material for

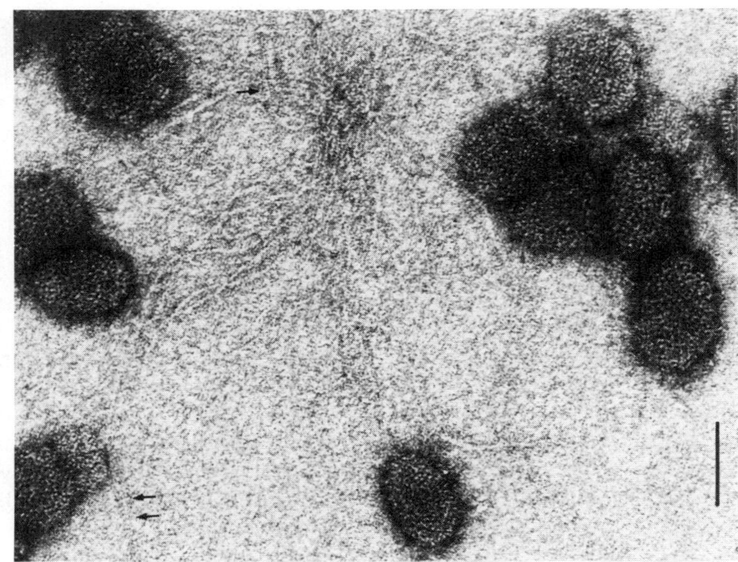

Fig. 15-3. Molecular morphology of *Physarum* titin-like protein purified under denaturing conditions. The titin-like protein was purified by SDS gel filtration, aggregated into a gel by SDS removal, and negatively stained with uranyl acetate. Note rather uniform globular structures (glycogen?) and linear superthin filaments (2-3 nm in diameter) with an occasional beaded appearance (arrows). Bar, 50 nm. (Reprinted with permission from Wissenschaftliche Verlagsgesellschaft, Stuttgart)

purification of the titin-like protein under denaturing conditions; see above): (1) in the unextracted in situ stage by use of the replica technique (Fig. 15-4); and (2) after SDS/urea/DTT-extractions, followed by an electrophoretic purification step (covalently cross-linked matrices) using polyethylene glycol, embedded, thin-sectioned, dewaxed, and critical point-dried specimens (Fig. 15-5). Note the similarity of the extraction-resistant residue (amino acid composition identical with the titin-like protein purified by chromatography) with a filament system present in the unextracted stage.

Critical-point drying, in combination with the dry-cleaving technique and rotary shadowing, indicates the overall spatial organization of the superthin cytoskeletal filament system in *Physarum* and gives us a first idea of the architecture of the filament network in situ with the giant titin-like protein as a major component. Other observations, such as variability of morphological appearance, elasticity, and tension response (based on volume and structural

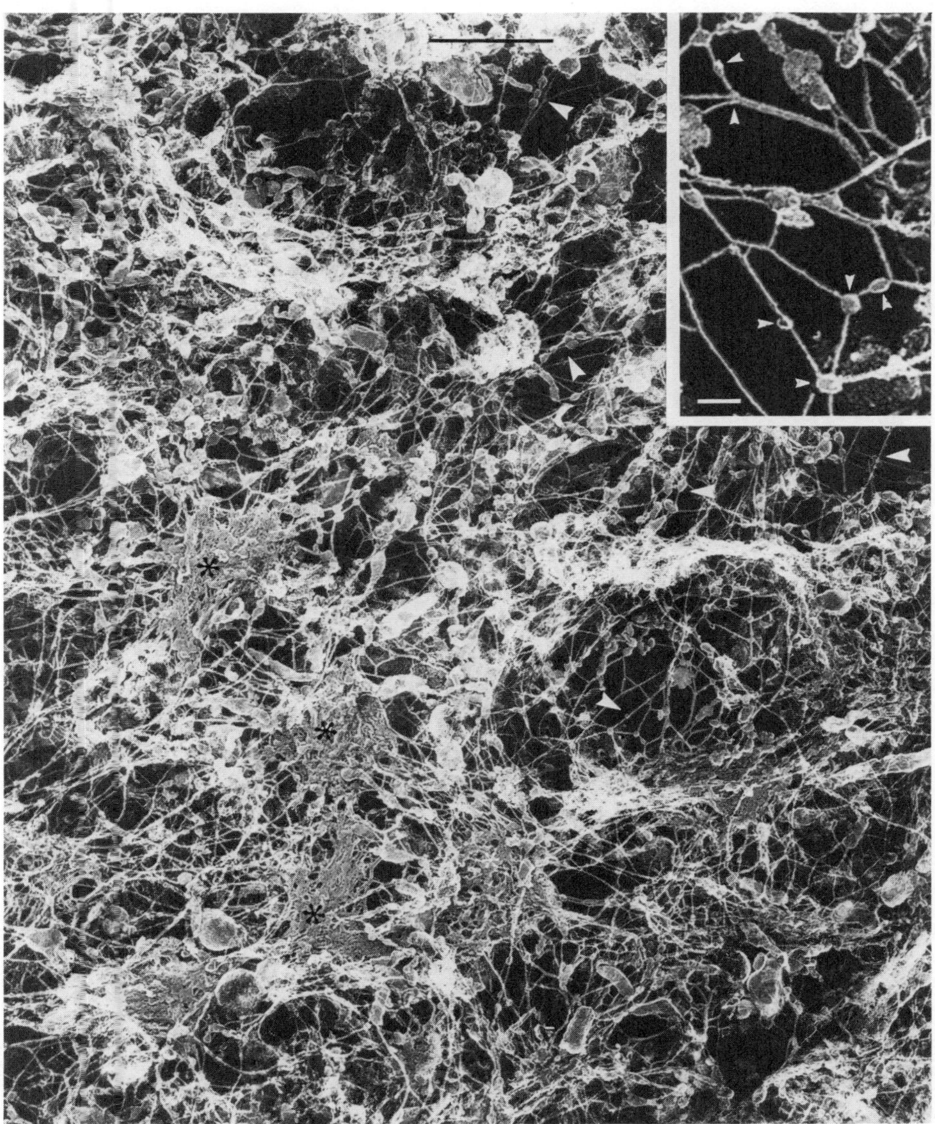

Fig. 15-4. Pt/C replica of an osmium/$HgCl_2$-fixed, critical-point-dried, and dry-cleaved 0-min endoplasmic drop. Note the anastomosing network of superthin filaments, mostly of uniform thickness and frequently associated with globular domains either along the filaments or at their nodal points (arrowheads). More condensed states of the lattice are marked with asterisks. Bar, 1 μm. Inset: Selected part of the lattice at higher magnification. Bar, 100 nm. (Reprinted with permission from Wissenschaftliche Verlagsgesellschaft, Stuttgart)

changes induced by mM Ca^{++}) possibly could mean that this network is a dynamic system capable of reversibly forming and unraveling distinct globular domains. Transformation into linear filaments, bundles, or coiled structures may result in changes of local or overall mesh densities.

IS NATIVE PHYSARUM TITIN-LIKE PROTEIN ESTABLISHED?

The availability of a procedure for purification of the *Physarum* titin-like protein in the native state without any denaturants would allow numerous techniques to be applied for examination of its molecular morphology, conformation, and interaction with other proteins. Previously, three groups (Maruyama et al., 1984; Trinick et al., 1984; Wang et al., 1984) reported on the successful purifi-

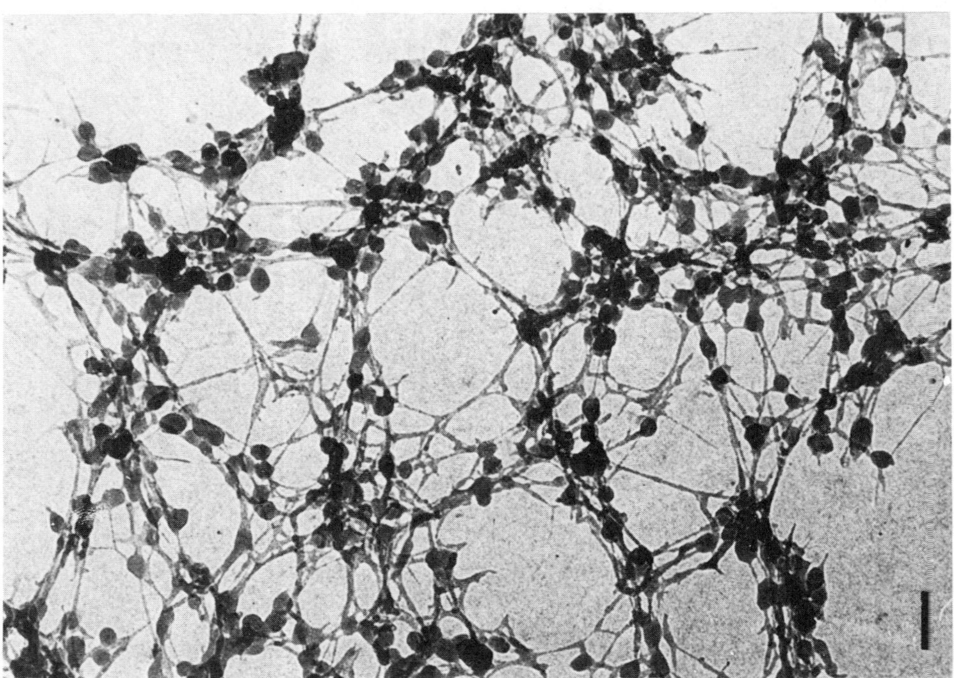

Fig. 15-5. Ultrastructure of covalently cross-linked matrices of 0-min endoplasmic drops after SDS-urea/DTT extraction followed by an electrophoretic purification step. Dewaxed, critical-point-dried thin section (free of embedding medium) after polyethylene glycol embedding. Note the anastomosing network of filaments associated with globular structures. Bar, 100 nm (From Gassner, unpublished, 1985)

cation of native titin from muscle, probably a proteolytic cleavage product of the genuine, but so far salt-insoluble, protein. Electron microscopy revealed that native titin chains self-assembled into long, flexible, and very slender strands with a width of 3-4 nm, sometimes with an axial 4-nm periodicity.

Moreover, the strands tended to associate into a number of morphological features ranging from thin strands of different thickness to nodules of various size, fiber bundles, and reticular networks.

Concerning native titin-like protein from *Physarum*, Maruyama's group (see Chapter 16) have reported on the isolation of a native connectin (titin)-like protein preparation starting from the whole plasmodium. By use of a bicarbonate extraction followed by a single gel filtration step, they reported that "fractions following the void volume *mainly* consisted of high molecular weight protein", which did not move into SDS gels at all. Moreover, their preparations were "*more* or *less* contaminated with polysaccharides" of the slime. From these statements, it seems clear that a successful isolation of pure titin-like protein from *Physarum* must be the aim of future work. Similar efforts for purification of native *Physarum* titin-like protein are being performed in our laboratory. However, it should be mentioned that, so far, we have no exact purification strategy available to obtain native titin-like protein in a highly purified form. This holds true even when starting with endoplasmic drops to avoid most of the carbohydrates of the plasmodial slime and using a sequence of purification steps including salt precipitation, gel filtration, and ionic exchange chromatography sufficient for isolation of native titin from myofibrils (Trinick et al., 1984; Gassner, 1985). Additional experiments to isolate the native molecule in highly purified form are currently being performed.

ACKNOWLEDGMENT

This investigation was supported by the Deutsche Forschungsgemeinschaft (Wo 20-22/2).

REFERENCES

Gassner, D., 1985, Myofibrillar interaction of blot immunoaffinity purified antibodies against native titin as studied by direct immunofluorescence and immunogold staining, *Eur. J. Cell. Biol.*, submitted.

Gassner, D., Shraideh, Z., and Wohlfarth-Bottermann, K.-E., 1983, An extraction resistant tensile protein in the protoplasmic matrix of *Physarum*, *Cell Biol. Int. Rep.*, 7:905.

Gassner, D., Shraideh, Z., and Wohlfarth-Bottermann, K.-E., 1985, A giant titin-like protein in *Physarum polycephalum*: evidence for its candidacy as a major component of an elastic cytoskeletal superthin filament lattice, *Eur. J. Cell Biol.*, 37:44.

Maruyama, K., Kimura, S., Yoshidomi, H., Sawada, H., and Kikuchi, M., 1984, Molecular size and shape of β-connectin, an elastic protein of striated muscle, *J. Biochem.*, 95:1423.

Maruyama, K., Matsubara, S., Natore, R., Nonomura, Y., Kimura, S., Ohashi, K., Murokami, F., Handa, S., and Eguchi, G., 1977, Connectin, an elastic protein of muscle. Characterization and function, *J. Biochem.*, 82:317.

Ozaki, K., and Maruyama, K., 1980, A connectin-like protein from the plasmodium *Physarum polycephalum*, *J. Biochem.*, 88:883.

Trinick, J. A., Knight, P., and Whiting, A., 1984, Purification and properties of native titin, *J. Mol. Biol.*, 180:331.

Wang, K., 1982, Purification of titin and nebulin, *Methods Enzymol.*, 85: 264.

Wang, K., 1984, Cytoskeletal matrix in striated muscle. The role of titin, nebulin and intermediate filaments, *in*: "Contractile Mechanisms in Muscle", G. H. Pollack and H. Sugi, eds., p. 285, Plenum Press, New York.

Wang, K., 1985, Sarcomere-associated cytoskeletal lattices in striated muscle. Review and hypothesis, *in*: "Cell and Muscle Motility", J. W. Shay, ed., p. 315, Plenum Press, New York.

Wang, K., Ramirez-Mitchell, R., and Palter, D., 1984, Titin is an extraordinarily long, flexible, and slender myofibrillar protein, *Proc. Natl. Acad. Sci., U.S.A.*, 81:3685.

Wohlfarth-Bottermann, K. E., 1983, Dynamic cellular phenomena in *Physarum* accessible to laser techniques, *in*: "The Application of Laser Light Scattering to the Study of Biological Motion", J. C. Earnshaw and M. W. Steer, eds., NATO ASI Series, Series A: Life Sciences, Vol. 59, p. 501, Plenum Press, New York and London.

Chapter 16: ISOLATION OF A NATIVE CONNECTIN-LIKE PROTEIN FROM THE PLASMODIUM OF PHYSARUM POLYCEPHALUM AND ITS INTERACTION WITH MYOSIN AND ACTIN

Di Hua Hu, Sumiko Kimura, Tsuneo Suzuki, and Koscak Maruyama

Department of Biology
Chiba University
Chiba, JAPAN

INTRODUCTION

In 1980 we reported that there is a sodium dodecyl sulfate (SDS)-insoluble filamentous protein in *Physarum*, and this peculiar protein is very similar in filamentous structure and amino acid composition to connectin, an elastic protein of striated muscle (Ozaki and Maruyama, 1980). Recently, Gassner and his colleagues at the University of Bonn have been able to isolate a similar protein in an SDS-soluble form and have characterized it from the viewpoint of a new cytoskeletal matrix in the plasmodium (Gassner et al., 1985; see Chapter 15). We have also separated a similar connectin-like protein in the presence of SDS; however, all the preparations mentioned above were denatured proteins, and, therefore, its physiological function was not studied at all except for its electron microscopic structure.

The emphasis of our current work is to isolate the protein in question in the native form and to study its interactions with myosin and actin filaments in vitro, following our previous work with chicken breast muscle connectin (Kimura et al., 1984).

ISOLATION OF NATIVE PROTEIN FROM THE PLASMODIUM

A plasmodium was gently homogenized in 10 vol of 1 mM $NaHCO_3$ in the cold and centrifuged for 5 min at 7000 X *g*. The sediment was stirred in 10 vol of 1 mM $NaHCO_3$ for 15 min at 4°C. The mixture was centrifuged again, and the supernatant was filtered through a sheet of filter paper. The filtrate containing the high molecular weight protein was directly subjected to gel filtration in a

Toyopearl HW 65 column (2.4 X 80 cm). The flow rate was about 17 ml/hr. The fractions (100-120 ml) following the void volume consisted mainly of the high molecular weight protein as revealed by SDS gel electrophoresis. The solution collected was concentrated by dialysis against aquacide (Calbiochem).

The molecular weight of the native protein was even larger than those of chicken muscle connectins (2-3 million) (see Maruyama et al., 1984), since the mobility was slower than the connectin bands in SDS gel electrophoresis with 1.8% polyacrylamide gels. Coelectrophoresis of both muscle and _Physarum_ proteins showed that muscle connectin migrated at the same mobility in mixtures as when alone, suggesting that the _Physarum_ preparation did not contain any substance interfering with the electrophoretic mobility. Samples negatively stained by uranyl acetate showed entanglements of very thin filaments and globular domains just as in SDS-denatured preparations (Gassner et al., 1985). The amino acid composition was almost the same as those reported before (Ozaki and Maruyama, 1980; Gassner et al., 1985).

INTERACTION WITH MYOSIN AND ACTIN

It has been reported that native muscle connectin interacts with both myosin and actin at low ionic strengths (Kimura et al., 1984). Also native connectin-like protein from _Physarum_ plasmodia interacted with myosin and actin from rabbit skeletal muscle under conditions in which the proteins exist as filaments. The results are shown in Figs. 16-1 and 16-2. The turbidity of myosin suspension increased by adding connectin-like protein at 50-80 mM. This effect was small in comparison with muscle connectin (Kimura et al., 1984). At 50 mM KCl, the aggregates of myosin filaments were formed in the presence of the native protein from _Physarum_ plasmodia (Fig. 16-1).

By contrast, the turbidity increase of an F-actin solution induced by the connectin-like protein was more remarkable than that caused by muscle connectin, as seen in Fig. 16-2. The effect was evident even at 0.15 M KCl, where muscle connectin was barely active. Entanglements of actin filaments were observed in the presence of the _Physarum_ protein (Fig. 16-2). It is to be noted that bundles of actin filaments were not formed by the connectin-like protein, in contradiction to the action of muscle connectin (Kimura et al., 1984).

It is highly desirable to reinvestigate the interactions of the _Physarum_ connectin-like protein with both myosin and actin prepared from _Physarum_ plasmodia. A possible anchoring role of these superthin filaments for myosin and actin filaments may be important from the cytoskeletal matrix point of view in the plasmodium.

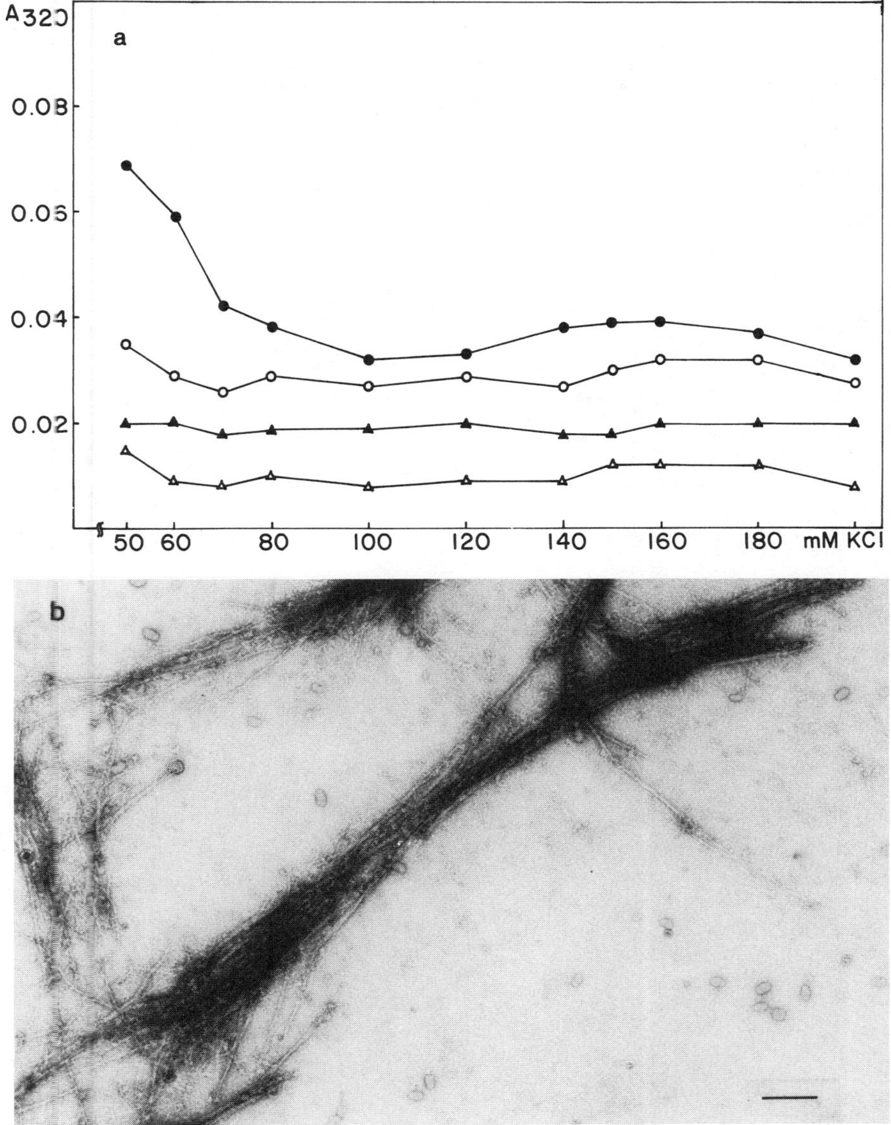

Fig. 16-1. a. Effect of <u>Physarum</u> connectin-like protein on the turbidity of a myosin suspension at various KCl concentrations. Connectin-like protein, 0.01 mg/ml, was mixed with rabbit skeletal muscle myosin, 0.05 mg/ml, in the presence of 5 mM phosphate buffer, pH 7.0, and KCl concentrations as shown in the abscissa. After 12 hr at room temperature, absorbance at 320 nm was measured. △, myosin; ▲, connectin-like protein; ●, myosin and connectin-like protein; ○, sum of myosin alone and connectin-like protein alone. b. Electron micrograph of a mixture of myosin and connectin-like protein at 50 mM KCl. Bar, 0.2 μm.

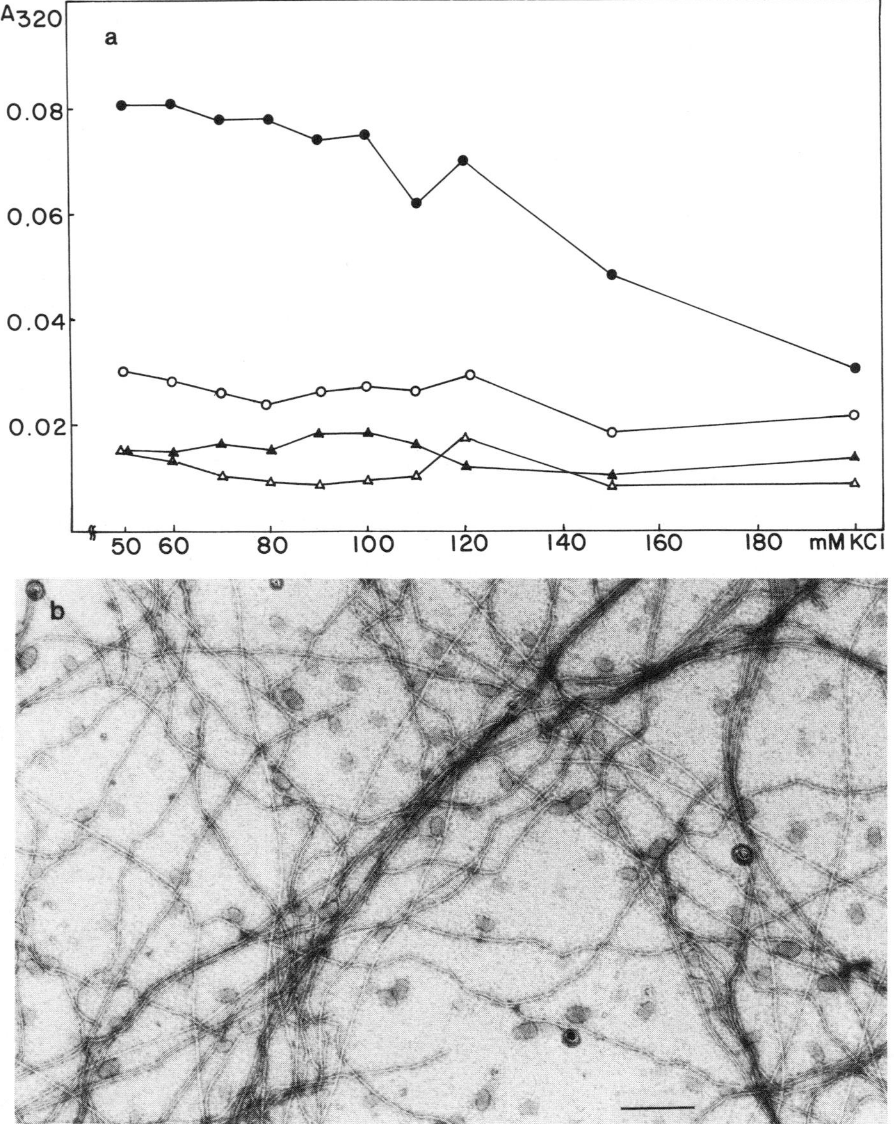

Fig. 16-2. a. Effect of Physarum connectin-like protein on the turbidity of F-actin solution at various KCl concentrations. Conditions as in Fig. 16-1 except that connectin-like protein (0.012 mg/ml) and F-actin (0.06 mg/ml) were used. △, F-actin; ▲, connectin-like protein; ●, F-actin and connectin-like protein; ○, sum of F-actin and connectin-like protein alone. b. Electron micrograph of a mixture of F-actin and connectin-like protein at 50 mM KCl. Bar, 0.2 μm.

Finally, it should be mentioned that the present native connectin-like protein preparations are partially contaminated with acid polysaccharides (slime), although the latter filamentous structure is distinguishable from that of connectin filaments (Ozaki and Maruyama, 1980). It is to be noted that, when fractionated material in 1 mM $NaHCO_3$ was centrifuged at 200,000 X *g* for 1 hr, the connectin-like protein was sedimented onto a carbohydrate layer. The former could be taken out and suspended in 1 mM $NaHCO_3$. This carbohydrate-free suspension also interacted with myosin and actin filaments.

REFERENCES

Ozaki, K., and Maruyama, K., 1980, A connectin-like protein from the plasmodium *Physarum plasmodium*, *J. Biochem.*, 88:883.

Gassner, D., Shraideh, Z., and Wohlfarth-Bottermann, K.-E., 1985, A giant titin-like protein in *Physarum polycephalum*: evidence for its candidacy as a major component of an elastic cytoskeletal superthin filament lattice, *Eur. J. Cell Biol.*, 37:44.

Kimura, S., Maruyama, K., and Huang, Y. P., 1984, Interactions of muscle β-connectin with myosin, actin, and actomyosin at low ionic strengths, *J. Biochem.*, 96:499.

Maruyama, K., Kimura, S., Yoshidomi, H., Sawada, H., and Kikuchi, M., 1984, Molecular size and shape of β-connectin, an elastic protein of striated muscle, *J. Biochem.*, 95:1423.

Editorial footnote: We recognize the transient problem in communication created by the use of two designations--titin and connectin--for what may be a single entity. We hope this subject is young enough that it can evolve to a consensus term, on the basis of function rather than publication priority.

Chapter 17: TO BRING MOLECULAR GENETICS TO THE STUDY OF BIOLOGICAL MOTILITY IN <u>PHYSARUM</u>

William F. Dove

McArdle Laboratory for Cancer Research and
Laboratory of Genetics
University of Wisconsin, Madison, WI, USA

MOTILITY IS A FUNDAMENTAL BIOLOGICAL PROPERTY

Back in the Golden Age of Molecular Biology, the 1960s, hordes were occupied with unraveling the mysteries of protein structure, self-assembly, the genetic code, protein synthesis, DNA replication, regulation, and the set of problems that we now call macromolecular metabolism (Stent, 1968). On the side, a small cohort including Norman Horowitz, Joshua Lederberg, Carl Sagan, and Wolf Vishniac was concerned with designing experiments to detect life on the moon and on nearby planets as there emerged a capability to place landing vehicles on these celestial objects (Horowitz, 1966). Tests for numerous metabolic pathways diagnostic of terrestrial life were adapted to the harsh conditions of outer space, and culture media to support growth were formulated from various aerobic and anaerobic cuisines. Finally, it was necessary to cover the range of possibilities in which extraterrestrial life would employ a metabolism different from that of terrestrial life ("silicon life"). Two phenomenological criteria were added to the repertoire of the landing vehicle laboratory--irritability and motility. Indeed, it seems that all terrestrial forms higher than viruses display integrated motility systems. These translocate materials within cells, or translocate the organism along surfaces, or propel it through three dimensions.

This anecdote contrasts the current status of experimental analysis of metabolism on the one hand, and of biological motility on the other. The problem of metabolism, including such complex processes as DNA replication, protein synthesis, and respiration, is apparently succumbing to the combined assault of the biochemist, working with soluble extracts, and the geneticist, isolating or even designing mutants affected in key genes. By contrast, the study of motility (and, even more, of irritability) still is shrouded in mystery. It may require integration of molecular information over the entire cell. Perhaps by the 21st century that integration will emerge from the fusion of numerous biochemical studies on isolated subcellular complexes. Alternatively, it may emerge only after the geneticist has identified by mutational subtraction the crucial genes at work and the logical pathways by which they operate.

In practice, the lines of progress that will lead to a molecular understanding of biological motility may involve fruitful unions between biochemical analysis and genetic analysis, involving either of two styles of molecular genetics marriage. Style B starts with a molecule identified by the biochemist and thought to be crucial in the process of interest on the basis of reactions in vitro--for example, myosin (Sheetz and Spudich, 1983) or kinesin (Vale et al., 1985). The marriage is consummated when mutations are found that affect the structural gene for that molecule, and the phenotypic consequences are assessed in vivo (MacLeod et al., 1977). With _Physarum_ this process may be accomplished over the next decade.

Linking of genetic and biochemical analysis can alternatively arise from a marriage of Style G, which starts with the geneticist finding a mutant gene, with unknown product but affected in the process of interest. (Style G has previously been called Plan A by Thomas et al., 1984.) Here, the marriage is consummated when the gene product and its biochemical activity are identified. Again, we shall talk about the emergent feasibilities for this scientific process with _Physarum_.

The apparent complexity of a responsive motility system seems to dictate that significant progress involve at least one style of molecular genetics. It seems unlikely that the system can be assembled directly from known purified components. (The same may well be said also for cell cycling [see Chapter 5] or for differentiation [see Chapters 6 and 7].) For those of us who work on the genetics of _Physarum_, the challenge is clear!

PHYSARUM DISPLAYS MOST KNOWN FORMS OF BIOLOGICAL MOTILITY

The plasmodium displays an amplified form of protoplasmic streaming, offers adequate quantities of structural and regulatory molecules that may participate in this process and, crucially, gives the opportunity to re-introduce quantities of material adequate for a biochemical analysis of their fate (see Chapters 20 and 21). Research papers in this volume indicate biological and biochemical studies that have long been carried out on this form of motility.

The plasmodium also displays surface translocation, including chemotaxis (Coman, 1940; Carlile, 1970). It seems possible that this process shares elements with the surface translocation of myxamoebae, though the latter has not yet been shown to include chemotaxis (Jacobson and Dove, 1975).

Propulsion through liquids is displayed by the flagellate, and conditions now exist for the conversion of more than 90% of myxamoebae to flagellates within 1 hour (Pagh and Adelman, 1982).

Descriptive molecular biology has already joined the descriptive biology summarized above. For cytoskeletal macromolecules likely to be involved in these motile processes, it is now known that:

* From the three-member β-tubulin gene family, one member is expressed in both plasmodia and myxamoebae, a second only in plasmodia, and a third in both proliferative forms (Burland et al., 1984);
* An α-tubulin is modified posttranslationally when the myxamoeba converts to the flagellate (Green and Dove 1984);
* Plasmodia and myxamoebae differ in the myosin heavy chain and in a myosin light chain, as judged by peptide mapping, antigenicity, and electrophoretic mobility (Chapter 10).

We can expect this catalog to grow rapidly with recent advances in the purification of macromolecules from <u>Physarum</u>, in developing specific antibodies (Chapter 6), in gene cloning (Chapters 22-24), and in gene mapping (Chapter 2). Soon enough, it may reach the magnitude of the Manhattan telephone directory! This <u>Physarum</u> catalog may on its own be useful in deducing rules of function. Because the organism can carry out so many different motility functions, we shall come to know which molecular parts are used universally and which ones are modified to construct particular motile systems. We can expect to learn much from the results of the evolutionary process when the genes for different motility processes have cohabited the same genome. This exemplifies the benefit of studying an organism that can "do it all".

<u>Physarum</u> has a singular position for the development of a molecular genetic analysis of eukaryotic motility. The microbes <u>Chlamydomonas</u>, <u>Dictyostelium</u>, and <u>Aspergillus</u>, and cultured animal cells, each display a subset of the motility functions. (The yeasts display rather little motility!) Of these experimental organisms, <u>Dictyostelium</u> is currently severely hampered for genetic manipulation by lacking a meiotic process in laboratory strains, making gene mapping very difficult and the construction of multiply mutated strains prohibitive. These limitations are shared by cultured

animal cells, which not only lack meiotic genetics, but also lack a haploid phase in which recessive loss-of-function mutations affect the phenotype. Much has been written about ways in which studies of Physarum can shed some light on the dark mysteries surrounding the biology of malignant neoplasia (Rusch, 1980). Because the malignant potential of a neoplasm is entwined in its capacity for metastasis, and because there is so little knowledge of the fundamental mechanisms involved in this process (Alby and Auerbach, 1984), work on Physarum biology may have a crucial contribution to make in the early explorations of this domain.

FUNCTIONAL ANALYSIS

Style B Molecular Genetics

Feasibilities to inactivate genes for known cytoskeletal macromolecules. In the case of tubulins expressed in the myxamoebal phase, it is possible to select directly for mutations that alter the sensitivity to anti-mitotic benzimidazoles (see Chapter 2). Because such resistance mutations are recessive, it seems possible now to select in a second step for loss-of-function mutations at the benzimidazole-resistance loci. This selection starts with a diploid myxamoebal strain heterozygous for a recessive drug-resistance mutation. Loss of function of the wild type allele in such a strain would then lead to re-emergence of the resistant phenotype. Already the use of drug-resistance mutations in identifying the active genes for tubulin shows that much can be gained if other drugs can be identified with specific action against particular components of the motility systems of Physarum. Phalloidin or cytochalasin may offer such an opportunity for actin, though they are not active when added to plasmodial cultures (M. Adelman, personal communication).

A more general way to inactivate genes encoding known cytoskeletal macromolecules will depend upon DNA transformation. If one has a cloned DNA sequence for a portion of the coding region of the molecule in question, and if DNA transformation includes events occurring by homologous recombination, the structural gene can be "disrupted" (see Chapter 25 and Smithies et al., 1985). A less complete inactivation of a gene for a known cytoskeletal macromolecule could also emerge from a workable DNA transformation system by introducing cloned DNA for the molecule in question joined to a promoter that directs synthesis of the anti-sense RNA (Izant and Weintraub, 1985). For motility functions that are essential for viability, such as plasmodial streaming, this approach would seem best implemented were the anti-sense promoter an inducible one, such as that for metallothionein (Butt et al., 1984).

These projections may become reality in a rather short period of time. For those working on the biochemistry of macromolecules likely to be involved in Physarum motility, it is important to aim directly for a dowry of information that will permit a Style B marriage with genetics: drugs with known molecular targets; antibodies to the molecule of interest, preferably several to guard against spurious cross-reactions, permitting DNA cloning from expression libraries (see Chapter 23); or amino acid sequence information, by microchemistry if necessary (Hunkapiller et al., 1984), to permit the construction of oligonucleotide probes with which to screen any of the Physarum libraries (see Chapters 22-24).

Style G Molecular Genetics

The search for motility phenotypes that permit isolation of mutants affected in molecules relevant to motility. The isolation of mutants requires the use of clonal growth systems, preferably haploid. Otherwise, in general a mutation in a gene will not affect the cellular phenotype to permit mutant isolation. The myxamoeba is therefore naturally suited to this enterprise and can be grown into large populations. In selfing strains, isolated plasmodia are clonal and can be screened for mutants by the determined investigator. In mass cultures of plasmodia, however, the rapid somatic fusion breaks down clonal isolation, preventing selectional genetics on mass populations of microplasmodia. (It would be invaluable if it were possible first to isolate a mutant strain that had lost the somatic plasmodial fusion system. It is interesting to wonder if such nonfusing plasmodia would be viable under proper care.)

Screening of isolated plasmodial clones for mutants is laborious. The frequency of loss-of-function alleles for a single gene after optimal mutagenesis may be in the order of 10^{-4}. Thus, to screen isolated clonal plasmodia for a defect in a particular motility gene would require in excess of 10^4 separate plasmodial microcultures. Were the motility system vital for growth--e.g., plasmodial protoplasmic streaming--then only conditional alleles such as thermosensitives could be sought. Such alleles would be more rare, by perhaps an order of magnitude, making plasmodial screening prohibitive. For the foreseeable future, then, the isolation of motility mutants is restricted to those based upon phenotypes displayed by the myxamoeba.

Two myxamoebal motile behaviors form the basis for mutant isolation. One is migration on surfaces, as in colony formation. Jacobson and Dove (1975) described colony-morphology variants in which the migratory behavior is altered. Some showed Mendelian inheritance. Jacobson (1979, 1980) went on to study myxamoebal migratory behavior by microcinematography. He showed that the speed of individual cells in contact with bacteria is affected by the mov-1 mutation. (These films are available from M. Adelman, personal communication.)

Pure chemotactic substances have not yet been defined for Physarum myxamoebae, to the best of the author's knowledge. As such single substances become known, strong selection would become feasible for mutants in chemotaxis specifically or, more generally, in cellular translocation. An example of such a selection for the myxamoeba of Dictyostelium is reported by Barclay and Henderson (1982), who found mutants that fail to migrate through filters toward cyclic AMP. Note here that the great facility for meiotic mapping in Physarum (see Chapter 2) permits tests for linkage between mutations selected for aberrant migratory behavior and structural genes for known cytoskeletal or receptor macromolecules, identified by cloned DNAs.

A second myxamoebal phenotype that permits mutant isolation is the motility of the flagellate. Ability to form the flagellate is open to selection because myxamoebae adhere more strongly than do flagellates to glass beads in columns (Jacobson et al., 1976; Blindt, Dee, and Gull--see Chapter 6). Migration of the flagellate is in principle open to selection in diffusion chambers, in analogy to the selection used by Barclay and Henderson (1982).

There is clearly much to do for those engaging a Style G molecular genetic analysis of motility in Physarum. The biology of the myxamoeba and of the flagellate need far more study to expand the range of phenotypes for which mutant selection can be devised. The laboratory of Michel Wright has made interesting contributions in this area. For example, Mir et al. (1979) report that of 76 myxamoebal variant isolates thermosensitive for growth, 16 were also defective in the formation of flagellates. Of these, five isolates showed a defect in flagellate formation that was not conditional on the temperature, unlike the defective growth phenotype. In one such case, an explicit test was made to show that the growth defect was genetically separable from the flagellation defect.

An effort to develop sets of motility mutants will contribute to an understanding of motility in a way that cannot be matched by other approaches. In principle, Style G permits one to identify molecules that are crucial to a motility system but that are not yet in the biochemistry textbooks. Regulatory molecules of low concentration may fall into this category. The consummation of a Style G marriage, of course, will depend on finding effective ways to identify the products of genes known only by their mutations. Elsewhere in this volume, glimmers of hope for such ways can be found--DNA transformation (see Chapter 25), were it to reach efficiencies high enough to permit cloning by mutant-rescue; and transposon mutagenesis mediated by known, cloned transposons (Bingham et al., 1981; see Chapter 3). Identification of the product of a gene affecting a plasmodial motility function has a further experimental option unique to *Physarum*--transfer of microliter volumes of cytoplasm into the plasmodium (Chapters 20 and 21; Ueda et al., 1979). In principle, any mutant plasmodium provides the substrate for a one-component bioassay for the missing gene product. Such an assay would be linear (see, by contrast, Loidl and Gröbner, 1984) and can guide the purification of active single gene products.

SUMMARY

Physarum displays a broad range of motility behaviors within a single organism. The capability for molecular genetic analysis in the organism is emerging, resting on the fundamental power of haploid clonal myxamoebal populations, clonal plasmodia, and a facile meiotic system for gene mapping and for construction of informative recombinant genotypes. Style B molecular genetics seems to be as imminent as a tractable DNA transformation system, even one of low efficiency (Chapter 25). Technical progress described in this volume, combined with an energetic biological investigation of the motile behaviors of the myxamoeba and the flagellate to open up Style G molecular genetics, can surely permit *Physarum* to make unique contributions to the experimental biology of eukaryotic motility.

ACKNOWLEDGMENTS

The long-term dialogs that I have enjoyed with David Jacobson, Mark Adelman, Stephen Barclay, Gary Borisy, and Tim Burland have been distilled into this essay. My research is supported by Core Grant CA-07175 and Program-Project Grant CA-23076 from the National Cancer Institute.

REFERENCES

Alby, L., and Auerbach, R., 1984, Differential adhesion of tumor cells to capillary endothelial cells in vitro, *Proc. Natl. Acad. Sci., U.S.A.*, 81:5739.

Barclay, S. L., and Henderson, E. J., 1982, Thermosensitive development and tip regeneration in a mutant of *Dictyostelium discoideum*, *Proc. Natl. Acad. Sci., U.S.A.*, 79:505.

Bingham, P., Levis, R., and Rubin, G. M., 1981, Cloning of DNA sequences from the *white* locus of *D. melanogaster*, *Cell*, 25:693.

Burland, T. G., Schedl, T., Gull, D., and Dove, W. F., 1984, Genetic analysis of resistance to benzimidazoles in *Physarum*: differential expression of β-tubulin genes, *Genetics*, 108:123.

Butt, T. R., Sternberg, E. J., Borman, J. A., Clark, P., Hamer, D., Rosenberg, M., and Crooke, S. T., 1984, Copper metallothionein of yeast: structure of the gene and regulation of expression, Proc. Natl. Acad. Sci., U.S.A., 81:3332.

Carlile, M. J., 1970, Nutrition and chemotaxis in the myxomycete Physarum polycephalum: the effect of carbohydrates on the plasmodium, J. Gen. Microbiol., 63:221.

Coman, D. R., 1940, Additional observations on positive and negative chemotaxis. Experiments with a myxomycete, Arch. Pathol., 29:220.

Green, L. L., and Dove, W. F., 1984, Tubulin proteins and RNA during the myxamoeba-flagellate transformation of Physarum polycephalum, Mol. Cell Biol., 4:1706.

Horowitz, N. H., 1966, The biological significance of the search for extraterrestrial life. Technical Report No. 32-1000, Jet Propulsion Laboratory, California Institute of Technology, Pasadena.

Hunkapiller, M., Kent, S., Caruthers, M., Dreyer, W., Firca, J., Giffin, C., Horvath, S., Hunkapiller, T., Tempst, P., and Hood, L., 1984, A microchemical facility for the analysis and synthesis of genes and proteins, Nature, 310:105.

Izant, J. G., and Weintraub, H., 1985, Constitutive and conditional suppression of exogenous and endogenous genes by anti-sense RNA, Science, 229:345.

Jacobson, D. N., 1979, The role of regulation of cell speed in the behavior of Physarum polycephalum amoebae, Exp. Cell Res., 122:219.

Jacobson, D. N., 1980, Locomotion of Physarum polycephalum amoebae is guided by a short range interaction with E. coli, Exp. Cell Res., 125:441.

Jacobson, D. N., and Dove, W. F., 1975, The amoebal cell of Physarum polycephalum: colony formation and growth, Devel. Biol., 47:97.

Jacobson, D. N., Johnke, R. M., and Adelman, M. R., 1976, Cell motility in Physarum polycephalum, in: "Cell Motility", R. Golman, T. Pollard, and J. Rosenbaum, eds., p. 749, Cold Spring Harbor Laboratory, Cold Spring Harbor.

Loidl, P., and Gröbner, P., 1982, Acceleration of mitosis induced by mitotic stimulators of Physarum polycephalum, Exp. Cell Res., 137:469.

MacLeod, A. R., Waterston, R. H., Fishpool, R. M., and Brenner, S., 1977, Identification of the structural gene for a myosin heavy-chain in Caenorhabditis elegans, J. Mol. Biol., 114:133.

Mir, L., Del Castillo, L., and Wright, M., 1979, Isolation of Physarum amoebal mutants defective in flagellation and associated morphogenetic processes, FEMS Microbiol. Lett., 5:43.

Pagh, K., and Adelman, M. R., 1982, Identification of a microfilament-enriched, motile domain in amoeboflagellates of Physarum polycephalum, J. Cell Sci., 54:1.

Rusch, H. P., 1980, Introduction, *in*: "Growth and Differentiation in *Physarum polycephalum*," W. F. Dove and H. P. Rusch, eds., 250 pp. Princeton University Press, Princeton.

Sheetz, M. P., and Spudich, J. A., 1983, Movement of myosin-coated fluorescent beads on actin cables *in vitro*, *Nature*, 303:31.

Smithies, O., Gregg, R. G., Boggs, S. S., Koralewski, M. A., and Kucherlapati, R. S., 1985, Insertion of DNA sequence into the human chromosomal β-globin locus by homologous recombination, *Nature*, 317:230.

Stent, G. S., 1968, That was the molecular biology that was, *Science*, 160:390.

Thomas, J. H., Novick, P., and Botstein, D., 1984, Genetics of the yeast cytoskeleton, p153ff, *in*: "Molecular Biology of the Cytoskeleton," G. G. Borisy, D. W. Cleveland, and D. B. Murphy, eds., 512 pp., Cold Spring Harbor Laboratory, Cold Spring Harbor.

Ueda, T., Gotz von Olenhusen, K., and Wohlfarth-Bottermann, K.-E., 1979, Reaction of the contractile apparatus in *Physarum* to injected Ca^{++}, ATP, ADP and 5'AMP, *in*: "Protides of the Biological Fluids," H. Peeters, ed., Proceedings of the 26th Colloquium, 1978, p. 607, Pergamon, Oxford.

Vale, R.D., Reese, T.S., and Sheetz, M.P., 1985, Identification of a novel force-generating protein, kinesin, involved in microtubule-based motility, *Cell*, 42:39.

Chapter 18: THE CULTURE OF *PHYSARUM* AMOEBAE IN AXENIC MEDIA

Jennifer Dee

Genetics Department
University of Leicester
Leicester, ENGLAND

INTRODUCTION

The defined and semi-defined liquid and agar-based media reported for the culture of *Physarum polycephalum* in the early 1960s (Daniel and Baldwin, 1964) were successful in supporting the growth of a wide range of plasmodial strains but apparently failed to support the growth of myxamoebae. By the 1970s, several liquid synthetic media had been devised for the culture of amoebae of various *Physarum* species (see Aldrich, 1982). Only a few strains of *Physarum polycephalum* amoebae were cultured in these media, and two of these, originally designated RSD4 and CLd, were investigated in detail (McCullough, et al., 1978). After numerous subcultures in liquid axenic media, during which amoebal growth apparently improved, it was found that the amoebae in liquid culture were genetically different from the strains originally inoculated, and the new sublines were therefore designated RSD4-AXE and CLd-AXE respectively. The genetic differences between the new and original strains were demonstrated by culturing them on bacterial lawns for a period that allowed at least 20 cell doublings -- when all four strains were then inoculated in liquid axenic media, RSD4-AXE and CLd-AXE grew immediately, whereas RSD4 and CLd failed to grow. Apart from such direct growth tests, no other methods have yet been devised for distinguishing between an amoebal strain and its 'AXE' derivative. From these and later experiments, it has now become clear that amoebae able to grow in axenic media carry rare ('*axe*') mutant genes. These genes have not yet been identified, and the symbol 'AXE' is used simply to denote strains that have the phenotype of ability to grow in axenic medium. The vast majority of amoebal strains used in Physarum research, however, are unable to grow in axenic media and are therefore grown in monoxenic culture on bacterial lawns. Many interesting genetic mutants have been isolated in such strains, and in order for them to be studied biochemically, it will be necessary to make derivative strains that have the ability to grow in axenic media. To avoid such labor in the future, it will be useful to construct AXE strains in which all types of mutants can be directly isolated.

Advances in Physarum research have led to a greatly increased interest in the amoebal stage of the life cycle and to an urgent need in some laboratories for amoebae that can be subjected to biochemical analysis or genetic manipulations under axenic conditions. Although the few AXE strains already mentioned have been cultured in quantities sufficient for some analyses, these strains lack some of the biological attributes required for certain lines of research. For example, it does not seem to be possible to induce either flagellation or the amoebal-plasmodial transition in these strains, although encystment apparently can occur successfully (F. Bernier, D. Pallotta, and G. Lemieux, personal communication). Since 'axenic' growth has a genetic basis, it should be possible to solve these problems by using genetic analysis to construct AXE amoebal strains with desirable combinations of characteristics.

The manipulations of classical genetics in Physarum, such as mutant isolation, cell-cloning, crossing, and progeny analysis, have been satisfactorily performed with amoebae cultured on bacterial lawns. Even when AXE strains are involved, monoxenic culture is used for all these basic procedures, since the amoebae grow very much faster on bacteria than in axenic medium and can be much more easily cloned. A major difficulty with monoxenic culture, however, is that nutritional mutants cannot be isolated, and recently the unanswered demands for such mutants have become more persistent, particularly from those laboratories most keen to develop DNA transformation methods. This problem may be harder to solve than those mentioned above, since good growth has not yet been achieved in fully defined minimal media nor on agar-based axenic media. Thus, improvements in axenic culture methods, as well as strain construction, are required.

To summarize, there is clearly an urgent need, strongly expressed at the Workshop, for Physarum amoebae that can be cultured and induced to differentiate in axenic media. This article will be divided into two major sections: the first on genetic aspects of axenic growth, and the second on culture methods. Both sections are of practical importance, since it is now clear that difficulties with axenic cultures in the past were due to both genetic and technical factors, some of which are now understood and can be avoided. Future progress in this area will also depend on an understanding of both genetic and technical aspects.

GENETIC ASPECTS

Background

McCullough et al. (1978) obtained evidence that the amoebal strain CLd-AXE had acquired the ability to grow in axenic media as the result of mutant alleles (axe alleles) that could be transmitted to subsequent generations of amoebae. Like CLd, from which it had been derived, CLd-AXE carried matAh (from CL), a mutation that allows apogamic development, and also npfC, a closely-linked mutation that blocks apogamic development. In CLd-AXE, as in CLd, npfC occasionally reverted to the wild-type npf^+ allele, and the amoebae then formed a haploid plasmodium which completed the apogamic life cycle, forming spores containing a new generation of amoebae. McCullough et al. (1978) found that CLd-AXE progeny amoebae formed in this way had inherited the ability to grow in axenic medium; however, these revertant amoebal strains were difficult to maintain because of their readiness to form plasmodia in clonal cultures. CLd-AXE amoebae were also crossed with heterothallic amoebal strains, for example LU862, carrying matA3. The ability to grow in axenic media was inherited by a high proportion (20-30%) of amoebal progeny clones of both matA types, and at least some of these clones grew as well as the parent strain CLd-AXE. The results suggested that the ability to grow in axenic medium had a simple genetic basis and that it would be a straightforward matter to construct a range of AXE strains with various desirable attributes obtained from the genetically characterized strains commonly used for Physarum genetics. Initial attempts to repeat these results and to carry out similar crosses in Madison, Wisconsin, were successful (J. A. Gorman, personal communication), and at least one of the progeny strains of a CLd-AXE cross, MA185, is still being used in axenic cultures. A few years later, however, attempts to culture AXE strains stored in Leicester were unsuccessful, and crosses using CLd-AXE or its derivatives, LU320 and MA185 (Table 18-1), as parents not only failed to give the expected segregation of AXE progeny, but often produced results that suggested that these strains were no longer haploid clones (Chainey, 1981; T. G. Burland, personal communication). A further difficulty, which became apparent at that time, was that CLd-AXE amoebal cultures in several laboratories gave rise to npf^+ revertant plasmodia at a much lower frequency than formerly, and when such plasmodia did arise, they were morphologically abnormal and failed to sporulate. It was, therefore, no longer possible to derive npf^+ AXE progeny amoebal strains from the apogamic life cycle. These difficulties not only delayed AXE strain construction, but also led to suspicions that the genetic basis of ability to grow in axenic medium might be more complex than initially supposed (Dee, 1982) and that strains able to grow in such media might inevitably suffer from other abnormalities. McCullough et al. (1978) had already reported some apparent defects in the growth of CLd-AXE amoebae on

TABLE 18-1. Strains Able to Grow in Axenic Medium ('AXE' Strains)

Strain	Genotype	Derivation	Reference
RSD4-AXE	matA1	AXE mutant of RSD4, a Wis 1 amoebal clone	McCullough et al., 1978 Dee, 1982
CLd-AXE	matAh npfC matB1 fusA2 (npf?)*	AXE mutant of CLd	McCullough et. al., 1978 Dee, 1982
MA185	matA3 matB3 fusA1	AXE progeny clone of CLd-AXE x LU862	McCullough et al., 1978 J. Gorman, unpublished
LU320	matAh $npfC^+$ matB1 fusA2 (npf?)*	Amoebal clone from spores formed by a "CLd-AXE plasmodium" ($npfC^+$ revertant)	Chainey, 1981 J. Dee, unpublished

* There is evidence that LU320 contains a npf mutation unlinked to matA, which reduces and delays plasmodium formation (on bacterial lawns), particularly at high temperature. (Plasmodium formation is quite good at 22°C, infrequent at 26°C and above. This mutation probably arose in CLd-AXE but was not expressed because plasmodium formation was prevented at all temperatures by npfC, which reverted to $npfC^+$ during formation of LU320.

bacterial lawns, compared with CLd amoebae, and slow growth of CLd-AXE plasmodia. For some years, therefore, the prognosis for the construction of a useful range of AXE amoebal strains was poor, and the majority of laboratories requiring axenic cultures for their work were forced to use CLd-AXE with all its defects, or to shelve their amoebal projects for a time.

Recent AXE Strain Construction

Various abnormal results obtained during attempts at AXE strain construction, considered against an increasing background of genetic knowledge in *Physarum* in the 1980s, suggested that some cultures of CLd-AXE and its derivative strains might contain a large proportion of diploid amoebae. The CLd-AXE amoebae originally cultured by McCullough et al. (1978) had haploid DNA content, which was consistent with the normal genetic ratios found among the progeny of crosses performed at this time. If diploid cells were present, many aneuploid progeny clones would be expected from crosses with normal haploid strains, and characteristics due to recessive markers might not be detected. On this assumption, and because there were other reasons to suspect that diploid cells might be present in AXE strains (see Methods section), a new program of AXE strain construction was recently attempted in Leicester in which various measures were introduced to circumvent the problem. A culture of CLd-AXE, which had not been continuously maintained in liquid medium, was obtained from the McArdle Laboratory and subcultured on lawns of live bacteria. Crosses were set up between the AXE strains CLd-AXE, MA185, LU320 (Table 18-1), and various standard strains used for *Physarum* genetics. Crossed plasmodia were isolated from a number of replicate plates of each cross, and a large number of spore batches obtained from these were screened by spreading small samples of a fairly dense suspension of each spore batch on a few plates incubated at 26°C and 30°C. Spore batches giving rapid plasmodium formation in a high proportion of colonies at 30°C were discarded, because this is an indication of the presence of aneuploid or diploid cells heterozygous for *matA* (Adler and Holt, 1975). The spore viability was also estimated from these initial platings, and batches of spores with good viability, giving a high proportion of normal amoebal colonies, were then used to make fresh suspensions, which were diluted appropriately and spread on a large number of plates for the isolation of amoebal progeny clones. The results of analyzing the selected crosses between CLd-AXE and heterothallic non-AXE strains showed that the ability to grow in axenic medium segregated among the amoebal progeny and recombined freely with all the known genetic markers involved in the crosses. Some of the AXE progeny showed growth as good as that of the CLd-AXE parent strain, and the proportion of this type obtained in each cross was approximately 25%, similar to that reported by McCullough et al. (1978) for the cross CLd-AXE X LU862 and by J. Gorman (personal communication) for several crosses. These results supported the conclusions from earlier studies that ability to grow in axenic medium had a simple genetic basis. It would be premature to propose a specific genetic model, however, until further analysis in progress has been completed (J. Dee and J. L. Foxon, unpublished).

These recent crosses have produced a range of amoebal strains carrying various combinations of known genetic markers together with the ability to grow in axenic media. Some of these strains are now being used in further crosses to test whether they will reliably transmit their AXE characteristics; if they do, these crosses should also lead to the isolation of AXE strains carrying combinations of markers that could not be obtained from CLd-AXE in one step. Since progeny from earlier CLd-AXE crosses, such as MA185, have proved disappointing when used as parents for the construction of AXE strains, it is important that we should test the new strains for their stability and efficacy as parents before making them generally available to other laboratories.

Among the progeny of the recent crosses are some strains that combine the ability to grow in axenic medium with various desirable attributes of normal _Physarum_ amoebae. For example, some AXE strains of genotype _matAh npfC_ revert readily to give _npf_$^+$ plasmodia with normal morphology and sporulation. These plasmodia in turn produce AXE amoebal progeny, which show normal apogamic plasmodium formation on bacterial lawns. (Efficient apogamic development in liquid axenic culture has not yet been achieved, however.) At least some AXE progeny strains can flagellate normally and show high plating efficiency and normal colony morphology on bacterial lawns. Thus the results of recent analyses have dispelled our fears that the ability of amoebae to grow in axenic media is unavoidably associated with particular morphological or functional defects in amoebae or plasmodia and have suggested that the defects observed in CLd-AXE amoebae and plasmodia were due to mutations that had accumulated during long periods of culture. Methods for maintaining amoebal strains that will avoid such alterations in the future will be given in the section below on culture methods.

Genetic Basis of Growth in Axenic Media

In crosses between amoebal strains able and unable to grow in axenic media, there is a clear segregation of progeny clones resembling the two parental types in their response to axenic media. This segregation indicates that a small number of genes is involved in determining the ability to grow. However, a substantial proportion of the progeny clones show a response intermediate between those of the parents -- these clones neither grow well nor die quickly; the amoebae survive and grow slowly through several subcultures. These results suggest that more than one gene may be involved in the inheritance of the ability to grow well in axenic media. Until further growth tests have been carried out on larger numbers of progeny, it is premature to define classes of growth, or to deduce the ratios of progeny in these classes in order to propose a genetic model. Two points are important, however: first, care should be taken to select the best growers among the progeny of an AXE cross; and second, these strains may carry two or more <u>axe</u> mutant genes, all of which may be essential for good growth to continue.

A particular difficulty in analyzing the genetic basis of amoebal growth in liquid axenic media is that single cells cannot be tested. As explained more fully below, a large number of cells must be inoculated into each liquid culture for it to be successful. Although the amoebae may have been very recently re-cloned on a bacterial lawn, each clone must then be grown to high density for inoculation into liquid medium, and during this time changes in DNA content or mutations may occur, so that the cells may already differ in their ability to grow in the medium. During subcultures in axenic medium, in which the cells must always be kept at high density, unobserved selection may easily take place, gradually altering the genetic characteristics of the population. This may create problems when one attempts to categorize strains into different classes with respect to growth. One must always remember that a culture may no longer consist of a single clone.

Another problem is that amoebae cultured in liquid axenic media seem gradually to increase in volume and DNA content. Several different laboratories have reported at the Workshop that their cultures of CLd-AXE now apparently contain diploid amoebae, in confirmation of the suspicions we had already formed as a result of our difficulties with genetic analysis. The trend towards higher DNA content seems to be persistent in liquid cultures, perhaps indicating that increased cell size is advantageous to cells in liquid culture or that defects occur more frequently in cell or nuclear division in these conditions. However, an increased incidence of aneuploid and polyploid cells has also been observed in various non-AXE strains of amoebae cultured on bacteria (Adler and Holt, 1974) and may simply be the result of keeping cells in continuous culture. We intend to monitor the DNA content of AXE amoebal strains, cultured in liquid and on bacteria, and to investigate whether any changes in content are related to the growth characteristics of the strains. Meanwhile, however, it is important for everyone concerned to realize that such changes occur, that they may cause difficulties, and that they can probably be avoided by frequently re-establishing liquid cultures from stored stocks, as explained in the section on Methods.

The cellular or molecular basis of ability to grow in axenic media is as yet unknown. Tests carried out in a range of defined and semidefined media by McCullough and Dee (1976) suggested that the AXE strains had acquired a general ability to survive and grow in liquid axenic media, rather than having changed in any specific nutritional requirements. However, only a few AXE strains have been tested in several different media, since the normal practice now is simply to use the same semi-defined medium for plasmodial and amoebal cultures. When a wider range of AXE strains becomes available, further investigations of nutritional abilities may be worthwhile.

When AXE and non-AXE amoebae are inoculated in axenic media, differences in cell morphology become observable quite rapidly. AXE amoebae have an irregular, slightly rounded form similar to that of amoebae actively growing on a bacterial lawn, whereas non-AXE amoebae become flagellated or form small, spherical cysts. However, it seems unlikely that these differences in response are the cause of ability or inability to grow in axenic medium, since no differences have been detected between AXE and non-AXE amoebae in their ability to flagellate or to encyst when they are exposed to the conditions normally used to induce these transformations. So far, comparisons of AXE and non-AXE amoebae plated on bacterial lawns have also failed to reveal any consistent difference between them -- for example, in plating efficiency, growth rate, or colony morphology -- that might be correlated with the AXE phenotype. Apart from the intrinsic interest of identifying a cellular characteristic associated with the ability to grow in liquid media, it would be of great practical value to find a feature by which AXE strains could be rapidly identified without the lengthy tests in axenic medium necessary at present.

CULTURE METHODS

Maintenance of Strains

Stocks of amoebal strains of defined genotype should always be kept in a dormant state so that cultures can be reinoculated at intervals from these stocks. This must be done to avoid the changes in DNA content and the accumulation of unwanted mutations that may occur in cells maintained in continuous culture. As discussed in the previous section, it now seems likely that CLd-AXE amoebae have repeatedly become diploid during culture in liquid media, and also that various unfavorable properties such as slow growth and abnormal morphology arose in CLd-AXE cultures in the past. Several methods have been described for freezing or drying *Physarum* amoebae in a viable condition (Evans, 1982; Anderson et al., 1983). Details of the method successfully used at the McArdle Laboratory by T. G. Burland are given in the Appendix. Alternatively, cultures of amoebae grown on bacterial lawns and allowed to encyst can be stored for long periods at 4°C and will remain viable so long as the agar does not dry out.

All AXE strains of amoebae grow more rapidly on bacterial lawns than in axenic media, and for all routine genetic manipulations these strains are cultured on live E. coli by the standard published methods (Burland et al., 1981). Stocks for storage should also be cultured on live E. coli. Before inoculation into liquid media, the amoebae are transferred to formalin-killed bacteria (FKB). It is important to note that cultures on FKB should not be stored in the cold, since they apparently lose viability rapidly; this factor was probably responsible for some of our early difficulties with AXE strain construction.

Cultures in Liquid Axenic Media

As originally reported by McCullough and Dee (1976), an initial cell density of at least 5 X 10^5 cells/ml seems to be necessary for a liquid culture of AXE cells to grow satisfactorily. With a good strain, a doubling time of about 24 hr may then be expected during logarithmic growth, and a maximum yield of 3 X 10^7 cells/ml. Such growth may be achieved in a variety of culture vessels and at a range of temperatures between 26° and 30°C, but vigorous agitation on a reciprocating shaker is desirable, and the volume of the culture should not exceed about 1/5 the volume of the vessel (Plate 18-2). As mentioned in the previous section, good growth has been achieved in a variety of media, and the simplest method is to use the same medium for amoebae and plasmodia. The semi-defined medium used in Leicester is given in Table 18-2. When the stationary phase of growth begins, which may occur after a few days and at a cell density of about 10^7 cells/ml, the cells may begin to form large clumps and to encyst. It is advisable to subculture before this stage, since encysted cells will not grow immediately if transferred to fresh medium and may not grow at all if incubated at 30°C. In axenic medium, as in monoxenic cultures, a period of about 24 hr at a lower temperature (e.g., 26°C) seems to be necessary for excystment; cultures can then be transferred to 30°C if desired. In subcultures, as in the initial culture, the cell density must be at least 5 X 10^5 cells/ml. The time taken for changes in DNA content to occur in cultured cells is not yet known, but in Leicester we usually now discard cultures after about ten transfers.

Reports from various laboratories indicate that few difficulties are experienced when cultures of AXE cells have become established in liquid media, and that the details of culture conditions are probably unimportant; however, problems seem to arise when attempts are made to introduce stocks of amoebae from bacterial cultures into axenic liquid media. More precise details will therefore be given of the methods that have been found successful in Leicester with many different AXE strains. Stocks of amoebae are freed of live E. coli by a single transfer to formalin-killed bacteria on medium containing penicillin and streptomycin (Tables 18-3, 18-4). These cultures may be incubated at 26°C for up to a

Table 18-2. Semi-defined Medium for Culture of Amoebae and Plasmodia

Constituents	Concentration (g/liter)
Glucose	10.0
Soytone (Difco-Bacto)	10.0
KH_2PO_4	2.0
$CaCl_2.6H_2O$	1.35
$MgSO_4.7H_2O$	0.6
$FeCl_2.4H_2O$	0.039
$ZnSO_4.7H_2O$	0.034
Citric acid	3.54
EDTA disodium salt	0.224
Biotin	0.005
Thiamine	0.04

Add constituents to distilled water and make up to 1 liter.

Adjust pH to 4.6 with 20% NaOH.

Sterilize at 15 lb for 15 min in autoclave.

Add Hematin* (1 ml/100 ml medium) immediately before use.

* Hematin stock solution: 0.05 g Hematin, 1 g NaOH, 100 ml distilled H_2O; sterilize at 15 lb for 15 min; store in refrigerator.

Table 18-3. Preparation of Formalin-killed Bacteria (FKB)

Aseptic technique must be observed throughout.

Day 1: Inoculate 1 liter Oxoid Nutrient Broth with 1 ml dense suspension (10^9-10^{10} cells/ml) of *Escherichia coli*.

Incubate on rotary shaker at 37°C overnight.

Day 2: Spin down overnight culture of *E. coli*.

Resuspend pellet in 50 ml buffer (7 g Na_2HPO_4, 3 g KH_2PO_4, 4 g NaCl, 0.1 g $MgSO_4.7H_2O$ per liter water, sterilized at 15 lb for 15 min). Transfer suspension to sterile flask and add 4 ml formaldehyde solution (AR).

Stand suspension at 4°C overnight.

Day 3: Spin down bacterial suspension and discard supernatant.

Resuspend pellet in 50 ml buffer (see above). Add 5 ml 1 M glycine. Agitate on a rotary or reciprocating shaker for 1 hr at 30-37°C.

Wash twice in 50 ml buffer.

Resuspend pellet in 20 ml sterile distilled water.

Dispense 1-ml aliquots in screw-cap bottles and store at 4°C.

This should give a suspension of FKB containing 10^9-10^{10} cells/ml. For standardization check the optical density of diluted suspension.

To check that no live bacteria are present, streak some FKB suspension on nutrient agar plates, and incubate overnight at 37°C.

Table 18-4. Procedure for Culturing Amoebae on FKB for Inoculation into Axenic Medium

a) Medium:

Constituent	ml/100 ml water agar
Liver infusion stock solution*	4.0
Sodium phosphate buffer (0.5 M; pH 6.8)	1.0
Semi-defined medium with hematin (Table 18-1)	6.25
Penicillin (stock solution, 10 units/ml)	2.5
Streptomycin (stock solution, 10 mg/ml)	2.5

The constituents are added to sterilized, molten 1.2% water agar. The medium is poured into 50-mm Sterilin petri dishes (approx. 10 ml/dish) at least 3 days before use so that the agar surface is dry before FKB is added. Plates stored longer than 3 days should be kept in sealed bags.

* Liver infusion stock solution: 100 g desiccated liver (Oxoid Product no. L26), 1 liter distilled water. See Oxoid Manual for method of preparation.

b) Each 50-mm plate is inoculated with 0.04 ml FKB suspension (Table 18-3), which is spread with the tip of the pipette to form a patch approximately 10-15 mm in diameter. When the liquid has soaked into the agar, amoebae are inoculated at the center of the FKB patch with a toothpick. The plates are incubated at 26°C until the patch of amoebal growth is large enough for inoculation into axenic medium (see text).

week, but longer incubation seems to cause some deterioration; incubation at higher temperatures is also deleterious. When a single patch of dense amoebal growth is present in the FKB lawn (Plate 1), a block of the medium carrying these amoebae is cut out with a spatula and transferred to a small volume of liquid medium. It is probably important to avoid transferring FKB on this block, since this may inhibit growth in the liquid culture. Agitation for a few seconds on a vortex mixer frees the amoebae from the agar, and the suspension is then transferred to a fresh tube with a pasteur pipette, which is also used to inoculate a small sample in a hemacytometer. After a few trials with a batch of FKB cultures set up at the same time, it is possible to cut a standard-size block and to inoculate it in a standard volume of liquid in order to achieve the required starting density of 5 X 10^5 cells/ml in each liquid culture. Since the density of amoebae in the FKB cultures depends on the thickness and quality of the FKB lawn, it is not possible to specify the required block size in advance; however, as a rough guide, we have often found that a block 5 mm X 5 mm is

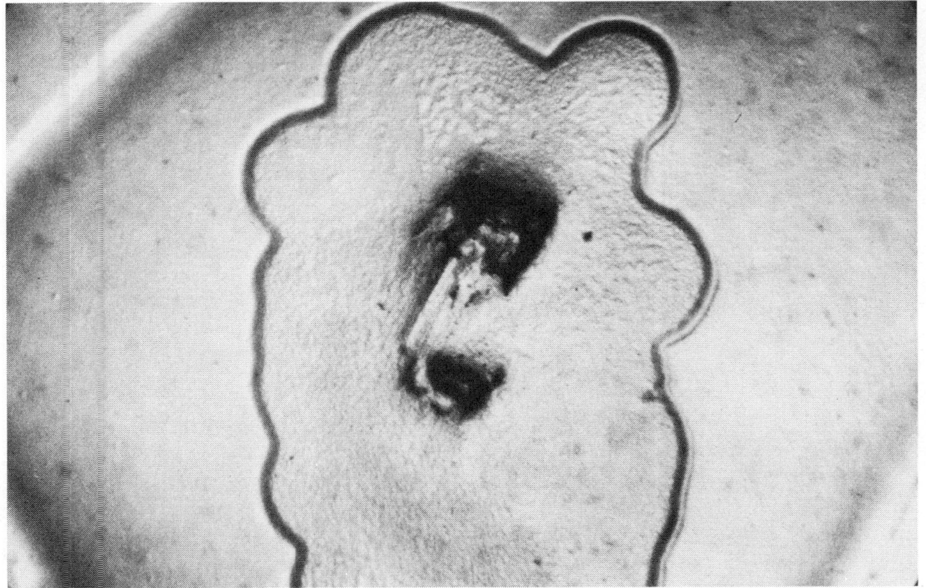

Plate 18-1. Amoebae growing on FKB. The impression of the toothpick with which the amoebae were inoculated can be clearly seen. The amoebae have grown out from the inoculum to cover an area of about 5 mm X 10 mm and are heaped up around the edge of this area, where they are still feeding on the FKB lawn. A block could now be cut from the center of the amoebal patch for inoculation into axenic medium.

sufficient in 1 ml of medium. For various reasons, it is probably best to inoculate small-volume cultures initially (Plate 2), and to scale up the culture size gradually when growth in liquid medium has become established.

Cultures on Axenic Agar Media

A number of experiments have shown that the growth of amoebae in axenic media is inhibited by the presence of agar (C. H. R. McCullough and J. Dee, unpublished). The reasons for this inhibition are not yet clear, but it is certainly not due to impurities, and it occurs even when pure agarose is added to liquid cultures. Some results initially suggested that the inhibition of an extracellular aminopeptidase was responsible (C. H. R. McCullough, J. Dee, A. Haars, and A. Hüttermann, unpublished). Further experiments to test this conclusion have been carried out since the Workshop, and these are still in progress (C. H. R. McCullough, personal communication). Discussion at the Workshop, involving a number of participants, indicated that the problem of growing amoebae on axenic agar media is one which many scientists would like to solve, but no very promising solutions were suggested. Very slow growth has been observed on fully defined media in the presence of agar (C. H. R. McCullough, personal communication) and on semi-defined medium containing a low concentration (0.7%) of agar (T. G. Burland,

Plate 18-2. Tubes for small-volume axenic cultures of amoebae. 16 X 100 mm disposable, conical-base test tubes supplied sterile with screw-caps in place (Sterilin product no. 144AS), arranged on a purpose-built deck for a reciprocating shaker. The arrows show the direction of movement on the shaker. Each tube contains 1 ml culture medium.

personal communication) but, since in both cases the doubling time was several days, these observations seem to be of little practical value. It is possible that membrane filters supported over liquid media might provide a solid substrate on which amoebal colonies could be cultured on defined or semi-defined media, but this has not yet been tested, and no one seemed certain whether such conditions would be satisfactory for experiments such as the isolation of nutritional mutants. The suggestion that received the most favorable response was that Caragheenan should be investigated as an alternative to agar, since preliminary results obtained by C. H. R. McCullough (personal communication) suggested that this substance may be a satisfactory solidifying agent with no inhibitory effects.

SUMMARY

1. In Physarum polycephalum, the few amoebal strains that have been found to grow in liquid axenic medium differ genetically from the majority of amoebal strains, which are unable to grow in these conditions.

2. Recent analysis of crosses between strains able and unable to grow in liquid axenic medium has suggested that this ability has a simple genetic basis; it is too soon to propose a specific genetic model, however.

3. Some strains of amoebae cultured in liquid axenic medium for long periods have increased in DNA content, and as a result they do not give normal genetic ratios among their progeny when crossed.

4. If amoebal strains are stored in dormant conditions, their properties remain stable. Strains able to grow in axenic medium should therefore be maintained in dormant form, and liquid cultures should be reinoculated from stored stocks at frequent intervals.

5. Some of the strains that have been cultured for long periods in liquid medium have acquired various abnormalities and can no longer complete normal developmental transitions.

6. Recent crosses have produced strains that are able to grow in liquid axenic medium and that also have all the normal properties of the Physarum life cycle, such as ability to flagellate and to develop into plasmodia. They can also be cultured efficiently on bacterial lawns. No association has been detected between ability to grow in axenic medium and any other characteristic of the cells.

7. No amoebal strain has yet been cultured satisfactorily on axenic agar medium.

ACKNOWLEDGMENT

Much of the work in Leicester discussed in this paper has been supported by grants from the Science and Engineering Research Council.

REFERENCES

Adler, P. N., and Holt, C. E., 1975, Mating type and the differentiated state in Physarum polycephalum, Dev. Biol., 43:240.

Adler, P. N., and Holt, C. E., 1974, Change in properties of Physarum polycephalum amoebae during extended culture, J. Bacteriol., 120:532.

Aldrich H. C., 1982, Culture methods, in: "Cell Biology of Physarum and Didymium", H. C. Aldrich and J. W. Daniel, eds., Vol II, p. 361, Academic Press, New York.

Anderson, R. W., Gray, A., Hutchins, G., and Price, J., 1983, Preservation of Physarum polycephalum amoebae with anhydrous silica gel, Physarum Newsletter, 15:3.

Burland, T. G., Chainey, A. M., Dee, J., and Foxon, J. L., 1981, Analysis of development and growth in a mutant of Physarum polycephalum with defective cytokinesis, Dev. Biol., 85:26.

Chainey, A. M., 1981, Studies on the Myxomycete, Physarum polycephalum, M. Phil. Thesis, University of Leicester.

Daniel, J. W., and Baldwin, H. H., 1964, Methods of culture for plasmodial myxomycetes, in: "Methods in Cell Physiology", D. M. Prescott, ed., Vol. 1, p. 9, Academic Press, New York.

Dee, J., 1982, Genetics of Physarum polycephalum, in: "Cell Biology of Physarum and Didymium", H. C. Aldrich and J. W. Daniel, eds., Vol. I, p. 211, Academic Press, New York.

Evans, T. E., 1982, Low-temperature preservation of amoebae, in: "Cell Biology of Physarum and Didymium", H. C. Aldrich and J. W. Daniel, eds., Vol. II, p. 244, Academic Press, New York.

McCullough, C. H. R., and Dee, J., 1976, Defined and semi-defined media for the growth of amoebae of Physarum polycephalum, J. Gen. Microbiol., 95:151.

McCullough, C. H. R., Dee, J., and Foxon, J. L., 1978, Genetic factors determining the growth of Physarum polycephalum amoebae in axenic medium, J. Gen. Microbiol., 106:297.

APPENDIX

PREPARATION OF FREEZER STOCKS OF PHYSARUM MYXAMOEBAE

USE ASEPTIC ROUTINE THROUGHOUT

1. Pre-spread two SM plates with 0.1 ml live bacteria each.

2. Pick amoebae from agar plate with a toothpick, and transfer to a pre-spread SM plate by making 5-6 parallel streaks. Repeat for the second pre-spread SM plate (spare plate in case of contamination etc). Incubate plates at 26°C unless otherwise indicated. (Exceptions include matAh strains, which must be incubated at 30°C to prevent selfing, and temperature-sensitive mutants).

Although plates will become confluent in 5-6 days, incubate plates for 10 days to allow plenty of time for encystment.

3. Examine the plates. The amoebae should be confluent. Check that the plates are not contaminated, then flood one of the SM plates with 7 ml 15% glycerol. Scrape off the myxamoebae with a glass spreader, and recover the suspension from the plate with a pipette. Transfer 3 ml of suspension to each of two 1-dram glass vials (Fisher Scientific Co. Cat No. 3 338A) on ice.

4. Transfer the vials to a freezer (-70° to -80°C). Check viability every 5-10 years. (Note that freezer stocks can be thawed at room temperature and re-frozen many times without substantially reduced viability.)

Reagents:

SM plates: To 1000 ml distilled water, add 0.7 g DifCo Bactotryptone, 0.2 g DifCo yeast extract, 0.6 g D-glucose, 0.8 g sodium phosphate monobasic dihydrate, 0.7 g sodium phosphate dibasic, and 15 g agar. Autoclave for 20 min at 15 lb/sq in. Pour plates.

15% Glycerol: Make 15 g glycerol (reagent grade) to 100 ml with distilled water. Autoclave 20 min.

Preparation of "Standard Bacterial Suspension", or "SBS":

1. Streak E. coli KB over a Luria broth (LB) plate and incubate at 37°C overnight or at 30°C for 2-4 days.

2. Flood plate with 10-12 ml distilled water, then scrape off bacteria with a glass spreader. Transfer suspension to test tube, vortex to remove clumps. Use 0.1 ml suspension per plate. Use same day.

Chapter 19: A NEW METHOD FOR THE PREPARATION OF HIGHLY PURIFIED AND MORE NATIVE NUCLEI SHOWING STAGE-SPECIFIC TRANSCRIPTION OF ACTIN AND TUBULIN GENES

Klaus-Dieter Nothacker[+], Anne-K. Werenskiold*, Thomas Schreckenbach*, and Armin Hildebrandt[+]

[+]University of Bremen
Bremen, FRG

*Max Planck Institute for Biochemistry
Martinsried, FRG

Physarum nuclei usually have been isolated by the standard procedure of Mohberg and Rusch (1971). This Triton-sucrose method (TS method) yields nuclei containing up to seven times more polysaccharides (mainly slime) than DNA (Hall et al., 1975). Some have presumed that the slime is located inside the nucleus (Farr and Horrisberger, 1978); it has been shown to have deleterious effects on enzymatic treatments of chromatin (Staron et al., 1980). Therefore, we have developed a new method of nuclear isolation. It was recently demonstrated that Triton X-100 dissolves nuclear membranes (Frederiks et al., 1978). We chose to use Surfynol as detergent because it keeps the nuclear membranes widely intact (Fig. 19-1). In order to reduce leakage during the isolation procedure, we used hexylene glycol instead of sucrose and, for higher purification, a Percoll density gradient instead of a sucrose cushion. For this reason, the new method is called the SHP-method (Nothacker and Hildebrandt, 1985, 1986). Percoll density gradients were also used earlier for the purification of nucleoli from contaminating slime (Seebeck et al., 1979). Recently, nuclei were prepared by being pelleted through a cushion of Percoll (28% v/v) instead of 1 M sucrose (Künzler et al., 1984).

MATERIALS USED FOR THE SHP-METHOD

Percoll (Pharmacia), Surfynol 485 (2,4,7,9-tetramethyl-5-decin-4,7-diol with associated ethyleneoxide chains, is produced by Air Products, P.O. Box 538, Allentown, PA 18105, USA, marketed in FRG by Biesterfeld, 2000 Hamburg 1, Ferdinandstr. 41, but only in industrial quantities; the quality of another Surfynol from Serva changes from batch to batch). Hexylene glycol, or 2-methylpentanediol-2,4, is from Merck.

THE SHP-METHOD

Nuclear medium (100-ml) (15 mM $MgCl_2$, 0.5 mM $CaCl_2$, 15 mM Tris-HCl pH 7.3, 5 mM EGTA, 0.5 M hexylene glycol, 0.8% Surfynol, 3 mM dithioerythritol (DTE) was added to 4 g *Physarum* plasmodial pellet. The suspension was then homogenized for 30 sec at position "high" in a Waring blendor (60 volt) and centrifuged for 5 min at 700 X *g*. The pellet was resuspended in 3 ml nuclear medium with a Pasteur pipette and then diluted to 70 ml of nuclear medium containing 25% Percoll. This suspension was centrifuged in a fixed-angle rotor for 35 min at 48,000 X *g* at 4°C. The nuclear band (close to the bottom) was aspirated by a Pasteur pipette, resuspended in 80-ml nuclear medium, and centrifuged for 5 min at 1100 X *g* to remove the Percoll.

COMPARISON OF THE TS-METHOD WITH THE SHP-METHOD

Biochemical Composition of Nuclei

SHP-nuclei contained almost twice as much protein as TS-nuclei (Table 19-1). As an example of the preservation of nuclear proteins, we have found by electron microscopy intact microtubules in SHP nuclei, but not in TS nuclei. Roobol and her colleagues (1984) have modified the Mohberg and Rusch (1971) protocol to include 5M glycerol and 5 mM EGTA in the isolation medium and have also observed intact microtubules. We have not investigated the use of spermine (1 mM) and EDTA (5 mM) reported by Grainger et al. (1980) to give a high recovery of chromatin proteins in isolated nuclei of *Physarum*. Equal amounts of DNA and RNA were detected in TS-nuclei and in SHP nuclei (Table 19-1). However, SHP nuclei contained one-third the amount of cytoplasmic protein contaminants of TS-nuclei, measured by the activity of malate dehydrogenase (MDH) as a marker enzyme of cytoplasm (Table 19-1). The use of Percoll dramatically decreased polysaccharide contamination to 1/100 of the amount detected with TS nuclei. We conclude that this contamination in the TS nuclear preparation is mainly slime, which, however, is not located inside the nuclei.

Preservation of RNA Synthesis Capacity

The decrease in contamination of the SHP nuclei has marked effect on RNA synthesis. SHP nuclei showed a nearly linear dependence on the nuclear concentration of transcriptional activity and

Table 19-1.

Biochemical Characterization of Nuclei Prepared by the SHP Method or by the TS Method

Biochemical composition of nuclei	SHP nuclei	TS nuclei	SHP/TS ratio
Protein/nucleus (pg)	9.6	5.3	1.8
DNA/nucleus (pg)	1.2	1.2	1.0
RNA/nucleus (pg)	0.9	0.9	1.0
Polysaccharides/ nucleus (pg)	0.15	15.7	0.01
Purification factors:			
MDH activity	174 X	57 X	3
Polysaccharides	436 X	4.2 X	104
Recovery of nuclei from 1 g plasmodium	47%	23%	2

transcriptional activity up to 8 times higher than that of TS nuclei, tested according to published assay conditions (Hildebrandt and Sauer, 1977; Schicker et al., 1979). Synthesis of high molecular weight transcripts ranging up to 4 kb was detected when freshly prepared SHP nuclei were incubated in the presence of the RNase-inhibitor RNasin (0.5 U/μl) (Fig. 19-2). No information is available as to whether these RNA chains were run-off products or the result of in vitro initiations (see Conclusions).

Using high concentrations (100 μg/ml) of the RNA polymerase inhibitor α-amanitin, we were able to detect RNA polymerase III activity in SHP nuclei (10% of the entire transcription rate). This activity was not detectable in TS nuclei, possibly owing to high protein leakage. To our knowledge this is the first report on RNA polymerase III activity in isolated nuclei of *Physarum*.

SHP nuclei displayed a steady increase in RNA yield for a period of 120 min, in contrast to TS nuclei, which showed a plateau value after the initial 40 min of incubation. We interpret this as an indication of reinitiation of RNA synthesis. This view is supported by the observation that the initiation inhibitor 5,6-dichloro-1-β-D-ribofuranosyl-benzimidazole (DRB) (Zendomeni et al., 1983) suppresses net amanitin-sensitive RNA synthesis to a significant extent (Fig. 19-3).

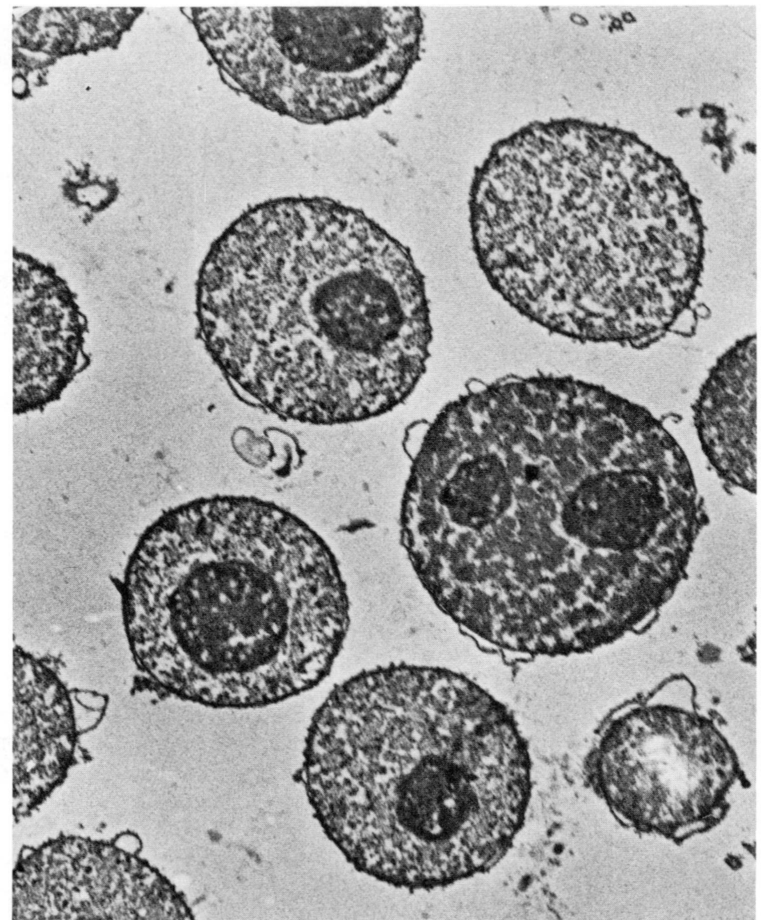

Fig. 19-1. Morphological appearance of SHP nuclei in the electron microscope. Crosssection through a pellet of SHP nuclei embedded in Araldite after fixation in 6% glutaraldehyde and 2% OsO_4.

When RNA synthesis was carried out with exogenously added RNA polymerase from E. coli, the incorporation of ^3H UTP was found to be elevated 10 times with SHP nuclei over TS nuclei (not shown).

DEVELOPMENTAL STAGE-SPECIFIC VARIATIONS IN ACTIN- AND TUBULIN TRANSCRIPTION IN SHP-NUCLEI

To check for transcription of actin and tubulin genes compared with that of ribosomal RNA genes in SHP nuclei from growing, starv-

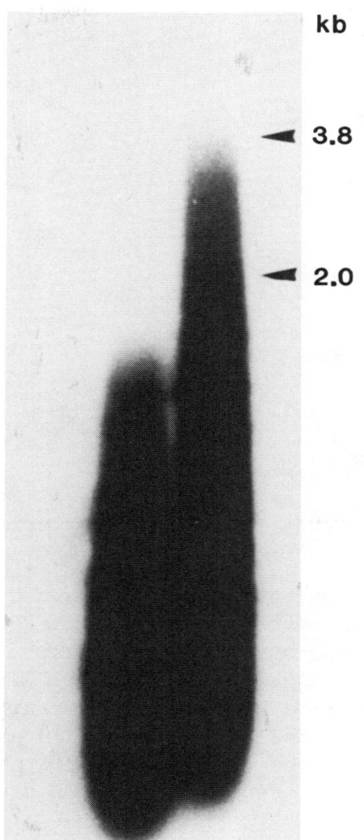

Fig. 19-2. Size of in vitro transcripts. ^{32}P-labeled RNA synthesized in isolated SHP-nuclei under high salt condition (250 mM KCl) in the presence or absence of RNAsin (0.5 U/µl) was isolated by phenol extraction, separated from unincorporated nucleotides by chromatography on a Sephadex G-50 column (1 ml), size-fractionated in a 1.5% denaturing agarose-formaldehyde gel (30,000 cpm per lane), and visualized by radioautography.

ing, or sporulating plasmodia, ^{32}P-labeled transcripts were analyzed by dot-hybridization to Drosophila actin DNA, Physarum tubulin cDNA, and rDNA as probes. The autoradiograph depicted in Fig. 19-4 reveals a four- and three-fold increase in the relative transcription of actin and tubulin sequences respectively during starvation. Tubulin transcription was inhibited in sporulating plasmodia, whereas actin-specific RNA was continuously synthesized at a higher rate. These results reveal that actin- and tubulin-specific transcripts are synthesized in SHP nuclei. α-Amanitin inhibition indicates that their synthesis depends on endogenous RNA polymerase II

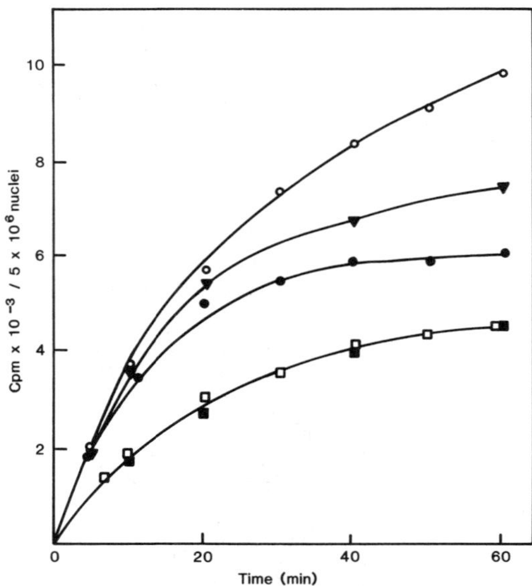

Fig. 19-3. Transcription in SHP nuclei in the presence of 90 mM KCl and the initiation inhibitor 5,6 dichloro-1-β-D-ribofuranosyl-benzimidazole (DRB). ○ total transcription without DRB; □ transcription in the presence of 1 μg/ml α amanitin; ▼ total transcription in the presence of 180 μM DRB; ● total transcription in the presence of 244 μM DRB; ■ transcription in the presence of 1 μg/ml α amanitin and 600 μM DRB.

activity (Werenskiold, 1985). However, the type of assay displayed in Fig. 19-4 is not adequate for an exact quantitation of mRNA concentration. We have found that only a small fraction of an individual mRNA is absorbed by the denatured double-stranded DNA probe on the filter. Therefore, the signals obtained are not strong enough to give reliable scintillation counting values when the samples are diluted to show a linear dependence on RNA concentrations. For this reason, we are working to achieve 100% absorption of an individual mRNA, using hybridization in solution with homologous single-stranded DNAs as probe.

CONCLUSIONS

Nuclear transcription systems are important tools for the investigation of transcriptional control in gene regulation. They allow labeling of nascent transcripts to high specific activity and have been used to determine in vitro the developmental control of transcription of individual genes (Keller et al., 1984; Miller

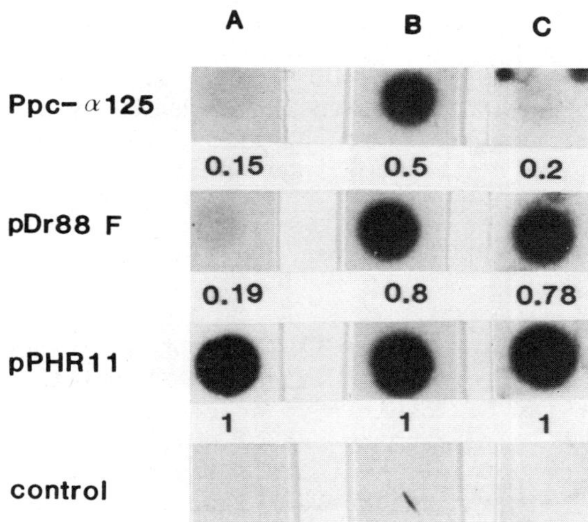

Fig. 19-4. Stage-specific variations in actin and α-tubulin-transcription. ^{32}P-labeled RNA synthesized in the presence of 250 mM KCl in SHP-nuclei--isolated from a growing (A), a starving (B), and a sporulating plasmodium (7 hr after photoinduction) (C)--was hybridized to 0.1 μg of immobilized DNA per dot and autoradiographed. Hybridization conditions: 5 X SSC, 50% formamide, 5 X Denhardt's medium, 0.1% SDS, 1% sarcosyl, 42°C, 48 hr. An 800-bp fragment of Ppc-α125 (α-tubulin) (Schedl et al., 1984), a 760-bp fragment of pDR88F (Drosophila actin), the plasmid pPHR11 (rDNA), and chromosomal DNA isolated from H. halobium (control) were used as probes. Dots were excised and quantitated by scintillation counting. The numbers represent activity relative to rDNA transcription (pPHR11, 160 cpm). (For further explanation, see text.)

and Elgin, 1981; Cook et al., 1985) and to identify transcription-regulating factors in higher eukaryotes (Wu, 1985). Nuclear transcription systems from Physarum reported so far have been used only to measure overall DNA synthesis and rRNA transcription (Davies and Walker, 1978; Mittermayer et al., 1966; Schicker et al., 1979). In the near future they may become important for an investigation of differential expression of RNA polymerase II-dependent genes during Physarum differentiation (also see Chapter 7).

The specific transcription of the abundantly transcribed actin and tubulin genes by RNA polymerase II has been described. So far one cannot exclude that this transcription represents elongation and possibly "read-through" of RNA molecules (250 mM KCl) initiated in vivo as a consequence of the high salt assay conditions in the experiments of Figs. 19-2 and 19-4. However, under lower salt conditions (90 mM KCl), the reinitiation capacity of nuclei, isolated by the new SHP method, could be demonstrated indirectly (Fig. 19-3). A direct proof for initiation by incorporation of β-^{35}S-ATP or β-^{35}S-GTP was not possible because of the high incorporation into nuclear protein. Using this approach, but with gamma-labeled nucleoside 5'-triphosphates, Sun et al. (1979) have demonstrated the initiation of a small percentage of all rRNA molecules transcribed in isolated TS nuclei.

PROSPECTS

Efforts to demonstrate initiation of RNA polymerase II transcripts in SHP nuclei will be continued. If initiation can be demonstrated, it may be possible to search for regulatory factors in this transcription system by use of complementation mixtures with nuclear or cytoplasmic extracts. Investigations are in progress aiming to adapt the nuclear isolation procedure to maintain undegraded RNA synthesized in vivo. Such nuclei can be tested for viability after reinjection (see Chapters 20 and 21).

REFERENCES

Cook, K. S., Hunt, C. R., and Spiegelman, B. M., 1985, Developmentally regulated mRNAs in 3T3-adipocytes: analysis of transcriptional control, J. Cell Biol., 100:514.

Davies, K. E., and Walker, I. O., 1978, Control of RNA transcription in nuclei and nucleoli of Physarum polycephalum, FEBS Lett., 86:303.

Farr, D. R., and Horrisberger, M., 1978, Structure of β-galactan isolated from the nuclei of Physarum polycephalum, Biochim. Biophys. Acta, 539:37.

Frederiks, W. M., James, J., Arnouts, C., Broekhoven, S., and Morreau, J., 1978, The influence of Triton X-100 on the nuclear envelope of the isolated liver cell nuclei, Eur. J. Cell Biol., 18:254.

Grainger, R. M., Ogle, R. C., and Keating, J. N., 1980, Purification of rDNA chromatin from Physarum polycephalum, Eur. J. Cell Biol., 22:94.

Hall, L., Turnock, G., and Cox, B. J., 1975, Ribosomal RNA genes in the amoebal and plasmodial forms of the slime mould Physarum polycephalum, Eur. J. Biochem., 51:459.

Hildebrandt, A., and Sauer, H. W., 1977, Discrimination of potential and actual RNA polymerase B activity in isolated nuclei during differentiation of Physarum polycephalum, W. Roux's Arch., 183:107.

Keller, L. R., Schloss, J. A., Silflow, C. D., and Rosenbaum, J. L., 1984, Transcription of α- and β-tubulin genes in vitro in isolated Chlamydomonas reinhdardi nuclei, J. Cell Biol., 98:1138.

Künzler, P., Pauli, U., and Braun, R., 1984, Regions in the ribosomal minichromosome of Physarum polycephalum are protected from restriction nucleases; protein is insensitive to high salt in the G phase and sensitive in the M phase of the cell cycle, J. Mol. Biol., 179:651.

Mohberg, J., and Rusch, H.-P., 1971, Isolation and DNA content of nuclei of Physarum polycephalum, Exp. Cell Res., 66:305.

Miller, D. W., and Elgin, S. C. R., 1981, Transcription of heat shock loci of Drosophila in a nuclear system, Biochemistry, 20:5033.

Mittermayer, C., Braun, R., and Rusch, H. P., 1966, Ribonucleic acid synthesis in vitro in nuclei isolated from the synchronously dividing Physarum polycephalum, Biochim. Biophys. Acta, 114:536.

Nothacker, K.-D., and Hildebrandt, A., 1985, Isolation of cell nuclei in a more native form using surfynol, hexylene glycol and Percoll-gradients, Eur. J. Cell Biol., 36 (Suppl. 7):48.

Nothacker, K.-D., and Hildebrandt, A., 1986, Isolation of highly purified and more native nuclei of Physarum polycephalum utilizing surfynol, hexylene glycol and Percoll, Eur. J. Cell Biol., in press.

Roobol, A., Paul, E. C. A., and Gull, K., 1984, The isolation of nuclei containing mitotic spindles from the plasmodium of the slime mould Physarum polycephalum, Eur. J. Cell Biol., 34:52.

Schedl, T., Burland, T. G., Gull, K., and Dove, W. F., 1984, Cell cycle regulation of tubulin RNA level, tubulin protein synthesis and assembly of microtubules in Physarum, J. Cell Biol., 99:155.

Schicker, C., Hildebrandt, A., and Sauer, H. W., 1979, RNA transcription of isolated nuclei and chromatin with exogenous RNA polymerases during mitotic cycle and encystment of Physarum polycephalum, W. Roux's Arch., 187:195.

Seebeck, T., Stalder, J., and Braun, R., 1979, Isolation of a minichromosome containing the ribosomal genes from Physarum polycephalum, Biochemistry, 18:484.

Staron, K., Jerzmanowski, A., Fronk, J., and Toczko, K., 1980, Some unusual features of Physarum polycephalum chromatin are due to the presence of slime, Acta Biochim. Polonica, 27:413.

Sun, J. Y.-C., Johnson, E. M., and Allfrey, V. G., 1979, Initiation of transcription of ribosomal deoxyribonucleic acid sequences in isolated nuclei of Physarum polycephalum: studies using nucleoside 5'-/gamma-S/ triphosphates and labeled precursors, Biochemistry, 18:4572.

Werenskiold, A. K., 1985, Light-dependent expression of actin and tubulins in Physarum polycephalum, Ph.D. Thesis, University of Munich.

Wu, C., 1985, An exonuclease protection assay reveals heat-shock element and TATA box DNA-binding proteins in crude nuclear extracts, Nature, 317:84.

Zandomeni, R., Bunick, D., Ackermann, S., Mittelman, B., and Weinmann, R., 1983, Mechanism of action of DRB. III: effect on specific in vitro initiation of transcription, J. Mol. Biol., 167:561.

Chapter 20: INCORPORATION OF SUBSTANCES INTO LIVING CELLS

Jörg Kukulies and Wilhelm Stockem

Institute of Cytology
University of Bonn
Bonn, FRG

A large variety of substances have been incorporated into living cells with the aim to induce and analyze specific interactions of these substances with different constituents of the cytoplasm. Three general ways to pass the plasmalemma as a barrier and to introduce substances into the cytoplasm (Chambers and Chambers, 1961; Celis et al., 1980) are:

1) <u>Cell membrane perforation</u> (Fig. 20-1). Treatment of the cell with detergents (e.g., Triton, saponin (Helenius and Simons, 1975)) or glycerol (Fig. 20-1a) induces both outflow of soluble cytoplasmic components and inflow of externally applied substances (Fig. 20-1b,c). Because some insoluble cytoplasmic constituents such as the actomyosin system are retained in the resulting cell model, this technique was applied to *Physarum polycephalum* (Achenbach and Wohlfarth-Bottermann, 1986) and has become of special interest for experiments to test the

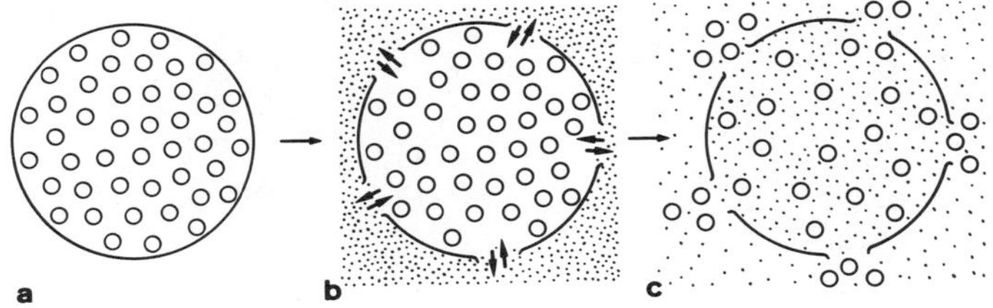

Fig. 20-1. Schematic illustration of transfer of substances into the cytoplasm of living cells by membrane perforation.

influence of regulating factors (e.g., Ca^{++} concentration) under defined conditions (Wohlfarth-Bottermann, 1985).

2) Substance transfer by liposome (or cell) fusion (Fig. 20-2). This technique includes both the incorporation of the liposome membrane into the plasmalemma and the discharge of the liposome content into the cytoplasm of the target cell (Fig. 20-2a-c). Instead of liposomes, cell ghosts (e.g., from red blood cells; Wojcieszyn et al., 1983) have also been used as vehicles to transfer substances into other cell types and to study the distribution and dynamics of labeled lipids and proteins within the plasmalemma and cytoplasm of the target cell respectively. However, inasmuch as defined differences in the surface properties of the vehicle and the target cell interfere with a successful fusion, this method must be specialized for each application.

3) Microinjection (Fig. 20-3). Microinjection by capillaries is generally preferred because this technique does not destroy the integrity of the target cell (Fig. 20-3a-c). Fast attachment of specimens to be microinjected is important and can be achieved by coating slides with poly-L-lysine, Alcian Blue, polysaccharides such as *Physarum* slime (Kukulies et al., 1984; Kukulies and Stockem, 1985a), or by mechanical devices.

The reliability of injection capillaries depends on the quality of the glass, the diameter of the tip (0.3-1 μm), and treatment of the tip by sharpening the edges with fluoric acid or by modulating the surface properties by siliconization. The injection pressure has to be carefully regulated by hand with a syringe or by automatic devices, and the properties of the injection medium should match as closely as possible those of the cytoplasm of the target cell. The volume of the injected medium should be controlled quantitatively during capillary filling. Generally it should be in the range of about 10% of the cell volume. Nearly all substances of interest for experimental cell research, such as ions, proteins, lipids,

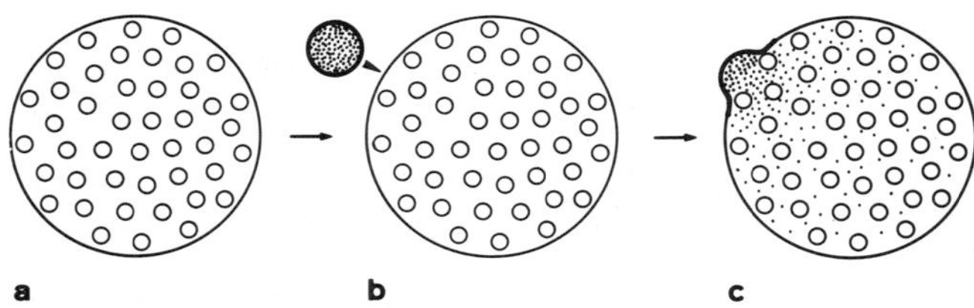

Fig. 20-2. Schematic illustration of liposome (or cell) fusion to transfer substances into the cytoplasm of living cells.

3. MICROINJECTION

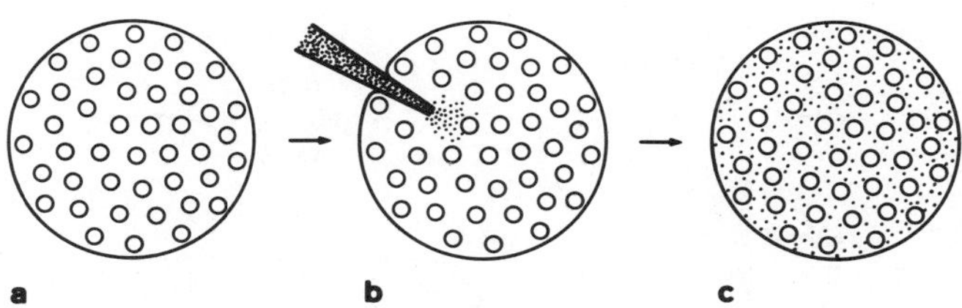

Fig. 20-3. Schematic illustration of transfer substances into the cytoplasm of living cells by microinjection.

sugars, nucleic acids, or drugs, have been microinjected into different cell types.

Our experiences with Physarum show that macroplasmodia (Fig. 20-4) are very difficult to microinject because rapid membrane flow and sealing phenomena prevent a transfer of the capillary content into the streaming protoplasm. Nevertheless, substances can be introduced into plasmodial strands by use of the endoplasm substitution technique. This technique is performed by injecting artificial medium by means of glass pipettes of a diameter of 50-100 μm into isolated plasmodial strands of at least 3 cm in length. The endoplasm is replaced by the injection solution over a length of 1-3 cm along the strand (for more details see Ueda et al., 1978). However, proteins such as fluorescently labeled actin or albumin are rapidly sequestered into numerous vacuoles after injection and are thus prevented from reaching the peripheral ectoplasm (Kukulies et al., 1984). Another disadvantage of macroplasmodia is their large extension and thickness, which impairs light microscopic investigations at high resolution.[1]

Physarum stages of appropriate size and nature for microinjection and subsequent high resolution light microscopic observation are cell fragments (Fig. 20-4) and microplasmodia from axenic shake cultures (Figs. 20-5, 20-6). Cell fragments can easily be microinjected because wound healing phenomena and sequestration of injected medium have rarely been encountered during injection (see also Chapter 14). The spherical cell fragments adhere strongly to glass slides covered with Physarum slime, exhibit a vigorous protoplasmic streaming activity, and incorporate injected proteins such as fluorescently labeled actin into dynamic, cytoplasmic structures

---

[1]During the Workshop discussion on microinjection, Kazarinoff reported successful microcapillary injection into Physarum plasmodia. His report and procedures are given in Chapter 21.

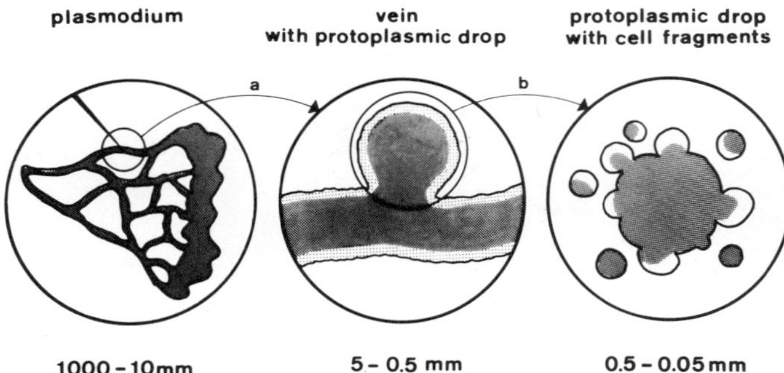

Fig. 20-4. Schematic illustration of different procedures to produce Physarum cell fragments. Puncturing a plasmodial strand generates a protoplasmic drop from which spherical cell fragments of appropriate size for light microscopic observation are pinched off after submersion in a caffeine solution. The ranges of diameters of the specimens are indicated below.

like the microfilament system (Kukulies and Stockem, 1985a). In contrast to such cell fragments, the axenically cultured microplasmodia are covered by a thick and rigid slime layer (Fig. 20-5a-c), which has to be removed by gentle washing and centrifugation before successful microinjection can be performed. Injected substances such as fluorescent actin or phalloidin first diffuse randomly in

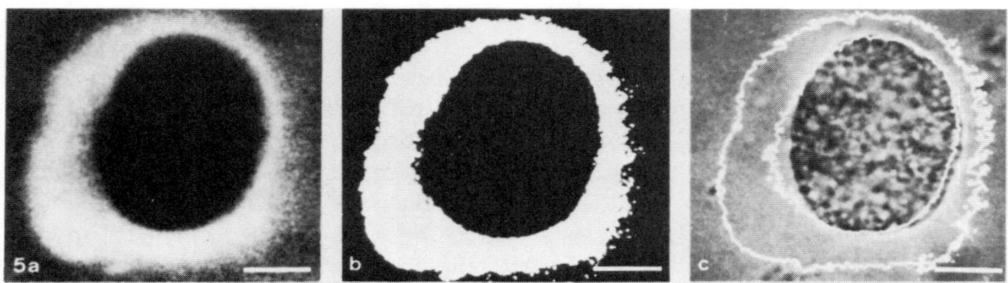

Fig. 20-5. Fluorescence micrograph (a), binary picture after digital image processing (b), and bright-field picture with computer overlay lines to indicate the countour of the external slime layer (c), after staining of a living microplasmodium with the rhodamine-labeled lectin Ricinus communis agglutinin I. - Bar 50 μm. (Reprinted with permission from Kukulies et al., 1985).

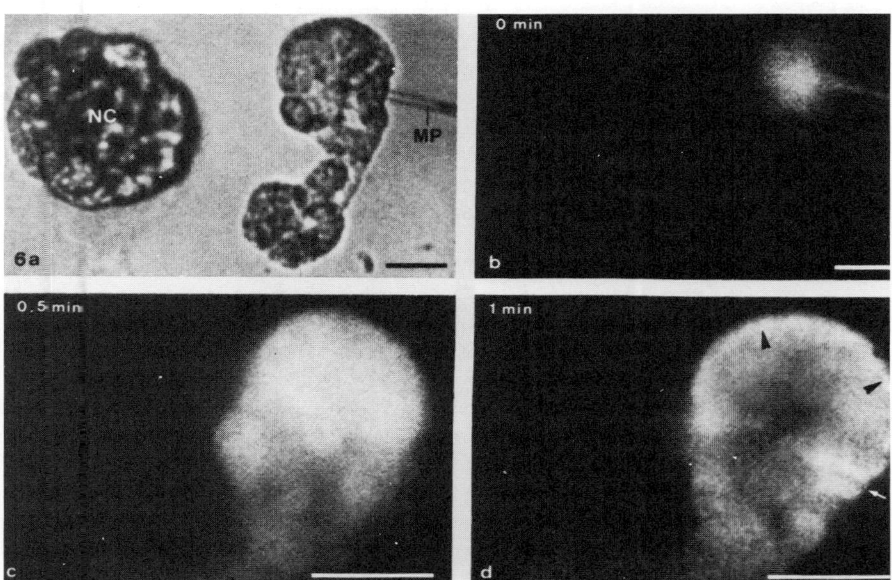

Fig. 20-6. Brightfield (a) and fluorescence micrographs (b-d) of a living dumbbell-shaped microplasmodium. The microinjected, rhodamine-labeled phalloidin first distributes randomly (c), and then distinctly stains the F-actin in the peripheral filament cortex (d, arrowheads). The arrow points to a plasmalemma invagination. MP, microcapillary; NC, non-injected cell. Bars, 50 µm. (Reprinted with permission from Kukulies et al., 1985).

the cytoplasm (Fig. 20-6a-c), but later show distinct concentration within the peripheral filament cortex delineating the cell surface and plasma membrane invagination system (Fig. 20-6d). So far no data have been gathered on microinjection of Physarum amoebae or flagellates.

Finally, it should be emphasized that new experimental possibilities are offered to cell biologists by the incorporation of labeled substances into living cells, especially for studies on the function and dynamic behavior of membrane and cytoskeletal components (Kukulies and Stockem, 1985b).

ACKNOWLEDGMENTS

This work was supported by a grant of the Deutsche Forschungsgemeinschaft (Sto. 126/4-2).

REFERENCES

Achenbach, F., and Wohlfarth-Bottermann, K. E., 1986, Reactivation of cell-free models of endoplasmic drops of *Physarum polycephalum* after glycerol extraction at low ionic strength, *Eur. J. Cell Biol.*, in press.

Celis, J. E., Graessmann, A., and Loyter, A., 1980, "Transfer of Cell Constituents into Eucaryotic Cells," Plenum Press, New York and London.

Chambers, R., and Chambers, E. L., 1961, "Exploration into the Nature of the Living Cell," Harvard University Press, Cambridge, Massachusetts.

Helenius, A., and Simons, K., 1975, Solubilization of membranes by detergents, *Biochim. Biophys. Acta*, 415:29.

Kukulies, J., Brix, K., and Stockem, W., 1985, Fluorescent analog cytochemistry of the actin system and cell surface morphology in *Physarum* microplasmodia, *Eur. J. Cell Biol.*, in press.

Kukulies, J., and Stockem, W., 1985a, Function of the microfilament system in living cell fragments of *Physarum polycephalum* as revealed by microinjection of fluorescent analogs, *Cell Tissue Res.*, 242:323.

Kukulies, J., Stockem, W., and Achenbach, F., 1984, Distribution and dynamics of fluorochromed actin in living stages of *Physarum polycephalum*, *Eur. J. Cell Biol.*, 35:235.

Kukulies, J., and Stockem, W., 1985b, Fluorescent analog cytochemistry of living cells, *Zeiss Inform.*, 98:in press.

Ueda, T., Götz von Olenhusen, K., and Wohlfarth-Bottermann, K. E., 1978, Reaction of the contractile apparatus in *Physarum* to injected Ca^{++}, ATP, ADP and 5' c AMP, *Cytobiologie*, 18:76.

Wohlfarth-Bottermann, K.-E., 1985, Cell-free models from *Physarum*--pitfalls and improvements, *in*: Proceedings of the 10th Yamada Conference 1984, "Cell Motility: Mechanism and Regulation", H. Ishikawa, H. Sato, and S. Hatano, eds., University of Tokyo Press, Tokyo, in press.

Wojcieszyn, J. W., Schlegel, R. A., Lumley-Sapanski, K., and Jacobson, K. E., 1983, Studies on the mechanism of polyethylene glycol-mediated cell fusion using fluorescent membrane and cytoplasmic probes, *J. Cell Biol.*, 96:151.

Chapter 21: PROTOCOL FOR MICROINJECTION OF MACROPLASMODIA OF *PHYSARUM POLYCEPHALUM*

Michael N. Kazarinoff and David C. Ruth

Division of Nutritional Sciences

Cornell University, Ithaca, NY

INTRODUCTION

We have used the following protocol over the last 18 months for the microinjection of ^{35}S-labeled proteins into macroplasmodia. Some patience and dexterity is required in filling the needles with material to be injected; considerably more is required during the injection. Uniform dispersion of injected material is verified by visual inspection in the stereo dissecting microscope with Blue Dextran used as a marker dye. Initially, the dye can be seen to be carried by the protoplasmic streaming. ^{35}S-Labeled material is also uniformly dispersed, as evidenced by constant specific activity in randomly chosen pieces of the plasmodium. Rupture of the plasmodium is rarely a problem when the injected material is equilibriated in the buffer described, but can occur if the microburet is dialed too rapidly. The use of black nitrocellulose for growing plasmodia is recommended to enhance the visual contrast. Standing liquid on the membrane or plasmodium causes problems and should be avoided.

PREPARATION OF NEEDLES FOR INJECTION

1. Wash SMI "green" capillaries (size "C") in chromic acid, to remove leached alkalis and nuclease activity, until the writing comes off (Scientific Manufacturing Industries Inc.; American Scientific Products No. P5070-902). Wash well with H_2O. Next, wash capillaries in 20 mM EDTA to prevent coagulation of plasmodial protoplasm. Then, rinse in distiled H_2O. Dry in oven.

2. Pull capillaries on Narishige PE-2 electrode puller at "8" on magnet, "15" on heater to make needles.

3. Store needles in plastic box taped to lid, since tips are very fragile.

4. Very gently touch the tip to a steel plate, holding the needle vertically to just barely break off end of needle, which tends to have a hole too tiny for successful manipulation of fluid.

BUFFER FOR INJECTION

Although we have no proof that the following buffer is essential to successful injection, it has been in use throughout successful injections (Taylor, 1977). The buffer is: 5 mM PIPES, pH 7.0; 5 mM K_2EGTA; 30 mM KCl; 1 mM $MgCl_2$; 1 mM Na_2ATP; 1 mM DTT; use Blue Dextran for dye.

PREPARATION OF MACROPLASMODIUM

1. Grow a macroplasmodium on a black nitrocellulose disc supported by 2-mm glass beads. The nitrocellulose filter used was Millipore Type HABP, 47-mm diameter, 0.45-μm pore size. To inoculate, pipette one drop of microplasmodia onto center of disc. Allow to dry or fuse for 1.5 hr at 26°C.

2. Add semidefined medium to the bottom of the petri dish to just cover bottom, or place the disc on a 1.5% agar plate with semidefined medium.

3. Grow the plasmodium for 24-48 hr prior to injection.

FILLING NEEDLES AND MANIPULATING FLUID

1. With a microburet, fill the syringe with mineral or paraffin oil.

2. Connect it to the microneedle with Tygon tubing, and fill the needle with oil from the syringe. Oil will fill to end of needle but will not leak out.

3. Place droplet of material to be injected on a silanized glass slide and, viewing it in a microscope, place the tip of the needle into the droplet and carefully withdraw syringe. With care, virtually all of the fluid can be withdrawn into the needle. Avoid breaking needle on glass. Note that the needle will bend before it breaks, so the glass can be touched.

INJECTION

Use a stereo dissecting microscope to view the plasmodium and needle during this procedure. Plasmodia to be injected are preferably grown on black nitrocellulose membranes.

1. Remove nitrocellulose from dish, and place on filter paper to "dry" for 15-30 min before injection. This can be omitted when plasmodia are grown on discs on agar. Choose translucent, yellow "rope-like arms" of macroplasmodia to inject; these are near the edges of the plasmodium.

2. Gently insert the needle. One can see the wall of the plasmodium yielding to the needle; patience is worthwhile. Touch wall, push needle gently into it; after needle penetrates, back it off a bit (so as not to jam it against the back wall). When injection is successful, dialing in of fluid from microburet goes quite well, and one can see fluid streaming down arms of plasmodia. Inject only a couple of hundred nanoliters at a time and allow fluid to disperse before dialing more; 1-2 µl can be injected in a single injection site. If the injection is done successfully, no liquid leaks upon withdrawal of needle.

ACKNOWLEDGMENT

This work was supported by USPHS Grant GM-28540.

REFERENCE

Taylor, D. L., 1977, Contractile basis of amoeboid movement. IV. The viscoelasticity and contractility of amoeba cytoplasm in vivo, <u>Exp. Cell. Res.</u>, 105:413.

Chapter 22: GENE CLONING AND CONSTRUCTION OF GENOMIC LIBRARIES IN PHYSARUM

Werner F. Nader

Dept. of Neurochemistry
Max-Planck-Institute for Psychiatry
Martinsried, FRG

With contributions from: T. Burland, R. A. Cox, W. F. Dove, M. J. Monteiro, and E. C. A. Paul

THE "UNCLONABLE" PHYSARUM DNA

Application of recombinant DNA techniques in cell biology has vastly increased our knowledge in this field during the last 10 years. Major breakthroughs have been achieved with different model organisms. Even the quite complicated differentiation mechanisms of mammals become more and more transparent. By contrast, the progress with *Physarum* has been comparatively slow. During the most recent *Physarum* meetings, it became obvious that a major disadvantage of that organism was a lack of a comprehensive genomic library. Genes like those for actin, tubulin, and amylase can be easily detected on Southern blots of restriction endonuclease-digested *Physarum* DNA with heterologous DNA probes, but could not be isolated from *Physarum* genomic libraries in lambda phages and cosmids (Knox et al., 1984). Until now, only one *Physarum* gene, that for histone H4, has been isolated and characterized (Wilhelm and Wilhelm, 1984). A second gene for H4 can be detected on Southern blots, but has not been isolated from the library (X. Wilhelm, personal communication).

Major progress on cloning of the *Physarum* genome and the successful isolation of genes for actin, α- and β-tubulin were reported during this Workshop. New cloning hosts were introduced, which will be helpful for other organisms as well, where similar problems are encountered (Murray et al., 1984; Wyman et al., 1985).

POSSIBLE CAUSES FOR INSTABILITY OF DNA IN *E. COLI*

Instability of cloned DNA was first reported by Lauer et al., (1980), who observed frequent deletions in their clones of human α-globin genes. Fairly common also are reports of sequences that cannot be found in eukaryotic libraries at all (Wyman et al., 1985). Possible causes for instability and vector inviability are direct or inverted DNA repeats. Direct repeats can generate unequal crossing over during the numerous recombination opportunities that accompany DNA replication of the phage. When those homologous regions are hyphenated by a spacer, this hyphen is deleted and, if the phage is shortened to a certain critical length, its DNA can no longer be packaged into virus particles. Inverted repeats, if they form a perfect DNA palindrome, frequently are deleted in plasmids (Collins et al., 1982) and cause complete inviability of phage lambda (Leach and Stahl, 1983). This instability is caused by the bacterial recombination exonucleases I and V, the products of the *sbcB* and *recBC* genes (for review, see Smith, 1983). But even on *recBC sbcB E. coli* mutant strains, frequent deletion of those palindromes can be observed, although phages carrying the palindromes do survive on those hosts.

CLONING IN PHAGE LAMBDA EMBL3 AND HOST *E. COLI* CES200

Direct and inverted repeats are quite common in eukaryotic DNA. Analysis of single-stranded and self-annealed *Physarum* DNA under the electron microscope reveals that foldback sequences occur regularly, interspersed within the genome at an average distance of 7 kb. In other organisms this distribution is more random or clustered (Hardman and Jack, 1979; see Chapter 3).

At Texas A & M University, we constructed a genomic library of *Physarum* in the replacement vector EMBL3. ("Replacement vector" means that a stuffer fragment of the phage, which is not essential for viral reproduction, can be replaced by foreign DNA.) DNA fragments between 9 and 22 kb can be cloned into EMBL3 (Frischauf et al., 1983). For library construction, genomic DNA is partially digested by restriction endonucleases MboI or Sau3a, which cut frequently within the genome and yield BamHI compatible cohesive ends. DNA fragments between 15 and 23 kb are then selected on a sucrose gradient or a preparative agarose gel, or the partial digest is dephosphorylated to avoid random religation and scrambling of the library. These fragments are then ligated to the BamHI sites at the vector arms.

Comparative plating of the library on *rec*$^{+}$, *recBC*$^{-}$, *sbcB*$^{-}$, *recBC*$^{-}$, *sbcA*$^{-}$, and *sbcB*$^{-}$ *E. coli* hosts revealed that on the strain without exonuclease I and V (*E. coli* CES200 from C. E. Shurvinton; see Nader et al., 1985 for details), the efficiency of plating was

5- to 10-fold higher than on the other mutant strains. Also, screening of the library for phages with actin sequences was successful only on CES200. Out of 90,000 clones, the equivalent of six *Physarum* genomes, 13 positive phages were selected and 10 could be plaque-purified.

HOMOPOLYMERS IN TWO INTRONS OF ACTIN GENE LOCUS *ardA*

The phage λPpA10 contains a complete EcoRI fragment of the actin gene locus *ardA* (Fig. 22-1) (Schedl and Dove, 1982). Frequently, deletion of a distinct 360-bp sequence from the 15-kb insert fragment of this phage can be observed. Its plating efficiency on CES200 is 5-fold higher than on the *rec*$^+$ host LE392. Phages which plate on the latter host no longer contain the 360-bp sequence. The EcoRI-HindIII fragment with the actin gene and the unstable sequence was further subcloned into plasmid pBR322 (Fig. 22-1c). Again, the 360-bp sequence was deleted in 50% of the plasmids, isolated from the *recA*$^-$ strain HB101, and in less than 5% of the plasmids from CES200. Electron microscopic analysis of single-stranded and self-annealed DNA from the EcoRI-HindIII fragment revealed secondary structures compatible with DNA "foldbacks", but it is not yet proven whether the observed inverted-repeat system is in the region of the 360-bp deletion (see the following sequencing results).

Not surprisingly, efforts to subclone this fragment into phage M13 failed. Luckily, the unstable 360-bp sequence contains a Sau3a restriction site in its center. Cloning of the region in halves as HaeIII-Sau3a fragments (Fig. 22-1c) yielded two stable M13 subclones. At present, we know approximately 80% of the sequence of *ardA* and have found that the coding region matches exactly the amino acid sequence published by Vanderkerckhove and Weber (1978). We have been able to locate five introns and a putative polyadenylation signal, an eight-times repeated AATAAA, at the 3' end. The unstable DNA extends from intron 4 through exon 5 into intron 5. Its most striking features are homopolymers, repeats with 30 or more $(GC)_n$ and $(AT)_n$ stretches. An exact sequence of the region could not be determined because GC-homopolymer sequences cause nonspecific termination of the sequencing reaction (Fig. 22-2). Efforts to improve the reaction by increasing the temperature to 50°C have failed (Gomer et al., 1985). (See note added in proof, at end).

This ^{32}P-labeled EcoRI-HindIII DNA fragment hybridized to a wide range of sequences in the *Physarum* genome, as was shown by Southern blot analysis (W. Knoerzer, unpublished). After unlabeled poly A and poly G RNA were added in high concentration (0.1 mg/ml) to the hybridization reaction, the probe hybridized specifically to restriction fragments known to carry actin sequences. These results indicate that homopolymers are broadly distributed within the Physarum genome.

ISOLATION OF PHYSARUM TUBULIN SEQUENCES AND THE ardC LOCUS

Major progress was also reported at the Workshop by Burland and Paul and by Cox and Monteiro. Burland constructed a genomic

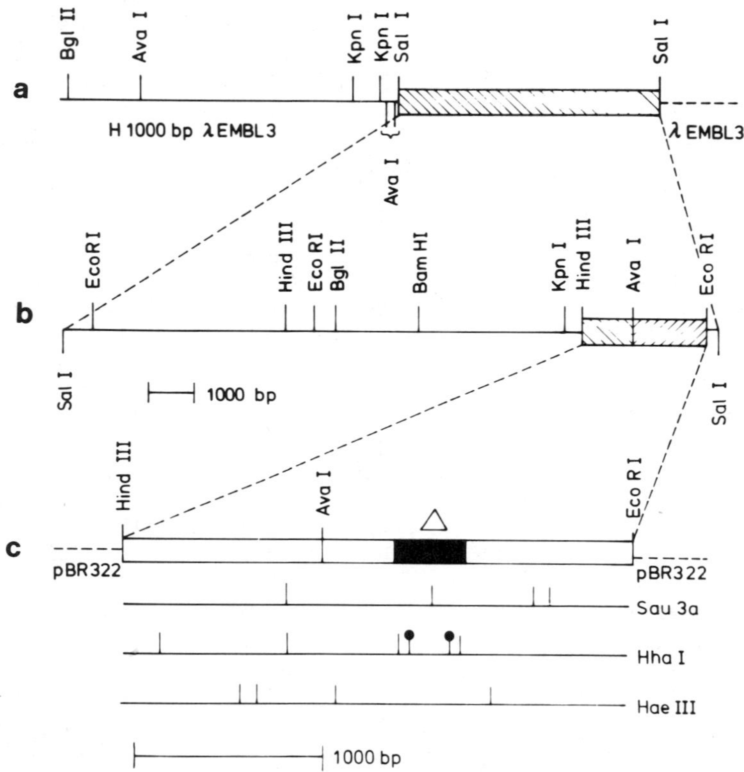

Fig. 22-1. Restriction map of phage λPpA10 with actin gene locus ardA. (a) The 15-kb insert fragment (hatched blocks) in phage EMBL3. (b) Restriction map of the insert with the actin sequence and the unstable DNA marked as hatched blocks. (c) Restriction map of the EcoRI-HindIII fragment from (b) subcloned into pBR322. The solid bar indicates the 360-bp sequence, which causes instability of phage lambda and frequently is deleted (Figure adapted from Nader et al., 1985, with modifications).

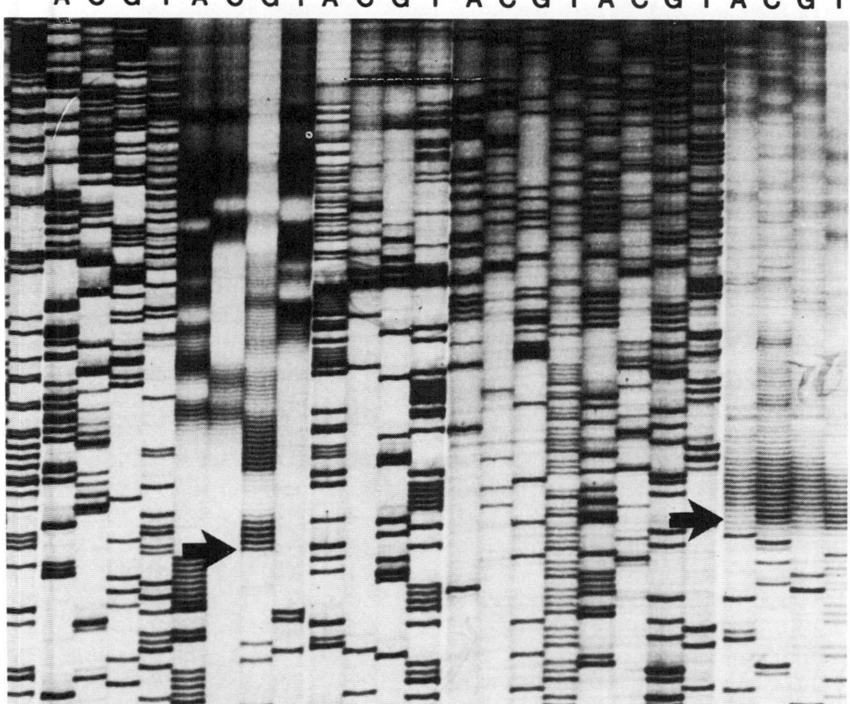

Fig. 22-2. Sequencing gel of M13 subclones from actin gene locus ardA. The sequencing reaction was performed as described by Biggin et al. (1983). Arrows indicate regions of polydG and polydC, where polymerization is nonspecifically terminated. The DNA for these reactions was from the two M13 subclones from the unstable region, which frequently is deleted (see Fig. 22-1).

library ML1 in the replacement vector Charon 35 (Loenen and Blattner, 1983), a phage that was successfully used to clone the wheat genome (Murray et al., 1984). In contrast to the red$^-$ gam$^-$ EMBL3 recombinants, Charon 35 phages with inserted DNA exhibit the red$^-$ gam$^+$ phenotype. The red gene products, lambda exonuclease and β protein, catalyze host-independent homologous recombination events, whereas the gam protein inhibits the bacterial exonuclease V, thus allowing phage growth on recA$^-$ hosts. Any recombination event is suppressed in such a vector-host system. Five phages, which hybridized to DNA probes with α- and β-tubulin sequences, were isolated from 15,000 recombinant phages, but only one could be plaque-purified. Because of poor growth of Charon 35 recombinants and their instability, Burland and Paul also screened a lambda EMBL3/ E. coli CES200 library, ML2, and found among 90,000 clones 60 phages with putative tubulin sequences. They also observed that phages with β-tubulin sequences plated 5- to 10-fold more efficiently on CES200 than on a *rec*+ host, whereas those with α-tubulin did not differ.

Cox and Monteiro reported the isolation of phages with tubulin and actin sequences from their libraries in lambda EMBL4 and NM1149 on standard *rec*+ hosts. Lambda EMBL4 is similar to EMBL3; the only difference is that the polylinkers between the stuffer fragment and phage arms are oriented in the opposite direction (Frischauf et al., 1983). Inserts of EMBL4 can be spliced out with EcoRI, whereas SalI can be used for EMBL3 inserts. Lambda NM1149 is a *cI* insertion vector with a unique HindIII site in that gene (Murray, 1983). Phages with inactivated *cI* gene can no longer lysogenize and therefore can form plaques on high frequency of lysogeny (*hfl*) *E. coli* mutant strains. This allows a powerful selection against *cI*+ nonrecombinant phages. The maximal size for the insert is 11 kb. Partial genomic libraries can be constructed by ligation of complete HindIII or EcoRI digests of DNA into the cloning sites of vector NM1149 (for HindIII) or NM641, a phage with a unique EcoRI site in *cI* (Murray et al., 1977). This technique can be applied only if the wanted gene is located on a EcoRI or HindIII fragment less than 11 kb in length. Furthermore, genes with a recognition site for one of those enzymes in their sequence (e.g., *Physarum ardA* with a HindIII site in codons 6 and 7) can be isolated only in pieces. A comprehensive library can be constructed in NM641 by cloning of blunt-ended DNA fragments, which can be randomly generated by the shearing of genomic DNA (Young and Davis, 1985). After filling the ends of those fragments with T4 DNA polymerase and size selection, the DNA is methylated with EcoRI methylase, linkers are attached, and EcoRI-compatible cohesive ends are produced by digestion with that enzyme.

With their cloning systems, Cox and Monteiro isolated phages for α-tubulin, actin gene locus *ardC*, and putative β-tubulin clones. The identity of the actin and α-tubulin clones was proven by a hybrid selection experiment. DNA from the lambda phages was used to isolate mRNA, which was found to yield peptides of the same molecular weight as for actin and α-tubulin after in vitro translation. All clones they characterized had relatively small inserts of 4.8 kb and 4.7 kb for actin and α-tubulin in NM1149, and 6 kbp for α-tubulin in EMBL4. By cloning smaller fragments of *Physarum* DNA, these researchers avoided those sequences which destabilize phage lambda and which were found in the actin *ardA* locus.

HOW CAN *PHYSARUM* GENOMIC CLONING BE FURTHER IMPROVED?

As was shown for λPpA10 (Nader et al., 1985) and for clones

with putative β-tubulin sequences (Burland and Paul, this Workshop), deletions are quite frequent in these phages even during propagation on CES200. This might be owing either to problems during DNA replication (e.g., "slippage replication") or to recF recombination, which is activated in this host. Also, plasmids with ColEl-derived replication origins such as pBR322 and the pUC-series are highly unstable on $recBC^-$ $sbcB^-$ E. coli strains (Bassett and Kushner, 1984), which might cause problems during further subcloning of Physarum genomic DNA. An additional introduction of a $recA^-$ or $recF^-$ mutation into CES200 might solve both problems. Experiments with CES201 (a $recA^-$ $recBC^-$ $sbcB^-$ construct by C.E. Shurvinton) show that this mutant is not very viable, highly light-sensitive, and thus problematic to work with.

A new set of host strains was introduced by Dove and McLeester during the Workshop. They introduced stepwise the $hsdR^-$ and either $recA^-$ or $recF^-$ mutations into the lambda host FS1578, which exhibits the hflA hflB and $recBC^-$ $sbcB^-$ phenotypes (F. W. Stahl, unpublished). They thereby gained a set of three new lambda hosts:

MB406 - recB21 recC22 sbcB15 hflA hflB $hsdR^-$ and tet^s;
MB407 - recA-del306 recB21 recC22 sbcB15 hflA hflB $hsdR^-$ and tet^r; and
MB408 - which differs from MB407 by carrying a $recF^-$ mutation instead of the $recA^-$.

Because of the two hfl mutations, these new strains are excellent selective hosts for cI insertion vectors like λgt10 (with an EcoRI insertion site; Huynh et al., 1984), NM641 (EcoRI insertion site; Murray et al., 1977) and NM728 (HindIII insertion site; ibid). In MB407 and MB408, recombination is totally suppressed, and exonucleases I and V are inactive, which will allow stable propagation of clones with direct and inverted DNA repeats in red^- phage lambda vectors.

An advantage of the cI insertion vectors over replacement vectors such as EMBL3 for cloning of DNA with extensive repeat systems is that smaller DNA fragments can be inserted. With this approach, systems of direct or inverted repeats can be broken into smaller nonrepetitive fragments, which can be stably propagated even on rec^+ hosts.

PROSPECTS

The results presented during this Workshop indicate that cloning of Physarum DNA is no longer impossible, but that new problems concerning stability of the cloned DNA and sequencing through homopolymer regions are still ahead. The sequencing problems might be overcome by using reverse transcriptase instead of DNA polymerase

Klenow fragment in the reaction, or the chemical sequencing method (Maxam and Gilbert, 1980). These efforts are worthwhile, because the arrangement of the homopolymers in the introns around an exon are quite interesting and might be of some biological function, e.g., during RNA splicing.

With exceptions (e.g., Physarum or wheat; Murray et al., 1984), recombinant DNA libraries of most organisms generally contain the majority of sequences from the genome from which they were derived. But, as has recently been shown for the human and Drosophila genomes (Wyman et al., 1985; Petri and Wyman, 1985), there might be regions in each genome that are refractory to cloning in common host strains. The experiences during the cloning of the Physarum genome will help to be aware of and to avoid those problems.

Genomic libraries, amplified on rec^{+} hosts and distributed to other laboratories, are now commonly used to isolate genes. These libraries not only lack certain regions of the genome (see above), but deletions in the cloned DNA because of recombination can also be observed quite frequently. As a precaution, several positive clones should be plaque-purified and analyzed for possible changes. These problems will be reduced if amplification is routinely done on nuclease or recombination deficient hosts such as CES200 or the MB-series. Genomic libraries of Physarum DNA in lambda EMBL3 and amplified on CES200 are now available from the author (Library TLC1 - Texas Library Colonia 1) and from Burland, Paul, and Dove (library ML2A - Madison Library 2, amplified).

REFERENCES

Bassett, C. L., and Kushner, S. R., 1984, Exonucleases I, III and V are required for stability of ColE1 related plasmids in Escherichia coli, J. Bacteriol., 157:661.

Biggin, M. D., Gibson, T. J., and Hong, G. F., 1983, Buffer gradient gels and ^{35}S label as an aid to rapid DNA sequence determination, Proc. Natl. Acad. Sci., U.S.A., 80:3963.

Collins, J., Volckaert, G., and Nevers, P., 1982, Precise and nearly precise excision of the symmetrical inverted repeats of TN5; common features of recA independent deletion events, Gene, 19:139.

Frischauf, A.-M., Lehrach, H., Poustka, A., and Murray, N., 1983, Lambda replacement vectors carrying polylinker sequences, J. Mol. Biol., 170:827.

Gomer, R., Datta, S., and Firtel, R., 1985, Sequencing homopolymer regions, FOCUS (BRL Comp., Gaithersburg, MD), 7:6.

Hardman, N., and Jack, P. L., 1979, Periodic organization of foldback sequences in Physarum polycephalum nuclear DNA, Nucl. Acids Res., 5:2415.

Huynh, T. V., Young, R. A., and Davis, R. W., 1984, Constructing and screening cDNA libraries in λgt10 and λgt11, in: "DNA Cloning Techniques: A Practical Approach", D. Glover, ed., IRL Press, Oxford.

Knox, C., Maher, M. J., and Marsh, R., 1984, Cloning Physarum DNA in Charon 4A and characterization of some clones, Texas J. Sci., 35:339.

Lauer, J., Shen, C.-K. J., and Maniatis, T., 1980, The arrangement of human α-like globin genes: sequence homology and α-globin gene deletions, Cell, 20:119.

Leach, D. R. F., and Stahl, F. W., 1983, Viability of λ phages carrying a perfect palindrome in absence of recombination nucleases, Nature, 305:448.

Loenen, W. A. M., and Blattner, F. R., 1983, Lambda Charon vectors (Ch32, 33, 34 and 35) adapted for DNA cloning in recombination deficient hosts, Gene, 26:171.

Maxam and Gilbert, 1980, Sequencing end-labelled DNA with base-specific chemical cleavages, Methods Enzymol., 65:499.

Murray, M. G., Kennard, W. C., Drong, R. F., and Slighton, J. L., 1984, Use of a recombination-deficient phage lambda system to construct wheat genomic libraries, Gene, 30:237.

Murray, N. E., Brammar, W. J., and Murray, K., (1977), Lambdoid recombinants, Mol. Gen. Genet., 150:53.

Murray, N. E., 1983, Phage lambda and molecular cloning, in: "Lambda II", R. W. Hendrix, J. W. Roberts, F. W. Stahl, and R. A. Weisenberg, eds., p. 395, Cold Spring Harbor Laboratory, Cold Spring Harbor.

Nader, W. N., Edlind, T. D., Huettermann, A., and Sauer, H. W., 1985, Cloning of Physarum actin sequences in an exonuclease-deficient bacterial host, Proc. Natl. Acad. Sci., U.S.A., 82:2698.

Petri, W. H., and Wyman, A. R., 1985, A significant fraction of the Drosophila genome cannot be cloned in rec^+ E. coli hosts, Genetics, 110:s59.

Schedl, T., and Dove, W. F., 1982, Mendelian analysis of the organization of actin sequences in Physarum polycephalum, J. Mol. Biol., 160:41.

Smith, G. R., 1983, General recombination, in: "Lambda II", R. W. Hendrix, J. W. Roberts, F. W. Stahl, and R. A. Weisberg, eds., p. 175, Cold Spring Harbor Laboratory, Cold Spring Harbor.

Vanderkerckhove, J., and Weber, K., 1978, The amino acid sequence of Physarum actin, Nature, 276:720.

Wilhelm, M. L., and Wilhelm, F.-X., 1984, A transposon-like DNA fragment interrupts a Physarum polycephalum histone H4 gene, FEBS Lett., 168:249.

Wyman, A. R., Wolfe, L. B., and Botstein, D., 1985, Propagation of some human DNA sequences in bacteriophage λ vectors requires mutant Escherichia coli hosts, Proc. Natl. Acad. Sci., U.S.A., 82:2880.

Young, R. A., and Davis, R. W., 1985, Immunoscreening λgt11 recombinant DNA expression libraries, in: "Genetic Engineering", J. K. Setlow and A. Hollaender, eds., p. 29, Plenum Press, New York.

NOTE ADDED IN PROOF

Recently the entire *ardA* gene locus has been sequenced in our laboratory, and all characteristics of a functional gene have been found. The deletion of 360 bp, which supposedly causes the unclonability of the gene locus in *rec*$^+$ *E. coli*, consists of homopolymers and exon 5, arranged in the fashion: poly T - poly C - exon 5 - poly C - poly T. We propose that the homopolymers themselves are responsible for the unclonability of this region.

Chapter 23: GENE CLONING AND CONSTRUCTION OF cDNA LIBRARIES IN PHYSARUM

Robert A. Cox
National Institute for Medical Research
London, UK

with contributions from: G. R. Barnett, F. Binette, F. Bernier, T. G. Burland, M. N. Kazarinoff, A. Laroche, G. Lemieux, M. Monteiro, D. Pallotta, E. C. A. Paul, T. Schedl, T. M. Shinnick, G. E. Sweeney, A. Tessier, G. Turnock, D. I. Watts, and M. L. and F. X. Wilhelm

INTRODUCTION

In principle, the construction of genomic DNA libraries with plasmid, cosmid, or bacteriophage as vectors is a matter of applying standard methods to *Physarum* DNA. In practice, more difficulties were encountered than were anticipated. The features of the nucleotide sequence of *Physarum* DNA that impede the construction of stable recombinants having a high plating efficiency are now beginning to emerge (Nader et al., 1985). These developments are welcome because we can now look forward to using cloned genes to investigate gene replication and expression during the mitotic cycle and cellular differentiation of *Physarum* and to capitalize on our knowledge of actin gene loci (Schedl and Dove, 1982) and tubulin gene loci (Schedl et al., 1984) derived from their characterization by Mendelian segregation of polymorphic restriction fragments of *Physarum* DNA. The construction of genomic libraries is dealt with extensively by Nader (Chapter 22), so only a few general points will be addressed here.

CONSTRUCTION OF GENOMIC LIBRARIES

Clones of rRNA genes have been available for some time mainly because rDNA can be isolated as a satellite band of high density in CsCl isopynic gradients, thus enabling restriction endonuclease fragments of DNA to be cloned directly without recourse to making a complete library of genomic DNA. *Physarum* rDNA comprises approximately 150 copies per haploid genome, each of which is a linear molecule of 60 kb in size. Each molecule comprises two sets of rRNA genes arranged in the form of a palindrome. Much progress has been made in sequencing rDNA (V. Vogt, personal communication).

The progress made in cloning genes other than rRNA genes has been slow. Clones of two other genes, histone H4 (Wilhelm and Wilhelm, 1984) and an *ardA* actin gene sequence (Nader et al., 1985), have been described. The limited progress made has not been the result of lack of effort. For example, at the Sixth European *Physarum* Meeting, April 24-28, 1984 in Font-Romeu, France, a group of investigators actively engaged in cloning experiments reported unsuccessful attempts to prepare cosmid clone banks of approximately 40-kb restriction fragments of *Physarum* DNA. No one was able to isolate more than approximately 100 recombinants, and these were not always stable. The discussion conveyed the impression that the construction of clone banks was difficult, possibly because of the nature of particular nucleotide sequences within the *Physarum* genome.

Features of the Physarum Genome

The organization of nucleotide sequences within the *Physarum* genome (approximately 2.7 X 10^8 bp) is very similar to that observed in many eukaryotes including man (Hardman et al., 1980; see Chapter 3). Single-copy DNA accounts for approximately 63% of the genome and is interspersed with short repetitive sequence elements. The repetitive sequences account for ca 31% of the genome and comprise ca 80 families, each containing ca 1800 repeats per family. The elements of repetitive sequences within a family are very similar but not identical. The size of the elements varies over a wide range, but the average is approximately 590 nucleotide residues. The remaining 6% of the genome is accounted for by fold-back sequences.

Choice of Vector

The choice of vector into which genomic DNA is cloned is generally either a cosmid (a plasmid containing the cohesive [cos] ends of phage-lambda DNA [Collins and Hohn, 1978]) or a phage-lambda (for review, see Kaiser and Murray, 1985). These vectors have gained favor because they can be introduced into *Escherichia coli* with high efficiency and reproducibility by in vitro packaging of cosmid DNA or phage lambda DNA (Hohn and Hohn, 1974). Earlier failures to clone *Physarum* DNA with cosmid vectors may have led to phage lambda emerging as the favored host.

The Advantages of Finding a Sympathetic Host

There are few difficulties in isolating, from *Physarum* nuclei, high molecular weight DNA which can be cleaved by restriction endonucleases and ligated. However, features of the nucleotide sequence of *Physarum* DNA (e.g., the foldback, possibly palindromic, sequences) may impede successful cloning because they are potentially unstable in the *recA*$^-$ strains of *E. coli* frequently used for recombinant DNA experiments. Nader (Chapter 22) discusses current improvements in *E. coli* hosts.

Influence of the Size of Insert on the Stability of the Clone

The presence of inverted repeat sequences, or direct repeat sequences within the fragment of cloned DNA, diminishes the stability of the recombinant in a *recA*$^-$ host. A smaller insert of genomic DNA would be less likely to have multiple repeat sequences and thus would be less likely to form cruciform structures; therefore, the recombinant would be expected to have higher stability. A genomic library with inserts of approximately 5 kb would be expected to be more stable than a library with inserts of approximately 20 kb. The disadvantages of working with recombinants with the smaller inserts are that four times the number of recombinants are needed for a complete library, and the chances of a single recombinant containing a complete gene are diminished.

Cloned Genes of Physarum

Progress was made in the isolation of genes for the highly conserved proteins histone H4, actin, and tubulin (for summary see Table 23-1). These gene sequences have been sufficiently conserved throughout eukaryotes so that it was feasible to use heterologous gene sequences as probes for the isolation of the *Physarum* gene. In each case the vector was phage lambda. Three of the four gene sequences that have been isolated contain repetitive sequence elements either within the gene or very close to it. The progress to date in characterizing the isolated gene sequences is summarized in Table 23-1. Since cloning of actin and tubulin genes is discussed fully in Chapter 22, only the case of histone H4 genes will be addressed here.

Histone H4 Gene

M. L. Wilhelm and F. X. Wilhelm (1984) constructed a genomic library by ligating *EcoRI* fragments of *Physarum* DNA strain M3 CVIII into phage λgtWES lambda B DNA. The phage plaques were screened in situ plaque hybridization. Putative histone clones were detected by using radioactive sea urchin histone H4 gene sequences. One histone recombinant was detected in 20,000 primary recombinant plaques. The recombinant contained an insert of 6.5 kb and the H4

Table 23-1.

Summary of Cloned Physarum Gene Sequences

Gene Sequences Isolated	Strain of *Physarum*	Presence of Repetitive Sequences Within or Near to Gene	Details of the Genomic Library[a]	Reference
Histone H4	M3CVIII	Yes	Complete EcoRI digests ligated into phage-lambda gtWES DNA. *E. coli* 1046 was the host.	Wilhelm and Wilhelm, 1984
Actin gene (*ard*A locus)	Cld-Axe (amoebae)	Yes	15-20 kb fragments of Mbol partial digest DNA were ligated with phage-lambda EMBL 3, *E. coli* CES200 was the host.	Nader et al., 1985
Actin gene (*ard*C2 locus)	M3CVIII (plasmodia)	Yes	Hind III DNA fragments ligated into phage-lambda NM1149 DNA. *E. coli* NM514 was the host.	Monteiro and Cox, unpublished; library gMHpl(NM)
α-Tubulin (*alt*B locus)	M3CVIII (plasmodia)	No	Hind III DNA fragments ligated into phage-lambda NM1149 DNA. *E. coli* NM514 was the host.	Monteiro and Cox, unpublished; library gMHpl(NM)
α-Tubulin	M3CVIII (plasmodia)	Yes	Sau3A DNA fragments ligated into phage-lambda EMBL 4. *E. coli* Q358 was the host.	Monteiro and Cox, unpublished; library gMHpl(NM)

Table 23-1. (Continued)

Gene Sequences Isolated	Strain of *Physarum*	Presence of Repetitive Sequences Within or Near to Gene	Details of the Genomic Library[a]	Reference
β-Tubulin	M3CVIII	Yes	Sau3A DNA fragments ligated into phage-lambda EMBL4. *E. coli* Q358 was the host.	Monteiro, Cox and Werenskiold, unpublished
α-Tubulin	M3CVIII (plasmodia)	--	System a. DNA cloned into phage-lambda Ch35/E using *E. coli* K802 recA as host.	Burland unpublished; library ML1
β-Tubulin		--	System b. DNA cloned into phage-lambda EMBL3. *E. coli* CES200 was the host.	Burland and Paul, unpublished; library ML2

[a], in each case the recombinant phage DNA was packaged in vitro.

gene coding sequences were located. The nucleotide sequence was established, and an 86-bp transposon-like insertion was found within the histone H4 coding sequence. The organization of the gene is summarized in Fig. 23-1.

The cloned histone gene sequences were used by Wilhelm and Wilhelm to investigate the time during the mitotic cycle when the histone H4 gene is transcribed. Histone H4 mRNA was found to accumulate as poly(A)$^{+}$mRNA during late G-2 and to be translated during S phase. They conclude that replication of the histone H4 gene and transcription are not linked (see Jalouzot et al., 1985). Similar results for the abundance of histone H4 mRNA were reported by V. Kung and R. Braun, personal communication.

Wilhelm and Wilhelm's colleagues also studied the organization of histone H4 gene sequences by hybridization of Southern blots of restriction endonuclease fragments of Physarum DNA with ^{32}P-labeled histone H4 probe. The presence of two types of H4 genes ($H4_1$ and $H4_2$), each present in one or two copies, was inferred from an analysis of the hybridization patterns. The cloned gene was identified as histone $H4_1$.

CONSTRUCTION OF cDNA CLONE BANKS

Techniques for the construction of cDNA libraries are well-established (for review see Glover, 1985). The libraries have been used as part of a successful strategy toward isolating a particular gene by first isolating a DNA copy of the mRNA encoded by that gene (a cDNA clone), then using this cDNA clone to isolate a genomic clone. Also, cDNA libraries have been used as a means of studying differential gene expression at the level of the transcription of the gene and the production of mature mRNA. In the

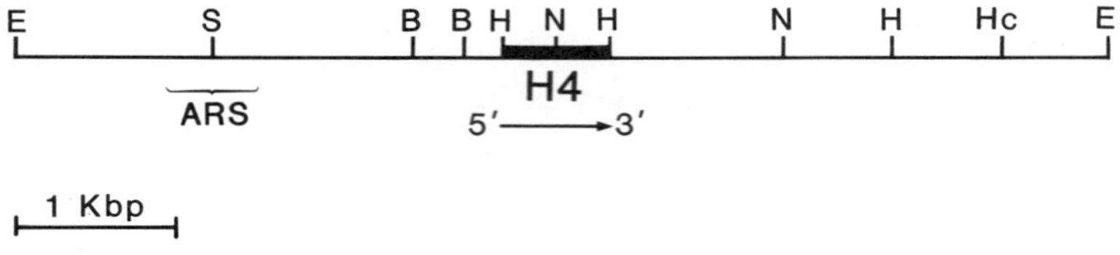

Fig. 23-1. Organization of a histone $H4_1$ gene cloned into phage-lambda gtWES lambda B (Wilhelm and Wilhelm, 1984). The approximate location of an autonomously replicating sequence (ARS) is indicated. The insert does not have recognition sites for BclI, HpaI, PstI, PvuII, Sall, and XhoI endonucleases. The recognition sites shown are E, EcoRI; B, BamHl; H, HindIII; N, NcoI; Hc, HincII; S, SmaI.

case of *Physarum*, knowledge of the differential expression of genes during the synchronous mitotic cycle or throughout the life cycle would be of great interest. Such studies should lead to valuable insights into the control of gene expression.

The features of the nucleotide sequence of *Physarum* genomic DNA that have impeded the construction of stable recombinants having a high plating efficiency (Nader et al., 1985) have not become apparent when cDNA copies from poly(A)$^+$mRNA were used as the starting point for the construction of a library. Moreover, the abundance of a particular mRNA species within the mRNA population is generally reflected in the abundance of the corresponding cDNA clone within the library. For this reason, the isolation of cDNA clones of rare mRNA species may require the construction of very large cDNA libraries representative of complex poly(A)$^+$mRNA populations, e.g., libraries of approximately 1 X 10^6 recombinants.

Double-stranded cDNA may be cloned into either a plasmid or a phage, such as phage lambda. A phage-lambda vector has the advantage that it may be introduced into *Escherichia coli* with high efficiency by the in vitro packaging technique (Hohn and Hohn, 1974). Insertion vectors such as phage-lambda NM1149, phage-lambda gt10, and phage-lambda gt11 are capable of accepting up to 7.6 kb of cDNA, which is adequate for a complete cDNA copy of mRNA with a coding region of 2.5 X 10^6 daltons. Phage-lambda gt10 was constructed to provide a vigorously growing phage-lambda vector, which can be packaged efficiently in vitro and from which recombinants can be grown selectively on a host such as *hflA150*.

Phage-lambda gt11 (*lac5* *cI857* *nin5* *S*am100) has a unique EcoRI site located within the *lacZ* gene, 53-bp upstream from the β-galactosidase translation termination codon. DNA sequences inserted into the *lacZ* gene have the potential to be expressed as fusion proteins with β-galactosidase. For this reason, cDNA libraries constructed in phage-lambda gt11 can be screened with antibody probes for antigen produced by specific recombinant clones.

Isolation of Particular cDNA Clones

Schedl and his colleagues at the McArdle Laboratory isolated a clone of cDNA copies from tubulin mRNA and ligated into the plasmid pBR322. The cDNA insert comprises the coding region of the gene for a plasmodial α-tubulin except for 25-30 C-terminal amino acid residues. The nucleotide sequence of this cDNA clone was established by Krämmer et al. (1985). A clone of actin cDNA was isolated, and the nucleotide sequence was determined by Pallotta and his colleagues (personal communication).

A phage λgt11 recombinant DNA expression library was constructed with cDNA synthesized from poly (A$^+$)RNA isolated by guanidinium thiocynate extraction of late log phase cultures of microplasmodia of *Physarum polycephalum*. The resulting library containing 7500 recombinants was amplified and screened for antigen-producing clones by use of rabbit antiserum raised against purified ornithine decarboxylase (Barnett et al., 1984). Among 40,000 recombinant phage screened, 12 were found to produce antigenic material recognized by the specific antiserum. Four recombinant clones were isolated by plaque purification from these 12. Three contained an insert of about 1 kb, and the fourth carried a 1.3 kb insert. ^{32}P-labeled clone hybridized to a single RNA species of about 1.5 kb on northern blots.

Kazarinoff, Barnett, and Ruth (personal communication) purified tyrosine aminotransferase (EC 2.6.1.5) from microplasmodia. The product was a single protein as judged by SDS-polyacrylamide gel electrophoresis and by HPLC, with a molecular weight of 37,000. Rabbit antiserum raised against the purified protein was used to screen the Cornell library described above. Twenty-five positives were found among 40,000 recombinant phage that were screened. One phage was purified and lysogenized into *Escherichia coli* Y1089. The lysogen was induced at 42°C with IPTG and was found to synthesize a new 155,000 dalton polypeptide, which reacted with both anti-aminotransferase and anti-β-galactosidase serum on Western blots. The recombinant carries a cDNA insert corresponding to *Physarum* tyrosine aminotransferase gene sequences.

Differential Gene Expression during the Life Cycle of Physarum

The life cycle of *Physarum* is notable because the organism proliferates in two vegetative phases, viz., the haploid amoebal phase and/or the diploid plasmodial phase. Moreover, starved amoebae form cysts, and starved plasmodia form spherules (sclerotia). On the addition of nutrients to the growth medium, cysts revert to amoebae, and spherules revert to plasmodia. Pallotta and his colleagues have investigated differential gene expression in all four phases, namely the two proliferative forms, amoebae and plasmodia, and the two dormant forms, cysts and spherules. For details of this strategy, see Chapter 24.

Changes in gene expression following the transition from the amoebal phase to the plasmodial phase were also studied by Sweeney, Watts, and Turnock, who made cDNA from G2 phase plasmodia of the apogamic strain, *Physarum* CL. The advantages of using this strain are that abundance of proteins of the amoebal and plasmodial phases were compared previously (Turnock et al., 1981) and that a second strain of amoebae having a different mating type is not required for the transition from the amoebal phase to the plasmodial phase. The plasmodial cDNA was ligated into phage M13. The recombinant phage were screened by a dot-blot procedure, in which phage DNA was immobilized onto nitrocellulose and probed with ^{32}P-labeled cDNA copied from either amoebal poly(A)$^+$mRNA or plasmodial poly(A)$^+$mRNA. Of the 250 clones analyzed, 10 were classified as being plasmodial-specific. DNA species isolated from several clones are currently being sequenced.

Monteiro and Cox used ^{32}P-labeled cDNA probes prepared from poly(A)+mRNA fractions isolated from exponentially growing amoebae or from exponentially growing microplasmodia to screen replica copies of a phage-lambda EMBL4 genomic library. A Sau3A partial digest of plasmodial nuclear DNA was fractionated by zone centrifugation. Fragments ranging in size from 5 kb to 20 kb were recovered and ligated into phage-lambda EMBL4. The library, gMHpl (EMBL), comprised more than 1 X 10^6 recombinants. After phage hybridization was carried out, 50 of the 7000 recombinants screened hybridized more strongly with the amoebal cDNA probe than with the plasmodial probe.

The 50 clones selected on the basis of visual inspection were further examined by use of prehybridized cDNA probes. The probes were made by hybridizing ^{32}P-labeled cDNA with heterologous mRNA to $Rot_{1/2}$ = 10, to diminish the concentration of radioactive cDNA species whose homologous mRNA was equally abundant in both amoebal and plasmodial phases of <u>Physarum</u>. The hybridization procedure was simplified by the use of crude cell lysates to which formamide was added (the final composition of the lysate was 4 M guanidinium chloride, 33% formamide, and 15 mM Tris pH 7.5 (Cox and Smulian, 1983).

Hybridization with prehybridized ^{32}P-labeled cDNA probes was studied quantitatively. DNA was isolated from those phage identified as hybridizing more extensively with amoebal cDNA than with plasmodial cDNA and immobilized on discs of aminothiophenol paper. Replicate discs were hybridized with plasmodial cDNA, or plasmodial cDNA prehybridized with plasmodial mRNA, or plasmodial cDNA prehybridized with plasmodial RNA, or amoebal cDNA hybridized with amoebal mRNA. The results reveal that mRNA complementary to immobilized DNA was approximately fivefold more abundant in the amoebal phase than in the plasmodial phase. Similar results were obtained by the conventional dot-blot procedure, in which unfractionated RNA was immobilized on aminothiophenol paper and then hybridized with radioactive phage-lambda DNA by nick-translation.

Prehybridized probes were also used to screen a cDNA library constructed by inserting amoebal or plasmodial derived cDNA into phage-lambda gt11. The library from amoebal cDNA was designated cMHam; that from plasmodial cDNA, cMHpl. In total, at least 12,000 cDNA recombinants were screened. No recombinant was identified as containing cDNA expressed exclusively in either the amoebal or the plasmodial phases. One recombinant was found to be at least tenfold more abundant in the amoebal than in the plasmodial phase, and two recombinants were found to be more abundant in the plasmodial than in the amoebal phase. The recombinants tentatively identified as being preferentially expressed in one phase or the other were confirmed as containing sequences expressed preferentially in either CLd AXE amoebae or in the isogenic strain CLd AXE plasmodia and so are phase-specific.

DISCUSSION

Further characterization of the genes expressed in the dormant forms of amoebae and plasmodia should be especially rewarding. The transitions from rapidly growing amoebae to cysts and from plasmodia to spherules appear to be readily amenable to detailed investigation by methods based on cDNA libraries, mainly because comparatively few genes appear to be expressed in the dormant forms.

Comparison of the genes expressed in proliferating amoebae with those expressed in proliferating plasmodia is inherently more complex. The three studies concerning differences in gene expression at the level of mRNA resulting from the transition from amoebae to plasmodia aim at identifying genes that are switched on at some stage in the cell cycle of one phase, but which are not expressed, or at least expressed to a much lesser extent, in the other phase. Having identified such a gene, the next step is to identify the protein for which it codes and then to establish the mechanism(s) that control gene expression.

The three studies are difficult to compare directly because of the absence of a common criterion for identifying differentially expressed genes. All three studies rely heavily on poly(A)$^{+}$mRNA or cDNA probes, but the experimental procedures differ significantly in some respects (see Table 23-2). Pallotta and his colleagues used homologous and heterologous ^{32}P-labeled poly(A)$^{+}$mRNA probes in their first screen and subsequently used cloned DNA made radioactive by nick-translation to screen a Northern transfer of amoebal and plasmodial poly(A)$^{+}$mRNA separated by agarose gel electrophoresis. Sweeney and his colleagues immobilized M13 DNA to nitrocellulose film and used as probes ^{32}P-labeled cDNA copies from poly(A)$^{+}$mRNA isolated from each of the two phases.

Monteiro and Cox used as a probe ^{32}P-labeled cDNA prehybridized with heterologous mRNA present in cell lysates, in order to neutralize cDNA species for which a complementary mRNA species was present, irrespective of the extent of polyadenylation of the mRNA species. The control was ^{32}P-labeled cDNA hybridized with homologous mRNA present in cell lysates. As a further check, dot blots of unfractionated RNA isolated from amoebae and from plasmodia were probed with recombinant DNA made radioactive by nick-translation. The proportion of differentially expressed cDNA clones revealed by

Table 23-2.

Summary of Results Obtained in Studies of Differential Gene Expression

Source of poly(A)$^+$mRNA for synthesis of cDNA	Vector	Number of Recombinants Screened	Screening procedure	Conclusions
Plasmodia (CLd AXE) Amoebae (CLd AXE) Spherules (CLd AXE)	pBR322 pBR322 pBR322	1344 800 --	Colony hybridization using ^{32}P-labeled plasmodial poly(A)$^+$mRNA and ^{32}P-labeled amoebal poly(A)$^+$ mRNA. Northern blots of poly(A)$^+$mRNA were probed with nick-translated recombinant DNA	160 recombinants, plasmodia-specific (See A) 20 recombinants, amoeba-specific, (See B) 247 recombinants, spherule-specific (See C)
G2 macroplasmodia (CL)	phage M13 (mp8)	500 (250 with DNA complementary to cDNA probe)	Immobilized M13 DNA hybridized with ^{32}P-labeled cDNA copied from plasmodial poly(A)$^+$mRNA or from amoebal poly(A)$^+$mRNA	10 recombinants, plasmodium-specific
Exponentially growing microplasmodia* (M3CVIII)	phage-lambda gt11	6000	Plaque hybridization with ^{32}P-labeled cDNA pre-hybridized with heterologous mRNA to $Rot_{1/2}$ 10	1 recombinant, amoeba-specific 2 recombinants, plasmodium-specific

Table 23-2. (Continued)

Source of poly(A)+mRNA for synthesis of cDNA	Vector	Number of Recombinants Screened	Screening procedure	Conclusions
Exponentially growing amoebae (CLd AXE)[+]	phage-lambda gt11	6000	(See above)	(See above)
Sau3A fragments 5 kb-20 kb of plasmodial (M3CVIII)* nuclear DNA	phage-lambda EMBL4	7000	Phage hybridization with ^{32}P-labeled cDNA copied from plasmodial poly(A)$^{+}$mRNA, or ^{32}P-labeled cDNA copied from amoebal cDNA, or dot blot of recombinant DNA probed with cDNA prehybridized with mRNA	25 clones, amoeba-specific

*Clones identified as being differentially expressed were also shown to differentially express in CLD AXE plasmodia.

A - 160 recombinants were identified as plasmodium-specific by differential hybridization; six different cDNA clones were confirmed as plasmodia-specific, four of which were found to code for abundant mRNA species.

B - 20 recombinants were found to be amoeba-specific by differential hybridization; seven different clones were confirmed as amoeba-specific, and one of these was found to code for an abundant mRNA species.

C - For spherules, 247 clones were identified as stage-specific by differential hybridization; eight different clones were confirmed as spherule-specific; of these, four were found to code for abundant mRNA and four for moderately abundant mRNA species.

these cloning and screening procedures was much smaller than the proportion obtained by the other two groups of investigators.

It is likely that the composition of the poly(A)+mRNA fraction will reflect the composition of the total mRNA fraction. However, this may not always apply to all mRNA species. If, for example, histone mRNA is polyadenylated in G2 phase but not in the later stages of S-phase, the profile of the abundance of histone-poly(A)+mRNA at various stages throughout the mitotic cycle would not be congruent with the profile of the abundance total histone-mRNA at various stages throughout the mitotic cycle. Uncertainties in the state of polyadenylation of mRNA are unlikely to be important when they are minimized by isolating the poly(A)+mRNA from exponentially growing cells.

CONCLUSIONS

The early attempts to clone Physarum genes gained encouragement from the successful cloning of fragments of rDNA. However, the earlier attempts to produce a library of genomic-containing inserts of 20 kb or more were not encouraging. The low yield of recombinants and their instability is now attributed to elements of repeated sequence and inverted repeat sequences in Physarum DNA. The development of suitable strains such as E. coli CES200 has allowed stable clone banks to be constructed. Cloning of histone H4, actin, and tubulin genes has been achieved. A histone H4 gene has been sequenced and the sequence of an actin gene of the ard A locus is well advanced.

Each of the cDNA studies has yielded clones that are of potential use as probes which are at least quantitatively specific for either the amoebal or plasmodial phases of the life cycle. Interest in these clones will increase as they are characterized further. The number of clones screened is probably too small to permit a realistic estimate of genes that are expressed preferentially in one phase or the other. This estimate will of course depend on the quantitative criterion chosen for phase specificity.

These lines of rapid progress will offer numerous molecular markers for particular phases of the interesting life cycle of Physarum.

REFERENCES

Barnett, G. R., and Kazarinoff, M. N., 1974, Purification and properties of ornithine decarboxylase from Physarum polycephalum, J. Biol. Chem., 259:179.

Collins, J., and Hohn, B., 1978, Cosmids: a type of plasmid gene-cloning vector that is packageable *in vitro* in bacteriophage lambda heads. *Proc. Natl. Acad. Sci., U.S.A.*, 75:4242.

Cox, R. A., and Smulian, N. J., 1983, A single-step procedure for the isolation of individual mRNA species from crude lysates of *Physarum polycephalum*, *FEBS Lett.*, 155:73.

Glover, D. M., 1985, "DNA Cloning", Vol. 1, IRL Press, Oxford and Washington.

Hardman, N., Jack, P. L., Fergie, R. C., and Gerrie, L. M., 1980, Sequence organization in nuclear DNA from *Physarum polycephalum*, *Eur. J. Biochem.*, 103:247.

Hohn, B., and Hohn, T., 1974, Activity of empty, headlike particles for packaging of DNA of bacteriophage lambda in vitro, *Proc. Natl. Acad. Sci., U.S.A.*, 71:2372.

Jalouzot, R., Toublan, M. L., Wilhelm, M. L., and Wilhelm, F. X., 1985, Replication timing of the histone H4 genes in *Physarum polycephalum*, *Proc. Natl. Acad. Sci., U.S.A.*, 82:6475.

Kaiser, K., and Murray, N. M., 1985, The use of phage lambda replacement vectors in the construction of representative genomic DNA libraries, *in*: DNA Cloning, Vol. 1, p. 1, by D. M. Glover, ed., IRL Press, Oxford and Washington.

Krämmer, G., Singhofer-Wowra, M., Seedorf, K., Little, M., and Schedl, T., 1985, A plasmodial alpha-tubulin cDNA from *Physarum polycephalum*, *J. Mol. Biol.*, in press.

Nader, W. F., Edlind, T. D., Huettermann, A., and Sauer, H. W., 1985, Cloning of *Physarum* actin sequences in an exonuclease-deficient bacterial host, *Proc. Natl. Acad. Sci., U.S.A.*, 82:2698.

Schedl, T., and Dove, W. F., 1982, Mendelian analysis of the organization of actin sequences in *Physarum polycephalum*, *J. Mol. Biol.*, 160:41.

Schedl, T., Owens, J., Dove, W. F., and Burland, T. G., 1984, Genetics of the tubulin gene families of *Physarum*, *Genetics*, 108:143.

Turnock, G., Morris, S. R., and Dee, J., 1981, A comparison of the proteins of the amoebal and plasmodial phases of the slime mould, *Physarum polycephalum*, *Eur. J. Biochem.*, 115:533.

Wilhelm, M. L., and Wilhelm, F. X., 1984, A transposon-like DNA fragment interrupts a *Physarum polycephalum* histone H4 gene, *FEBS Lett.*, 168:249.

Chapter 24: cDNA CLONING OF *PHYSARUM POLYCEPHALUM* STAGE-SPECIFIC mRNAs

Dominick Pallotta[1], François Bernier[2], Michel Hamelin[2], Rémi Martel[2], and Gérald Lemieux[2]

Departments of Biology[1] and Biochemistry[2]
Laval University
Québec, P.Q., CANADA

INTRODUCTION

A useful approach to the study of differentiation consists of isolating and characterizing the genes specifically expressed in the various stages and then studying their regulation. This chapter describes the methodology used in our laboratory to clone sequences from stage-specific mRNAs of *Physarum polycephalum*. It involved the preparation of intact poly $(A)^+$ RNA and the construction of cDNA libraries for four developmental stages (amoebae, plasmodia, spores, and spherules). These libraries were then screened by differential hybridization, and the specificity of the positive clones was confirmed by Northern blot hybridization. The quality of the results obtained at each step was controlled, and the methodology was found to be of general use in the identification of *Physarum* stage-specific mRNAs.

It is our opinion that this approach will facilitate the study of the mechanisms of cellular differentiation at the molecular level in *Physarum* and that this knowledge will contribute to a better understanding of this important and fundamental biological phenomenon.

CELL CULTURE

Amoebae of CLd-Axe, a strain capable of growing axenically in liquid medium, was used (McCullough et al., 1978). For the formation of plasmodia, CLd-Axe amoebae were allowed to differentiate asexually (Cooke and Dee, 1975). Amoebae and plasmodia were grown in 500-ml baffled flasks containing 200 ml of semi-defined medium supplemented with hematin (Daniel and Baldwin, 1964). The cultures were shaken constantly in the dark at 26°C. Only exponentially growing amoebae and plasmodia were used. Spherulating cells were prepared by transferring M3C microplasmodia to a non-nutrient salt medium (Daniel and Baldwin, 1964). Poly A^+ RNA was prepared after 12, 24, 36, 48, and 72 hr of spherulation. Our cDNA library was constructed from RNA isolated from cells spherulating for 24 hr, a period at which 10% of the cells were mature spherules, as judged from their viability (Bernier et al., 1985).

For sporulation studies, a diploid plasmodium was formed by crossing the amoebal strains LU648 X CH786 (Cooke and Dee, 1975; Youngman et al., 1979). Sporulation of macroplasmodia was induced by a starvation period of 4 days, followed by a 4-hr light pulse. Poly $(A)^+$ RNA was prepared at various times after illumination (0,4, 8, 12, and 16 hr). The mRNA prepared from cells collected 12 hr after the end of the light pulse was used to construct the library. At this time, the protoplasm accumulates in small nodules which, within 1 hr, became transformed into tiny pillars. The formation of the sporangia themselves starts about 1 hr later. The nodule formation, therefore, is the first macroscopic morphological change that takes place in sporulating plasmodia (Guttes et al., 1961).

Since the various Physarum strains do not differentiate equally well, it is important to check that the selected strain sporulates or spherulates efficiently.

POLY(A)$^+$ RNA PREPARATION

Physarum polycephalum contains very active RNases that can rapidly degrade the mRNA during the extraction procedure. Melera and Rusch (1973) devised a method for obtaining intact mRNA by rapid freezing of the intact cells followed by lyophilization. The dehydrated cells were then resuspended in a buffer containing EDTA, SDS, and Bentonite and were extracted with organic solvents. We have used this technique with two modifications. Bentonite was omitted, since it has been reported that this compound adsorbs some RNAs (Zellweger and Braun, 1971); instead, the cells were resuspended directly in a mixture of phenol and denaturing buffer. Under these conditions, the rehydration is accompanied by the simultaneous inactivation of enzymes, including RNases. The second modification was the addition of a LiCl precipitation before the chromatography on oligo-dT cellulose. This step eliminates much of the DNA, polysaccharides, and other impurities.

All material used in this procedure was RNase-free. Cells were collected, frozen immediately in liquid nitrogen, and lyophilized. Ten ml of dehydrated cells were suspended in 50 ml of homogenization medium which was a mixture of 35 ml of NES (100 mM sodium acetate, 10 mM EDTA, 1% SDS, pH 6.0) and 15 ml of freshly distilled phenol equilibrated with NES. The cells were homogenized with a Potter-Elvehjem tissue homogenizer and stirred for 15 min at room temperature. Fifteen ml of chloroform-isoamyl alcohol (24:1 v/v) was added, and the suspension was again stirred for 15 min. The two phases were separated by centrifugation and the aqueous phase was collected and reextracted with 30 ml of chloroform-isoamyl alcohol. The aqueous phase was collected and saved. The two organic phases were sequentially extracted with 35 ml of NES. The resulting aqueous phase was combined with the first one, and the high molecular weight RNA was precipitated overnight at 4°C by the addition of a 0.25 volume of 10 M LiCl. The RNA was collected by centrifugation, washed in 2 M LiCl, and recentrifuged. The pellet was dissolved in 8 ml NES, and the RNA was precipitated overnight at -20°C with 2.5 volumes of ethanol.

The phenol extraction method consistently gave intact RNA in good yields (about 3.5 mg of total RNA/ml of cells). It was successfully used with growing amoebae, plasmodia, encysted amoebae, young and mature spherules, and sporulating plasmodia. Other laboratories (Burland et al., 1983; Cox and Smulian, 1983; Putzer et al., 1983) have extracted RNA from _Physarum_ with guanidinium isothiocyanate (Chirgwin et al., 1979). We, however, have not used this method frequently, since our technique gave entirely satisfactory results.

The RNA was collected and dissolved in 25 ml of TS (10 mM Tris.HCl, 0.1% SDS, pH 7.4). The solution was heated for 10 min at 68°C, cooled rapidly, and adjusted to 0.5 M LiCl. The sample was then passed through an oligo-dT cellulose column. The column was washed with TS plus 0.5 M LiCl to remove the poly A^- fraction. The poly A^+ RNA was then eluted with TS and further purified by a second chromatography on an oligo dT column. The final poly A^+ fraction was adjusted to 0.3 M sodium acetate (pH 5.2), precipitated with ethanol, and stored at -70°C.

AGAROSE GEL ELECTROPHORESIS OF POLY $(A)^+$ RNA

The quality of the poly $(A)^+$ mRNA was first verified by electrophoresis on denaturing agarose gels (see below). After being stained with acridine orange, residual 26S and 19S rRNA bands were always seen. In addition, a smear of poly $(A)^+$ RNA was also present (Fig. 24-1). To further confirm the quality of our RNAs, we probed them by Northern blot hybridization. In all cases, sharp bands were seen. Thus, the electrophoretic analysis is, in itself, a good indication of the quality of the RNA.

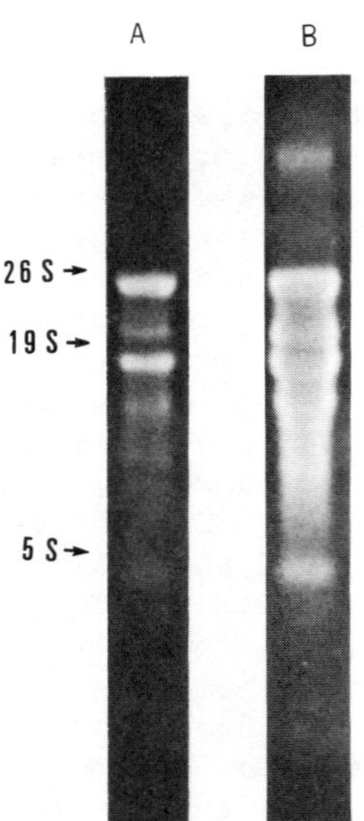

Fig. 24-1. Agarose gel electrophoresis of plasmodial RNA. (A) 2 μg of total RNA; (B) 5 μg of poly $(A)^+$ RNA. The gel was stained with acridine orange.

NORTHERN BLOTS

One to 5 μg of poly $(A)^+$ RNA was denatured with 2.2 M formaldehyde and fractionated on 1.5% agarose gels containing 1.1 M formaldehyde and 10 mM phosphate buffer, pH 7.4 (Maniatis et al., 1982a). After electrophoresis, the RNA was transferred to nitrocellulose filters. In some instances, glyoxylated RNA was employed (Thomas, 1983).

All preparations of poly $(A)^+$ RNA were tested in a Northern blot experiment before cDNA synthesis. In these experiments, we routinely used a Physarum actin probe labeled by nick translation (Rigby et al., 1977). Initially, the Drosophila actin clone DMA2 was chosen (Fyrberg et al., 1980), but more recently we changed to the Physarum actin cDNA clone PA35 (manuscript in preparation). The results of a typical experiment are seen in Fig. 24-2. The amoebal, plasmodial, and spherule poly A^+ RNA preparations gave a single sharp band that hybridized with the Physarum actin probe. Spore RNA could not be tested with this probe because actin mRNA is absent at this developmental stage. In this case, since the electrophoretic profile of spore poly $(A)^+$ RNA was similar to those obtained with other RNAs, we assumed that the spore RNA was undegraded. This was later confirmed by Northern analysis with spore-specific cDNA clones.

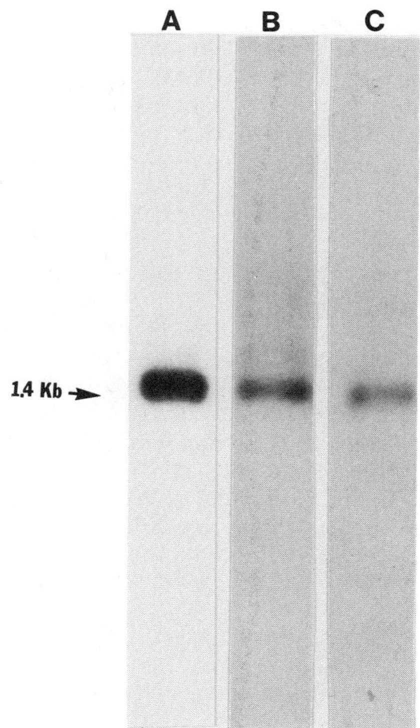

Fig. 24-2. Labeled Physarum actin cDNA hybridized to Northern blots of poly $(A)^+$ RNA. (A) amoebae; (B) plasmodia; (C) 24-hr spherules.

cDNA SYNTHESIS AND CLONING

Ten µg of poly $(A)^+$ RNA was used to synthesize the first strand of cDNA in a 50-µl reaction mix containing 28 mM β-mercaptoethanol, 25 units RNasin, 200 µg/ml oligo $(dT)_{12-18}$, 100 mM Tris.HCl (pH 8.3), 150 mM KCl, 10 mM $MgCl_2$, 1 mM of each dNTP, 20 µCi of [α-^{32}P] dCTP (800 Ci/mmole), and 200 units of AMV reverse transcriptase. The reaction was carried out at 43°C for 2 hr and stopped by the addition of EDTA to a final concentration of 20 mM (Maniatis et al., 1982b). The nucleic acids were then extracted with phenol-chloroform, precipitated twice with ethanol in the presence of 2 M ammonium acetate, and washed with 70% ethanol as described by Okayama and Berg (1982).

The second strand was synthesized with 1 µg of mRNA·cDNA hybrid in a volume of 100 µl containing 20 mM Tris.HCl (pH 7.5), 5 mM $MgCl_2$, 10 mM $(NH_4)_2SO_4$, 100 mM KCl, 50 µg/ml BSA, 40 µM of each dNTP, 0.5 units of RNase H, and 23 units of DNA polymerase 1 (Gubler and Hoffman, 1983). The reaction was carried out at 12°C for 1 hr, at 22°C for another hr, and then stopped by the addition of EDTA to a final concentration of 20 mM. The DNA solution was then extracted with phenol-chloroform and ethanol-precipitated, as described for the synthesis of the first strand. The lengths of the first and second strand cDNAs were determined by agarose gel electrophoresis. A typical result is shown in Fig. 24-3. The size of the duplexes ranged in length from 250 to 3000 bp.

Oligo (dC) tailing of the double-stranded cDNA was performed in 140 mM potassium cacodylate pH 7.2, 2 mM $CoCl_2$, 0.1 mM DTT, 500 μg/ml BSA, 5 μM dCTP, 0.4 μCi/μl [α-^{32}P]dCTP (800 Ci/mmole), and 640 units/ml terminal transferase, at a concentration of 3' ends of DNA fixed approximately at 30 nM (Michelson and Orkin, 1982). The reaction was incubated (about 20 min) at room temperature to give an average of 20 deoxycytidines per end. EDTA was then added to a final concentration of 20 mM. This cDNA was annealed to the oligo (dG)-tailed PstI site of pBR322 for 2 hr at 57°C, followed by 1 hr at room temperature in 10 mM Tris.HCl, 1 mM EDTA, 150 mM NaCl, pH 7.5. The total DNA concentration was 1 μg/ml, and the molar ratio of cDNA to vector was about 2:1.

Recombinant DNA clones were obtained by transformation of $CaCl_2$-treated *E. coli* MC 1061 cells (Mandel and Higa, 1970). Typically, about 5 X 10^4 clones were obtained per μg of double-stranded

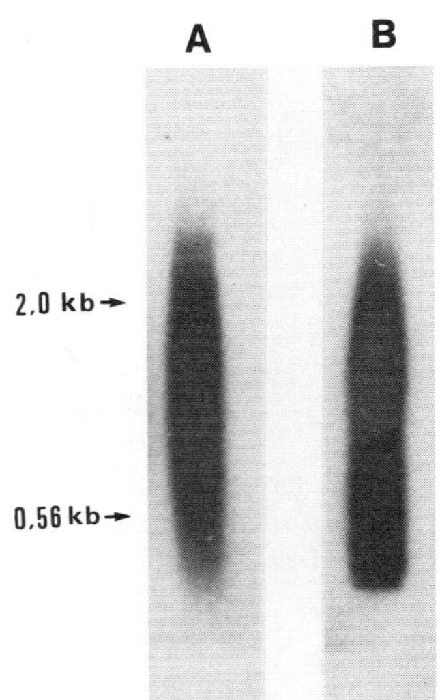

Fig. 24-3. Agarose gel electrophoresis of spore cDNA. (A) ^{32}P-labeled first strand cDNA run on an alkaline agarose gel; (B) ^{32}P-labeled second strand cDNA run on a non-denaturating agarose gel.

cDNA. All cDNA clones used carried a library identification code plus a clone number. For example, LAV1-1 is the first plasmodial-specific cDNA clone isolated from the plasmodium library LAV1 (Table 24-1).

LENGTH OF cDNA INSERTS

More than 99% of the transformed bacterial colonies were tetracycline-resistant and ampicillin-sensitive; this indicates that they contain pBR322 with a cDNA insert. To determine the length of these inserts, we randomly selected 20 clones from each of the amoebal, plasmodial, spherule, and spore cDNA libraries. Mini-preparations of plasmid DNA were made according to the boiling method of Holmes and Quigley (1981), except that the DNA was recovered by spermine precipitation (Hoopes and McClure, 1981). In some cases, the alkaline lysis method was used (Birnboim and Doly, 1979). For large-scale preparation, the plasmid DNA was isolated by alkaline lysis and further purified by CsCl density gradient centrifugation. The cDNA inserts were excised with PstI endonuclease and sized on agarose gels. They varied in length between 200 and 2200 bp (results not shown). This indicated that most of the clones in the libraries contained a cDNA insert of reasonable length that could be used in further experiments.

SCREENING OF cDNA LIBRARIES

The cDNA libraries were screened by differential hybridization. Clones from a given library were transferred in duplicate onto

Table 24-1. Nomenclature of Libraries and cDNA Clones

Name of library	Origin of poly $(A)^+$ RNA	Representative Stage-specific cDNA Clone*
LAV1	Plasmodium	LAV1-1
LAV2	Spherule	LAV2-1
LAV3	Amoeba	LAV3-3
LAV5	Spore	LAV5-4

* See Fig. 24-5 for the Northern blot analysis of the RNAs corresponding to these clones.

nitrocellulose filters. The filters were placed on LB agar plates containing 12.5 μg/ml tetracycline and grown overnight. The following day, the cells were lysed in 0.5 N NaOH, and the filters were prepared for hybridization by the usual methods (Hanahan and Meselson, 1983). The RNA used as the hybridization probe was labeled according to the method of Maizels (1976). Briefly, 3 μg of poly A^+ RNA was partially hydrolyzed in 15 μl of 50 mM Tris.HCl, pH 9.5, at 90°C for 20 min. The solution was placed on ice, and the following were added: 3 μl of 10X MDGT (100 mM $MgCl_2$, 50 mM DTT, 50% v/v glycerol, and 250 mM Tris.HCl, pH 9.5), 10.5 μl of γ-^{32}P ATP, 15 units of polynucleotide kinase, and water to give a final volume of 30 μl. The mixture was incubated at 37°C for 45 min, and the unincorporated nucleotides were removed by chromatography on G-50 Sephadex or by three precipitations with ethanol in the presence of 2 M ammonium acetate. The labeled RNA was heated to 65°C for 10 min, cooled on ice, and added to the hybridization solution (50% formamide, 5X SSPE, 2X Denhardt, 0.5% SDS, 20 μg/ml poly A, and 100-200 μg/ml denatured herring sperm DNA). For the amoebal and plasmodial libraries, one set of nitrocellulose filters was hybridized with labeled poly $(A)^+$ RNA from amoebae, and the other with labeled poly $(A)^+$ RNA from plasmodia. For the spherule library, one set of filters was hybridized with labeled plasmodial poly $(A)^+$ RNA, and the other one with spherule poly $(A)^+$ RNA. For the spore library, the RNAs from spores and plasmodia collected immediately after the light pulse (0 hr) were used as probes.

The filters were prewashed at 42°C in 50 mM Tris.HCl (pH 8.0), 1 M NaCl, 1 mM EDTA, 0.1% SDS for 2 hr and prehybridized for 3-4 hr at 42°C in the hybridization solution. All hybridizations were carried out for 1 to 3 days at 42°C. For RNA-DNA hybridizations, filters were washed twice for 10 min, each time in 2X SSC at room temperature, followed by two 30-min washes in 2X SSC at 50°C and by two 30-min washes in 0.1X SSC at 50°C. For DNA-DNA hybridizations, there were two 10-min washes in 2X SSC at room temperature, two 60-min washes in 0.2X SSC at 68°C, and a 30-min wash in 0.1X SSC at 68°C.

The clones that gave a positive signal with the poly $(A)^+$ RNA used to make the cDNA library, but no signal with the RNA from a different stage, were preliminarily classified as potentially stage-specific (Fig. 24-4). The definitive confirmation that these clones were stage-specific was obtained by Northern blot hybridization. The inserts were purified, labeled by nick translation, and hybridized to Northern blots of poly $(A)^+$ RNAs from various stages. For verification that the lack of a hybridization signal was not owing to the absence of RNA on the filter, an actin cDNA probe was used as a positive control (Fig. 24-2). Actin mRNA was observed in amoebae, plasmodia, and immature spherules.

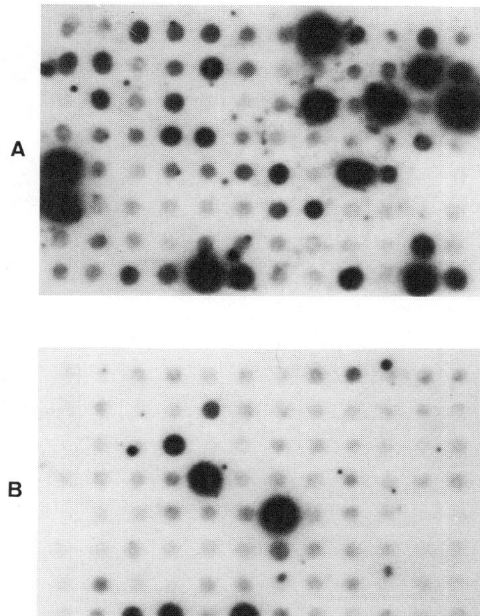

Fig. 24-4. Differential hybridization of 96 clones from the spherule cDNA library. The same clones were hybridized with ^{32}P-labeled poly (A)$^{+}$ RNA from (A) spherules and (B) plasmodia.

In most cases, the putative clones selected as stage-specific by differential hybridization were shown to be specific by Northern blot hybridizations. A few selected examples of these results are shown in Fig. 24-5. A single sharp band was seen with the homologous RNA, whereas no signal above background was detected with the RNA from another developmental stage. Even when the autoradiograms were greatly overexposed, the same results were obtained. This is clear evidence of the presence of stage-specific mRNAs.

It should be noted that for most of the mRNAs, the specificity was determined relative to one other stage. We are now determining whether this specificity can be generalized to all developmental stages.

FULL-LENGTH cDNA CLONE

Northern blot analysis showed that in some cases the cDNA inserts were shorter than their corresponding mRNAs. To determine

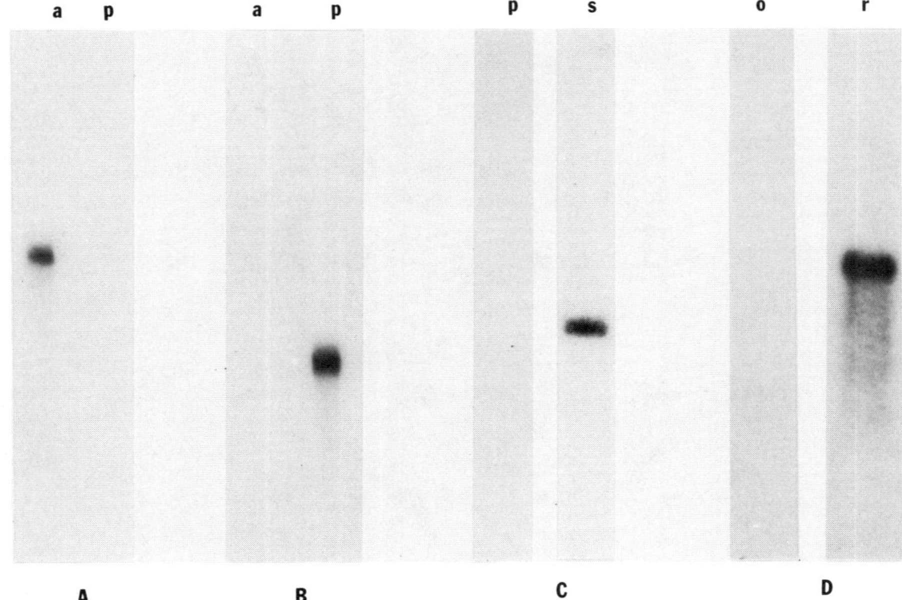

Fig. 24-5. Northern blots of poly $(A)^+$ RNA (a, amoebae; p, plasmodia; s, spherules; o, illuminated plasmodia; r, spores) hybridized with stage-specific cDNAs. (A) amoebal (LAV3-3); (B) plasmodial (LAV1-1); (C) spherules (LAV2-1); (D) spores (LAV5-4). The lengths of the mRNAs detected with the probes were: A, 1200; B, 860; C, 950; D, 1150 nucleotides.

whether full-length inserts could be found, we carried out a detailed experiment with actin. The lengths of the inserts in 19 amoebal and 89 plasmodial actin cDNA clones were measured (Fig. 24-6). Physarum actin mRNA contains about 1400 nucleotides, as determined by Northern blot hybridization. Seven percent of the plasmodial and 26% of the amoebal actin clones had cDNA inserts of about 1400 nucleotides; this indicates that full-length cDNAs are relatively abundant in our libraries.

CONCLUSION

Standard molecular biological techniques can be used to isolate stage-specific cDNA clones for each of the developmental forms of Physarum polycephalum. We have used the techniques presented in this chapter to isolate and characterize seven amoebal, seven plasmodial, nine spherule, and five spore-specific cDNA clones. Each clone corresponds to a different mRNA, and many

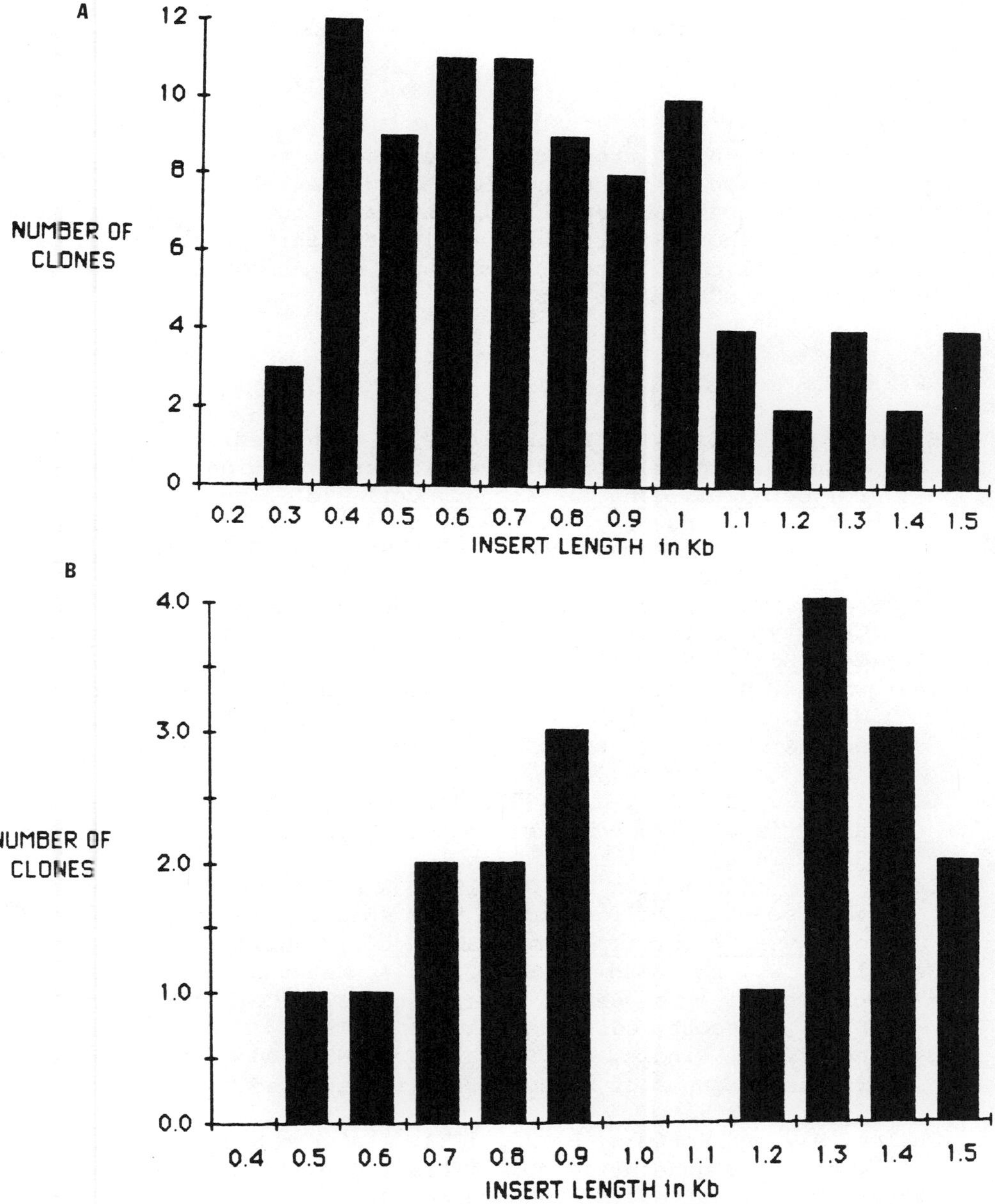

Fig. 24-6. Length of *Physarum* actin cDNA inserts in different clones isolated from (A) plasmodial and (B) amoebal libraries.

appear to contain a full-length cDNA insert. These clones will be useful in studying the molecular mechanisms of cell differentiation in Physarum.

ACKNOWLEDGMENTS

We thank André Laroche and Anne Tessier for the isolation and characterization of the clones LAV3-3 and LAV1-1 respectively. Luc Adam and Louise Savard kindly furnished the data on actin cDNA lengths. François Binette helped with the initial screening of the plasmodial cDNA library. This work was supported by the NSERC of Canada and FCAR of Québec.

REFERENCES

Bernier, F., Seligy, V. L., Pallotta, D., and Lemieux, G., 1985, Changes in gene expression during spherulation in Physarum polycephalum, Can. J. Biochem. Cell Biol., in press.
Birnboim H. C., and Doly, J., 1979, A rapid alkaline extraction procedure for screening recombinant plasmid DNA, Nucl. Acids Res., 7:1513.
Burland, T. G., Gull, K., Boston, A. S., and Dove, W. F., 1983, Cell type-dependent expression of tubulins in Physarum, J. Cell Biol., 97:1852.
Chirgwin, J. M., Przybyla, A. E., MacDonald, R. J., and Rutter, W. J., 1979, Isolation of biologically active ribonucleic acid from sources enriched in ribonuclease, Biochemistry, 18: 5294.
Cooke, D. J., and Dee, J., 1975, Methods for the isolation and analysis of plasmodial mutants in Physarum polycephalum, Genet. Res., 24:175.
Cox, R. A., and Smulian, N. J., 1983, A single step procedure for the isolation of individual mRNA species from crude lysates of Physarum polycephalum, FEBS Lett., 155:73.
Daniel, J. W., and Baldwin, H. H., 1964, Methods of culture for plasmodial myxomycetes, in: "Methods in Cell Physiology", D. M. Prescott, ed., p. 9, Academic Press, New York.
Fyrberg, E. A., Kindle, K. L., Davidson, N., and Sodja, A., (1980), The actin genes of Drosophila: a dispersed multigene family, Cell, 19:365.
Gubler, U., and Hoffman, B. J., 1983, A simple and very efficient method for generating cDNA libraries, Gene, 25:263.
Guttes, E. S., Guttes, S., and Rusch, H. P., 1961, Morphological observations on growth and differentiation of Physarum polycephalum in pure culture, Dev. Biol., 3:588.
Hanahan, D., and Meselson, M., 1983, Plasmid screening at high colony density, Methods Enzymol., 100:333.

Holmes, D. S., and Quigley, M., 1981, A rapid boiling method for the preparation of bacterial plasmids, Anal. Biochem., 114:193.

Hoopes, B. C., and McClure, W. R., 1981, Studies on the selectivity of DNA precipitation by spermine, Nucl. Acids Res., 9:5493.

Maizels, N., 1976, Dictyostelium 17S, 25S and 5S rDNAs lie within a 38,000 base pair repeated unit, Cell, 9:431.

Mandel, H., and Higa, A., 1970, Calcium dependent bacteriophage DNA infection, J. Mol. Biol., 53:154.

Maniatis, T., Fritsch, E. F., and Sambrook, J., 1982a, Electrophoresis of RNA through gels containing formaldehyde, in: "Molecular Cloning, A Laboratory Manual", p. 202, Cold Spring Harbor Laboratory, Cold Spring Harbor.

Maniatis, T., Fritsch, E. F., and Sambrook J., 1982b, Synthesis of double stranded DNA, in: "Molecular Cloning, A Laboratory Manual", p. 230, Cold Spring Harbor Laboratory, Cold Spring Harbor.

McCullough, C. H. R., Dee, J., and Foxon, J. L., 1978, Genetic factors determining the growth of Physarum polycephalum amoebae in axenic medium, J. Gen. Microbiol., 106:297.

Melera, P. W., and Rusch, H. P., 1973, A characterization of ribonucleic acid in the myxomycete Physarum polycephalum, Exp. Cell Res., 82:197.

Michelson, A. M., and Orkin, S. H., 1982, Characterization of the homopolymer tailing reaction catalyzed by terminal deoxynucleotidyl transferase. Implication for the cloning of cDNA, J. Biol. Chem., 257:14773.

Okayama, H., and Berg, P., 1982, High efficiency cloning of full length cDNA, Mol. Cell. Biol., 2:161.

Putzer, H., Werenskiold, K., Verfuerth, C., and Schreckenbach, T., 1983, Blue light inhibits slime mold differentiation at the mRNA level, EMBO J., 2:261.

Rigby, P. W. J., Dieckmann, M., Rhodes, C., and Berg, P., 1977, Labeling deoxyribonucleic acid to high specific activity in vitro by nick translation with DNA polymerase, J. Mol. Biol., 113:237.

Thomas, P. S., 1983, Hybridization of denatured RNA transferred or dotted to nitrocellulose paper, Methods Enzymol., 100:255.

Youngman, P. J., Pallotta, D., Hosler, B., Struhl, G., and Holt, C. E., 1979, A new mating compatibility locus in Physarum polycephalum, Genetics, 91:683.

Zellweger, A., and Braun, R., 1971, RNA of Physarum. I. Preparation and properties, Exp. Cell Res., 65:413.

Chapter 25: TOWARD A DNA TRANSFORMATION SYSTEM FOR *PHYSARUM POLYCEPHALUM*

Finn Haugli and Terje Johansen

Institute of Medical Biology
University of Tromso
Tromso, NORWAY

DNA TRANSFORMATION IS IMPORTANT

Physarum polycephalum is an organism with unusually flexible possibilities for genetic analysis of a variety of functions. One of the foundations for making such a claim is that through the life cycle the following nuclear conditions can be achieved; haploid uninucleate; diploid homo- or heterozygous uninucleate; haploid multinucleate; diploid homo- or heterozygous multinucleate; haploid-haploid heterokaryons; diploid-diploid (homo- or heterozygous) heterokaryons; diploid-diploid heterokaryons; and heterochronous heterokaryons of all the ploidy types listed. In addition, the organism has natural cell cycle synchrony, interesting differentiation processes, and motility systems.

As *Physarum polycephalum* has become a tool for molecular geneticists, it is clear that genetic manipulations at the molecular level are essential if the organism is to become an experimental system of central importance for the future. Thus a DNA transformation system must be developed. This is the background for our attempts to establish such a method. The rather preliminary data, to be viewed as a first step toward this goal, are reported here.

BASIC TECHNICAL REQUIREMENTS

Experiments toward achieving DNA transformation must establish the following bases: cell treatment methods to make cells competent for DNA uptake; availability of useful, preferably dominant, selectable markers; and vectors containing gene elements allowing establishment of foreign DNA as well as expression of the gene for the selectable marker, in the cells. The following briefly describes our present status with regard to these three basic requirements.

CELL COMPETENCY

Traditional ways of achieving DNA uptake in bacteria as well as in eukaryotic cells have included $CaCl_2$ treatment, polyethylene glycol (PEG) treatment, protoplast fusion, and induction of phagocytosis of DNA by DNA-calcium phosphate coprecipitation. Liposome-encapsulated DNA has also been introduced by fusion with cells. We have concentrated our attempts on $CaCl_2$ and PEG treatments and have had success (as defined by obtaining resistant clones) with both. With the $CaCl_2$ treatment procedure, 5 X 10^7 axenically growing myxamoebae of strain CLd AXE (Dee, 1982; see Chapter 18) are harvested, resuspended in 5 ml 0.05 M $CaCl_2$, and left on ice for 20 min. Cells are centrifuged, resuspended in 300 µl 0.1 M $CaCl_2$ containing 20-40 µg vector DNA, and incubated on ice for 60 min. Finally, cells are diluted with 9 ml SDM medium (see Chapter 18) and incubated in shake flasks with 5 µg per ml of G418 (containing 45% biologically active compound) for 24 hr. Another 5 µg G418/ml is added, and incubation is continued another 24 hr before plating on selective plates.

For the polyethylene glycol treatment, a similar number of cells are harvested and resuspended in 0.5 ml 1.25 M KCl, 100 mM N-tris [hydroxymethyl]methyl-2-aminoethane sulfonic acid (TES) (Sigma catalog no. T.1375), pH 7.5. 30 µg vector DNA is added, and the cells are left at room temperature for 10 min. Next, 2 ml 35% PEG-1000 in 100 mM TES, pH 7.5, is added, mixed, and left at room temperature for 2 min. Cells are centrifuged and resuspended in 10 ml SDM, shaken for 24 hr with 5 µg/ml G418, and for another 24 hr with 10 µg/ml G418, before plating on selective plates.

SELECTABLE MARKERS AND VECTORS

Reports on the use of kanamycin-resistance genes of bacterial transposons Tn5 and Tn601 to transform mammalian cells and yeast to resistance to the aminoglycoside G418 (geneticin) made this marker appear attractive (Jimenez and Davies, 1980; Colbere-Garapin et al., 1981).

The following vectors containing these aminoglycoside phosphotransferase genes are among those that have given rise to resistant clones either directly or by genetic modifications: pSV2 neo (Southern and Berg, 1982); pAG60 (Colbere-Garapin et al., 1981); and TCM12 (Cabezon et al., 1984).

Among the additional gene elements that have been introduced into these vectors are a putative origin of replication for the ribosomal DNA molecule from Physarum (Ferris and Vogt, 1982), a Physarum sequence that can function as an autonomously replicating sequence (ARS) in yeast (Gorman et al., 1981), as well as some random Physarum DNA sequences.

Since there is presently no reason to believe that these vectors, modified to contain the mentioned Physarum DNA, give transformation frequencies greater than those of the parent vector, these constructions will not be detailed.

PHYSARUM MYXAMOEBAE ARE SENSITIVE TO G418

In order to use G418 resistance as a selectable marker, we had to establish whether Physarum myxamoebae were sensitive to this drug at moderate levels. We tested the axenic strain CLd AXE in liquid culture in SDM and found cells to be markedy sensitive at 5-10 μg/ml. The same cells plated on agar plates with bacteria needed 20 μg/ml to barely prevent growth or to permit just a few divisions, and still allowed resistant clones to show up.

The non-axenic strain CLd, on the other hand, required 40 μg/ml of the commercially available G418 (GIBCO) to stop growth. This means that G418 is suitable as a selective agent. Note that the batches of G418 used had about 45% active compound; the amount given must be reduced or increased according to the amount of active compound in any given lot of G418 to achieve the same level of inhibition.

Another crucial question with regard to the usefulness of G418 as a selective agent is whether resistant clones arise spontaneously at a level that would interfere with the interpretation of resistant clones as transformants. We addressed this question by first plating 10^9 $CaCl_2$-treated cells on selective plates, obtaining no resistant clones. During the course of these experiments 2 X 10^9 cells have been plated in experiments yielding no resistant colonies. In comparison, as related below, in experiments that were successful in yielding resistant clones after treatment with DNA, frequencies were about 10^{-7}.

On the basis of this observation, we think it is unlikely that any of the resistant clones obtained were spontaneous.

SOME SUCCESSFUL EXPERIMENTS

Cells exposed to plasmid DNA harboring the G418 resistance gene gave resistant clones in only about 30% of the experiments.

This amounts to 100 putative transformants among 8 X 10^8 cells treated, or an average frequency in the successful experiments of about 10^{-7}.

This is a fairly low frequency, comparable to what is seen in yeast when successful transformation is dependent on general recombination into the genome. Thus, one might speculate that this is also the source of the putative transformants obtained here in *Physarum*.

SOME PROPERTIES OF RESISTANT CLONES

Among the questions that arise regarding the resistant clones are:

1. How resistant are the putative transformants?
2. Will the resistance be stably maintained in the absence of selective pressure?
3. Can foreign DNA sequences be shown to exist in the resistant cells, and if so, are they integrated in the genome or do they exist as plasmids?
4. Finally, can expression of the aminoglycoside phosphotransferase (APH) gene be detected?

With regard to the level of resistance, we have found that several clones selected at 20 μg/ml show resistance to 50, 100, and some even to 150 μg/ml and above. Thus the expression of the APH gene, at least in some clones, appears to be quite substantial, although as yet we have not been able to show convincingly that high levels of the enzyme exist in sonicates of resistant cells (see below).

We have addressed the second question by propagating six different clones (obtained with three different plasmids) for six generations without drugs and then asking what percentage of cells are still resistant to the level of G418 at which they were selected. Interestingly, four of the six clones appear perfectly stable, whereas two have lost resistance almost completely. This could reflect two alternative modes of establishment of the plasmid DNA (integrated in the genome or free plasmid), but we have not yet analyzed the condition of the DNA in these transformants by hybridization analysis and, thus, shall have to wait for an answer to this question.

Total DNA extracted from two different transformants has been analyzed by slot-blot membrane hybridization analysis with ^{32}P nick-translated total DNA of the plasmid used to induce the transformants, and by Riboprobe transcripts (Promega-Biotec, Madison,

WI) from the APH-gene labeled with photobiotin (BRESA, P.O. Box 498, Adelaide, South Australia, 5001).

In both instances, sequence homology has been shown to exist at high stringency. Although these experiments are preliminary, they do support the conclusion that the resistant cells have arisen as a consequence of incorporation of the gene from Tn5 giving resistance to G418.

A fourth "proof" one would like to have is the test for aminoglycoside phosphotransferase (APH) in the resistant cells. We have assayed APH by the method of Reiss et al. (1984). Although bacteria harboring plasmids carrying the APH-II gene (for which the assay was developed) show high levels of the enzyme activity, the putative transformed *Physarum* cells (extract of 10^8 cells) show only a very weak activity--barely above the background seen in wild-type extracts.

Presently, we do not know whether this reflects low enzyme activity, destruction of the enzyme by proteases in the sonicate, or the actual absence of the enzyme. It is necessary to continue work on this assay, as well as to develop alternative assays for expression of foreign genes in *Physarum*.

CONCLUSION

The data presented do not contain absolute and irrefutable evidence that DNA transformation in fact has been achieved. We can only suggest that the frequency data and the preliminary hybridization results indicate that the APH-gene sequences are present in DNA from transformants. We shall have to learn how to improve cell treatment methods to increase uptake of DNA, to search for additional dominant selectable markers, and to engineer new vectors containing *Physarum* gene control elements. The Workshop brought out very broad support for this transformation project. In practical terms this resulted in the offering of gene control elements--notably cloned *Physarum* genes containing promoters as well as putative origins of replication--from several laboratories involved with the molecular biology of *Physarum*. At the time of writing this report (September 1985), our group is eagerly awaiting input from other *Physarum* laboratories and are thus ready to proceed with what might be viewed as the second stage in the development of a DNA transformation system in *Physarum polycephalum*.

Discussion (Contributions from: S. Barclay, T. Burland, W. F. Dove, T. Laffler, W. Nader, and D. Pallotta).

The session for discussion of this report on DNA transformation in *Physarum* centered on G418 resistance as a selectable marker and on improved designs for the future. With regard to the former, Haugli reported that putative transformants grown to near encystment on G418-containing plates did not survive very well under prolonged storage in the cold.

Barclay agreed, adding that this might be a problem with expression of the APH-gene in Dictyostelium as well as in Physarum. If promoters are not functioning well and the gene is marginally expressed, there may be little room for additional stress, in addition to the exposure to G418, before the cells succumb. Barclay suggested that recovery of transformants might increase if initial exposure to the drug were delayed after treatment with DNA, plating selectively only after some growth in the absence of drug.

Barclay reported that another complication for Dictyostelium is that the presence of divalent cations (Ca^{++} and Mg^{++}) in millimolar amounts caused the usually G418-sensitive Dictyostelium cells to show resistance up to 100 μg/ml. Thus, whenever G418 is used as a selective agent, it would seem prudent to safeguard against higher concentrations of Ca^{++} and Mg^{++}, although we have no evidence with regard to this effect in Physarum.

Nader reported that he had observed a few spontaneously resistant mutants in Didymium iridis, and Barclay observed that occasional resistant mutants are seen in Dictyostelium. Their nature is not known, but some kind of membrane-uptake mutant would appear to be a likely mechanism.

With regard to spontaneous mutations to G418 resistance occurring in Physarum, Haugli and Johansen reported that none had been observed in up to about 3 X 10^9 cells plated. Dove reported that one resistant clone appeared to have arisen spontaneously in a transformation experiment in his laboratory, out of approximately 10^9 cells.

Finally, Pallotta reported on his attempts to develop a transformation system for Physarum amoebae growing in axenic liquid culture. He used a plasmid derived from pSV2-neo (Southern and Berg, 1982) by inserting a piece of Physarum r·DNA, possibly containing an origin of replication (Ferris and Vogt, 1982). This plasmid is similar to one construction used in successful transformation experiments by Haugli and Johansen. In Pallotta's experiments, no resistant cells developed out of approximately 10^{10} cells exposed. This could be taken as additional strong evidence that spontaneous resistance to G418 is not a problem in Physarum.

However, testing in liquid culture, as pointed out by Laffler, may have a weakness in that the lysing of sensitive cells could create a toxic environment for surviving, resistant cells. Still, in conclusion, no evidence was presented to suggest that the resistant colonies found in Haugli's and Johansen's transformation experiments could be explained as spontaneously arising mutants.

With regard to future strategies, Burland suggested that auxotrophic cell lines could be used as recipients in DNA transformation studies selecting for prototrophs. In light of the potential problems of G418 resistance as a selective marker discussed above, this is nice in principle. However, it was pointed out that the problems of a solid axenic minimal medium (see Chapter 18) would prevent its use at the present time, and even selection in liquid minimal medium could be problematic. Also, auxotroph caused by point mutations could revert at frequencies that could interfere. Thus, one would have to know that mutants carried deletions.

REFERENCES

Cabezon, T., Descurieux, M., Falque, J.-C., and deWilde, M., 1984, Abstract of the 12th International Conference of Yeast Genetics and Molecular Biology, p. 148, Edinburgh, Scotland, September, 1984.

Colbere-Garapin, F., Horodniceanu, F., Kourilsky, P., and Garapin, A.-C., 1981, A new dominant hybrid selective marker for higher eukaryotic cells, J. Mol. Biol., 150:1.

Dee, J., 1982, Genetics of Physarum polycephalum, in: "Cell Biology of Physarum and Didymium," H. C. Aldrich and J. W. Daniel, eds., Academic Press, New York.

Ferris, P. J., and Vogt, V. M., 1982, Structure of the central spacer region of extrachromosomal ribosomal DNA in Physarum polycephalum, J. Mol. Biol., 159:359.

Gorman, J. A., Dove, W. F., and Warren, N., 1981, Isolation of Physarum DNA segments that support autonomous replication in yeast, Mol. Gen. Genet., 183:306.

Jimenez, A., and Davies, J., 1980, Expression of a transposable antibiotic resistance element in Saccharomyces, Nature, 287:869.

Reiss, B., Sprengel, R., Will, H., and Schaller, H., 1984, A new sensitive method for qualitative and quantitative analysis of neomycin phosphotransferase in crude cell extracts, Gene, 30:211.

Southern, P. J., and Berg, P., 1982, Transformation of mammalian cells to antibiotic resistance with a bacterial gene under control of the SV40 early region promoter, J. Mol. Appl. Gen., 1:327.

Epilogue: PHYSARUM AS AN INTEGRATED EXPERIMENTAL ORGANISM

William F. Dove[1] and Keith Gull[2]

[1]McArdle Laboratory for Cancer Research and Laboratory of Genetics
University of Wisconsin
Madison, WI, USA

[2]The Biological Laboratory
University of Kent
Canterbury, UK

On the occasion of the 1985 Workshop on the Molecular Biology of *Physarum polycephalum*, literary contributions were gathered expressing views of the role of *Physarum* in experimental biology. The semi-anonymous winner of the first prize, kram namleda, offered the following view:

PHYSARUM is

the singular aroma of a cool damp forest
the vibrant yellow of a saucy lemon
the relentless rhythms of nature's multiple cycles
the storehouse of countless answers
the tantalizing challenge that lures us on

We recognize that the community of investigators who have been attracted to work on *Physarum* have been fascinated by one or another of a diverse set of features of the organism. Let us think about ways in which these researchers and their various interests can be fused into a whole. We believe firmly that the contributions that can be made to experimental biology by *Physarum* will flow most effectively if the organism develops an experimental integrity.

The Workshop allows us to focus on a range of experimental work which displays the richness of *Physarum* biology. There is, however, a danger: we must not be seduced to continue producing a multitude of descriptive projects. Rather, we must keep in mind the universal nature of biological problems and the consequent general application and excitement of solutions. If work with *Physarum* leads to important, generally applicable statements about how other cells go about organizing their lives, then *Physarum* biology will truly have come of age.

A number of key technical advances in molecular and genetic analysis have recently brought work on *Physarum* into the modern age of molecular biology. This creates the potential for a vast ouput of detailed molecular information. The biological richness of the organism brings that information to focus on several fundamental biological problems. To now, the molecular biology of *Physarum* has been limited on the molecular side. With facile molecular analysis, the molecular biology can become as interesting as the overall biology. Thus, we need to address how the biology of *Physarum* can be integrated into an active experimental enterprise.

It is informative to look in detail at the distribution of research grant support among experimental organisms. A recent informal scan of grants from the National Institutes of Health reveals that projects employing the laboratory mouse, *Drosophila*, and *S. cerevisiae* are far more numerous than those using other experimental eukaryotes. These three organisms have established a critical mass of investigators for several reasons, including the fact that each has a strong system of genetic analysis. The genetics creates a focus around which a number of positive scientific interactions revolve. For example, a person working with *Drosophila* will commonly attend a *Drosophila* meeting, whether he is studying neurobiology, embryology, or endocrinology. The vector for these positive interactions is the continual creation of useful strains. When one has the strain, one can do the experiment! It is noteworthy that, among geneticists working with the laboratory mouse, *Drosophila*, or *S. cerevisiae* (at least), it is traditional that any published strain becomes publicly available. Furthermore, it is important that, once a particular strain has been constructed or discovered, it is catalogued and maintained. These steps provide for a long-term view: a strain that appears to have little importance at one time may be of crucial importance at a later date in another role. Efficient cataloguing, storage, and distribution of strains are central to the efficiency of research using *Physarum*. Naturally, this concern for strains of the organism now extends in exactly the same way to the new probes that will become so useful for work on this organism--monoclonal antibodies and DNA clones.

Integration of a domain of experimental biology requires not only positive interactions, such as the continual creation and distribution of strains, but also strong negative interactions, such as the constructive exchange of critiques. Without this last component, the positive interactions serve to sweep up both good and bad science, creating a disorganized monster. In his excellent book, "Advice to a Young Scientist", Peter Medawar (1979) gives an account of the scientific process of debate, ending that particular chapter with these views:

> "Before he sets out to convince others of his observations or opinions, a scientist must first convince himself. Let this not be too easily achieved; it is better by far to have the reputation for being querulous and unwilling to be convinced than to give reason to be thought gullible. If a scientist asks a colleague's candid criticism of his work, give him the credit for meaning what he says. It is no kindness to a colleague--indeed, it might be the act of an enemy--to assure a scientist that his work is clear and convincing and that his opinions are really coherent when the experiments that profess to uphold them are slovenly in design and not well done. More generally, criticism is the most powerful weapon in any methodology of science; it is the scientist's only assurance that he need not persist in error. All experimentation is criticism. If an experiment does not hold out the possibility of causing one to revise one's views, it is hard to see why it should be done at all."

These positive and critical interactions must be fostered between workers who have been attracted from diverse directions into using the same experimental organism. It seems that the general health of a field of research is revealed by the nature and extent of its INTERACTIONS and CONNECTIONS. The established disciplines of Physics, Chemistry, Botany, and Zoology can be viewed as islands. The excitement in modern science is in the view from the bridges that link these islands--Microbiology, Biochemistry, Molecular Biology, Biophysics, etc. What we see are the connections. In looking at the work on Physarum one can now see these connections. The old islands of cell cycle, differentiation, motility, nuclear structure, etc., are being connected and cross-linked by current discoveries. Work involving cloning of specific DNA sequences, stage-specific myosins, tubulins, the transposon-like element, rDNA, transformation, novel proteins of the cytoskeleton, axenic strains, mutants, replication timing, periodic events, histone/chromatin organization, etc., has now started to produce a very fertile exchange of views. Each discovery provides some new insight into a rather different problem. This cross-fertilization becomes more intense as several laboratories with distinct styles focus on a small subset of entities, such as the tubulins. A good biological problem can easily stand the attention of several laboratories. It is this new sense of critical interaction and cooperation that we need to nurture into an "organism identity".

Over the past number of years the "*Physarum* Newsletter" has been circulated to inform investigators of recent publications in the field, upcoming conferences, and (occasionally) technical advances. Currently the "*Physarum* Newsletter" is circulated semi-annually by Dr. Henry C. Aldrich, Department of Microbiology and Cell Science, University of Florida, Gainesville, Florida 32611 USA.

As a consequence of discussions at the Workshop, Dr. Aldrich has added a set of technical co-editors who will communicate scientific advances germane to experimental work with *Physarum*. These co-editors include Roger Anderson, Norman Hardman, Finn Haugli, and Gerard Pierron. Within a month of each *Physarum* meeting, such co-editors will communicate a technical note, indicating changes in the status of position papers published in this volume and drawing attention to any technical advances crucial to further progress.

The constructive negative interactions of critique must of course often be anonymous. What each of us investigators in the field must accept is the responsibility to provide strong critiques whenever asked to serve as a referee for a submitted manuscript.

It will be interesting to see whether a balanced set of positive and negative interactions can lead over the next decade to a realization of the biological potential for experimental investigation with *Physarum polycephalum*.

ACKNOWLEDGMENT

Our joint research interactions are positively supported by grant INT-8501096 in the U.S.-West European Cooperative Science Program. We hope that our individual enthusiasms have been constructively tamed by mutual deletions from this epilogue.

REFERENCES

Medawar, P., 1979, "Advice to a Young Scientist," 94 pp, Harper and Row, New York.

ABBREVIATIONS

A: adenine

alc: locus that can mutate to deficiency in the amoebal stage of the life cycle

AMV: avian myeloblastosis virus

a-myosin: myosin of the *Physarum* amoeba

Ap_4A: diadenosine 5',5"-p1,p4-tetraphosphate

APH: aminoglycoside phosphotransferase

apt: locus that can mutate to deficiency in the amoebal-plasmodial transition

ardA, *B*, *C*, *D*: loci that contain DNA sequences related to actin

ARS: autonomously replicating sequence

AXE (or Axe): a myxamoebal strain capable of growth in axenic media

bp: nucleotide bases or basepairs

BSA: bovine serum albumin

BUDR: 5-bromodeoxyuridine

C value: the DNA content of an unreplicated haploid nucleus

C: cytosine

cDNA: complementary DNA

ChGAM: chicken gizzard actin modulator

CL: Colonia Leicester, selfing (apogamic) strain of *Physarum*

CLd: derivative of CL, delayed in plasmodium formation

C-terminal: carboxy-terminal segment of a polypeptide

dC: deoxycytidine

dNTP: deoxyribonucleotide triphosphate

DTE: dithioerythritol

DTT: dithiothreitol

EDTA: ethylenediamine tetraacetic acid

EGTA: ethyleneglycolbis(beta-aminoethylether)N,N,N',N'-tetraacetic acid

ETS: external transcribed spacer

EWAM: earthworm actin modulator

FAC: fluorescent analog cytochemistry

F-actin: the filamented polymeric form of actin

FKB: formalin-killed bacteria

G: guanine

_g_: the force of gravity

G-actin: the globular monomeric form of actin

gad: locus that can mutate to give greater asexual differentiation

GC-rich: greater than 50% guanine plus cytosine in DNA

G1: the phase in the cell cycle between mitosis and DNA synthesis

G2: the phase in the cell cycle between DNA synthesis and mitosis

HMW: high molecular weight

HPLC: high-performance liquid chromatography

HU: hydroxyurea

IEF: isoelectric focusing

imz: locus that can mutate to affect the ionic conditions for zygote formation

ind: locus that can mutate to affect the level of inducer of plasmodium formation

ITS: internal transcribed spacer

K: kilodaltons

kb: kilobases or kilobasepairs

K_D: equilibrium dissociation constant

lacZ: structure gene for the β-galactosidase of _E. coli_

LC-1: _Physarum_ myosin light chain-1

LC-2: _Physarum_ myosin light chain-2

matA: mating-type locus affecting development of zygotes into plasmodia

matB: mating-type locus affecting zygote formation

MBC: methyl benzimidazole carbamate

MDH: malate dehydrogenase

M_r: relative molecular weight, calculated from relative electrophoretic mobility

MI (MII, etc.) first (second, etc.) synchronous mitosis of the plasmodium after fusion of microplasmodia

MTOC: microtubule organizing center

NAD: nicotinamide adenine dinucleotide

npf: locus that can mutate to result in no plasmodium formation

oligo-dT: oligodeoxythymidylate polymer

PAGE: polacrylamide gel electrophoresis

PEG: polyethylene glycol

PIPES: piperazine n,n'-bis-[2-ethane sulfonic acid]

p-myosin: myosin of the *Physarum* plasmodium

poly(dG:dC): duplex DNA with homopolymer dG paired to homopolymer dC

poly (A^-) RNA: RNA lacking 3' polyadenylate residues

poly (A^+) RNA: RNA carrying 3' polyadenylate residues

PSAM: pig stomach actin modulator

recA: gene in *E. coli* that can mutate to a loss of homologous recombination

RITC: rhodamine isothiocyanate

$Rot_{1/2}$: initial RNA concentration multiplied by time required for 50% hybridization of single-copy sequences

RSAM: rabbit skeletal muscle actin modulator

rII: "rapid lysis" gene of bacteriophage T4

S: Svedberg unit of sedimentation

S: phase when DNA is replicated in the cell cycle

SDM: semidefined medium

SDS: sodium dodecyl (lauryl) sulfate

SDS PAGE: polyacrylamide gel electrophoresis in the presence of sodium dodecyl sulfate

SHP: nuclei purified in Surfonyl, hexylene glycol, and Percoll

SMP: sporangiophore morphogenetic protein

SSC: 180 mM NaCl, 10 mM NaH_2PO_4 (pH 7.4), 1 mM EDTA

SSPE: 150 mM NaCl, 15 mM trisodium citrate, pH 7.0

T: thymine

TAT: α-tubulin acetyltransferase

T-cell: thymus-derived lymphocyte

TDA: α-tubulin deacetylase

TES: N-tris-[hydroxymethyl]methyl-2-aminoethane sulfonic acid

TK: thymidine kinase

Tris: tris[hydroxymethyl]aminomethane

TRITC: tetramethylrhodamine isothiocyanate

tRNA: transfer RNA

TS: nuclei purified in Triton-X100 and sucrose

Wis 1: Wisconsin-1 isolate of the _Physarum_ plasmodium

Wis 2: Wisconsin-2 isolate of the _Physarum_ plasmodium

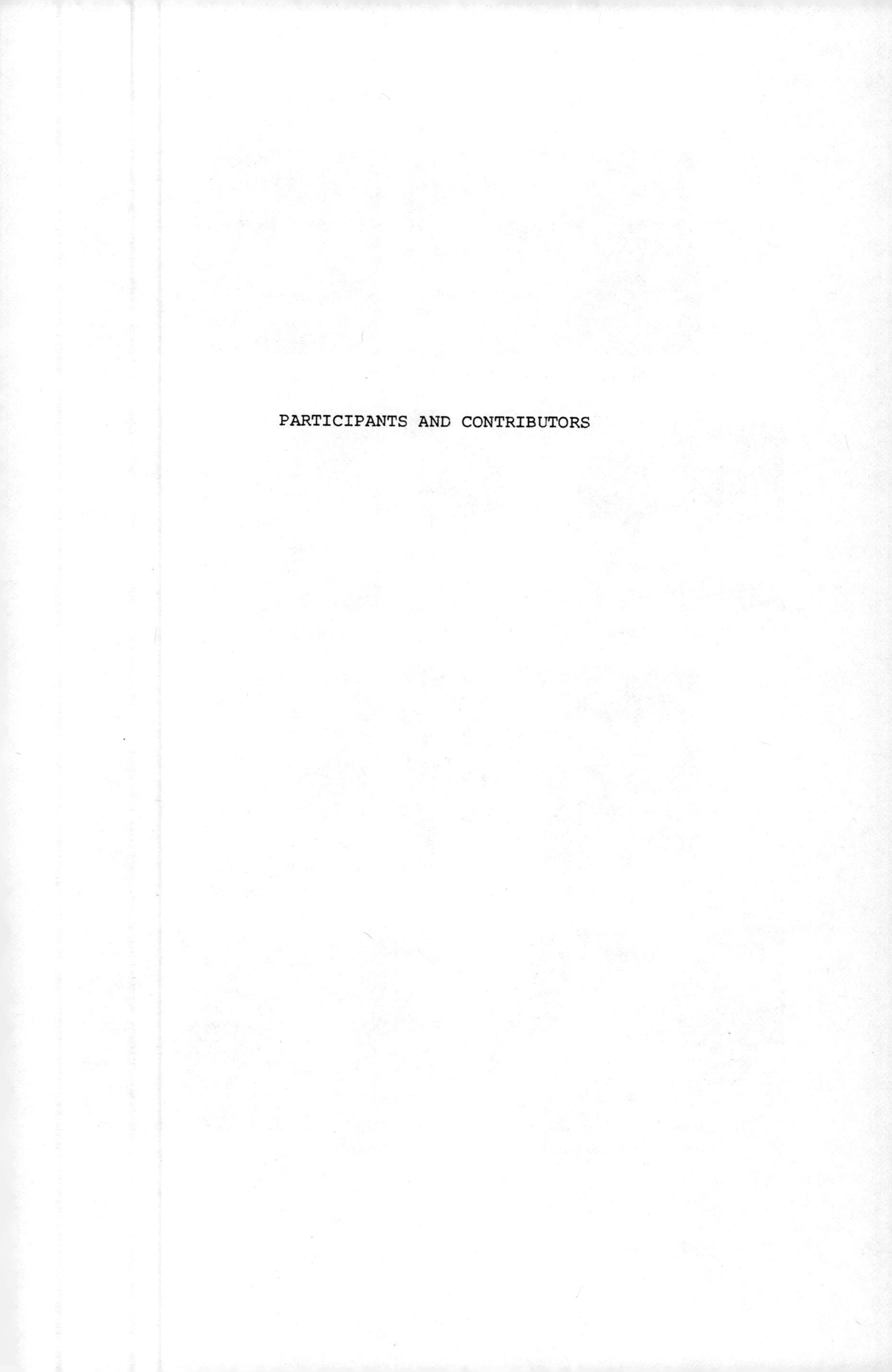

PARTICIPANTS AND CONTRIBUTORS

W. DOVE, H. RUSCH, T. BURLAND

T. JOHANSEN, F. HAUGLI

D. KESSLER, A. LOEWY

K. KOHAMA

M. WRIGHT

H. SAUER

G. PIERRON, K. GULL

M. ADELMAN, W. STOCKEM

K. MARUYAMA, S. HATANO

R. ANDERSON

T. LAFFLER

J. TYSON, J. M. MITCHISON

H. HINSSEN, W. STOCKEM

J. DEE

W. NADER, T. SCHRECKENBACH

R. COX, J. DANIEL

S. EBASHI, D. GASSNER

K.-E. WOHLFARTH-BOTTERMANN

V. VOGT, N. HARDMAN

T. SCHEDL, E. PAUL

D. PALLOTTA, F. BERNIER, W. DOVE

P. ANDERSON

PARTICIPANTS AND CONTRIBUTORS*

Friedhelm Achenbach
Institut für Cytologie
und Mikromorphologie
der Universität Bonn
53 Bonn 1
Ulrich-Haberland-Strasse 61a
FRG

Ulrike Achenbach
Institut für Cytologie
und Mikromorphologie
der Universität Bonn
53 Bonn 1
Ulrich-Haberland-Strasse 61a
FRG

Mark R. Adelman
Dept. of Anatomy
Uniformed Services University
of the Health Sciences
4301 Jones Bridge Road
Bethesda, MD 20814-4799
USA

Henry C. Aldrich
Dept. of Microbiology
and Cell Science
1059 McCarty Hall
University of Florida
Gainesville, FL 32611
USA

Robert G. Allen*
Max-Planck-Institut
für Biochemie
D-8033 Martinsried bei
München
FRG

Phil Anderson
Laboratory of Genetics
University of Wisconsin
Madison, WI 53706
USA

Roger W. Anderson
Dept. of Genetics
University of Sheffield
Sheffield S10 2TN
ENGLAND

Larry Barnes
Dept. of Biochemistry
University of Texas Health
Science Center
San Antonio, TX 78284
USA

Glenn R. Barnett
Division of Nutritional
Sciences
Cornell University
323 Savage Hall
Ithaca, NY 13853
USA

François Bernier
Dept. de Biochimie
Université Laval
Québec G1K 7P4
CANADA

F. Binette*
Dept. de Biologie
Faculte des Sciences
Université Laval
Québec, G1K 7P4
CANADA

A. B. Blindt*
Dept. of Genetics
University of Leicester
Leicester LE1 7RH
ENGLAND

E. Morton Bradbury
Dept. of Biological Chemistry
School of Medicine
University of California
Davis, CA 95616
USA

Richard Braun
University of Bern
Inst. of General
Microbiology
CH-3012 Bern,
Balter-Strasse 4
SWITZERLAND

Timothy Burland
McArdle Laboratory
for Cancer Research
University of Wisconsin
Madison, WI 53706
USA

John Carrino*
Dept. of Microbiology
and Immunology
Northwestern University
Ward Memorial Bldg.
303 E. Chicago Avenue
Chicago, IL 60611
USA

Mark E. Christensen
Dept. of Biology
Texas A & M University
College Station, TX 77843
USA

Jim Clark
School of Biological Sciences
University of Kentucky
Lexington, KY 40506
USA

Robert A. Cox
National Inst.
for Medical Research
The Ridgeway, Mill Hill
London NW7 1AA
ENGLAND

David Cunningham
McArdle Laboratory
for Cancer Research
University of Wisconsin
Madison, WI 53706
USA

Marta Czupryn
Institute of Biochemistry
Warsaw University
Al. Zwirki i Wigury 93
02-089 Warsaw
POLAND

John Daniel
Division of Radiation
Biology
Case Western Reserve
University
Cleveland, OH 44106
USA

Jennifer Dee
Dept. of Genetics
University of Leicester
Leicester LE1 7RH
ENGLAND

Jochen D'Haese*
Institute of Zoology II
University of Düsseldorf
Düsseldorf
FRG

Halil Demirtas
Medikal Bioloji
Tip Fakuiltesi
Erciyes Universitesi
Kayseri
TURKEY

Maureen Diggins
McArdle Laboratory for
Cancer Research
University of Wisconsin
Madison, WI 53706
USA

Beverly K. Dolberg*
Dept. of Biology
Haverford College
Haverford, PA 19041
USA

William F. Dove
McArdle Laboratory
for Cancer Research
University of Wisconsin
Madison, WI 53706
USA

Setsuro Ebashi
Natl. Inst. for Physio-
logical Sciences
Myodaiji, Okazaki
JAPAN

Frank E. Engels*
Institute of Molecular Biology
Austrian Academy of Sciences
Salzburg
AUSTRIA

Eileen Epstein
Section of Biochemistry,
Molecular & Cellular Biology
Wing Hall
Cornell University
Ithaca, NY 14853
USA

Helen Evans
Div. of Radiation Biology
Dept. of Radiology
Case Western Reserve University
Cleveland, OH 44106
USA

Thomas Evans
Div. of Radiation Biology
Dept. of Radiology
Case Western Reserve University
Cleveland, OH 44106
USA

Reinhard Exenberger
Institut für Exp.
Krebsforschung
Fritz-Pregl-Str. 3/VII
6020 Innsbruck
AUSTRIA

Alison M. Foote
Dept. of Neurogenetics
E.K. Shriver Center for
Mental Retardation
200 Trapelo Road
Waltham, MA 02254
USA

Kay Foster
Biological Laboratories
University of Kent, Canterbury
Kent, CT2 7NJ
ENGLAND

Gregorio Garcia-Herdugo
Facultad de Gencias
Universidad de Cordoba
14071-Cordoba
SPAIN

Dieter Gassner
Institut für Cytologie und
Mikromorphologie
Universität Bonn
Ulrich-Haberland-Strasse 61a
D-5300 Bonn 1
FRG

Larry Green
McArdle Laboratory
for Cancer Research
University of Wisconsin
Madison, WI 53706
USA

Keith Gull
Biological Laboratory
University of Kent,
Canterbury
Kent, CT2 7NJ
ENGLAND

Michel Hamelin*
Biology Department
Laval University
Quebec G1K 7P4
CANADA

Norman Hardman
Dept. of Biochemistry
Marischal College
University of Aberdeen
Aberdeen, AB9 1AS
SCOTLAND

Sadashi Hatano
Inst. of Molecular Biology
Faculty of Science
Nagoya University
Chikusa-ku, Nagoya
JAPAN 464

Finn Haugli
Inst. of Medical Biology
University of Tromsø
9000 Tromsø
NORWAY

Richard J. Heads
Biophysics Laboratory
Portsmouth Polytechnic
St. Michael's Bldg.
White Swan Road
Portsmouth, Hants PO1 2DT
UNITED KINGDOM

Armin Hildebrandt*
Faculte de Biologie
Universität Bremen
Fachber. Biologie NW2
Univ. Postf. 330440
280 Bremen 33
FRG

Horst Hinssen
California Institute
of Technology
Division of Biology 156-29
Pasadena, CA 91125
USA

Di Hua Hu*
Dept. of Biology
Chiba University
Yayoi-cho
Chiba, 260
JAPAN

Fu-Sen Hu
School of Biological Sciences
University of Kentucky
Lexington, KY 40506
USA

Raymond Jalouzot
Laboratoire de Biologie
Cellulaire
Faculte des Sciences
BP 347
51062 Reims Cedex
FRANCE

Terje Johansen
Inst. of Medical Biology
University of Tromsø
9000 Tromsø
NORWAY

Michael N. Kazarinoff
Division of Nutritional
Sciences
Cornell University
323 Savage Hall
Ithaca, NY 14853
USA

Norman Kerr
Dept. of Genetics & Cell
Biology
250 Bioscience Center
University of Minnesota
St. Paul, MN 55108
USA

Sylvia Kerr
Dept. of Biology
Hamline University
St. Paul, MN 55104
USA

Dietrich Kessler
Dept. of Biology
Colgate University
Hamilton, NY 13346
USA

Sumiko Kimura*
Dept. of Biology
Chiba University
Yayoi-cho
Chiba, 260
JAPAN

Kazuhiro Kohama
Dept. of Pharmacology
Faculty of Medicine
University of Tokyo
Bunkyo-ku
Tokyo 113
JAPAN

Sheri Kostelny
Section of Biochemistry,
Molecular & Cellular Biology
Wing Hall
Cornell University
Ithaca, NY 14853
USA

Jörg Kukules*
Institut für Cytologie
und Mikromorphologie
Universität Bonn
Ulrich-Haberland-Strasse 61a
5300 Bonn 1
FRG

Mario Lachapelle
Biology Department
Laval University
Quebec G1K 7P4
CANADA

Thomas Laffler
Dept. of Microbiology
and Immunology
Northwestern University
Ward Memorial Bldg.
303 E. Chicago Avenue
Chicago, IL 60611
USA

J. G. Lafontaine
Biology Department
Laval University
Quebec G1K 7P4
CANADA

A. Laroche*
Dept. de Biologie
Faculte des Sciences
Universite Laval
Quebec, G1K 7P4
CANADA

Gerald Lémieux*
Biology Department
Laval University
Quebec G1K 7P4
CANADA

Ariel Loewy
Dept. of Biology
Haverford College
Haverford, PA 19041
USA

Peter Loidl
Inst. für Medizinische
Chemie und Biochemie
Universität Innsbruck
Fritz-Pregl-Strasse 3
6020 Innsbruck
AUSTRIA

Angus M. MacNicol
National Institute for
Medical Research
Mill Hill
London NW7 1AA
ENGLAND

Rémi Martel*
Biology Department
Laval University
Québec G1K 7P4
CANADA

Hiroshi Maruta
Kline Biology Tower 310
Dept. of Biology
Yale University
219 Prospect Street
New Haven, CT 06511-8112
USA

Koscak Maruyama
Dept. of Biology
Faculty of Science
Chiba University
Yayoi-cho
Chiba, 260
JAPAN

Harry R. Matthews
Dept. of Biological Chemistry
School of Medicine
University of California
Davis, CA 95616
USA

Joan H. McCune
Dept. of Biological Sciences
Box 8007, Idaho State University
Pocatello, ID 83209-0009
USA

Ronald W. McCune
Dept. of Biological Sciences
Box 8007, Idaho State University
Pocatello, ID 83209-0009
USA

J. M. Mitchison
Dept. of Zoology
University of Edinburgh
West Mains Road
Edinburgh EH9 3FT
SCOTLAND

Joyce Mohberg
Division of Science, CAS
Governors State Univ.
Park Forest South, IL 60466
USA

Mervyn J. Monteiro
National Institute for Medical Research
Mill Hill
London NW7 1AA
ENGLAND

Werner F. Nader
Max-Planck-Institut für Psychiatrie
Abt. Neurochemie/Am Klopferspitz 18a
8033 Planegg-Martinsried
FRG

Claude Nations
Dept. of Biology
Southern Methodist University
Dallas, TX 75275
USA

Klaus-Dieter Nothacker*
Universität Bremen
Fachber. Biologie NW2
Univ. Postf. 330440
280 Bremen 33
FRG

Ron G. Opstelten
Dept. of Chemical Cytology
Faculty of Science
University of Nijmegen
Toernooiveld
6525 ED Nijmegen
THE NETHERLANDS

Dominick Pallotta
Dept. de Biologie
Faculte des Sciences
Université Laval
Québec, G1K 7P4
CANADA

Eileen Paul
Biological Laboratory
University of Kent, Canterbury
Kent, CT2 7NJ
ENGLAND

Douglas H. Pearston
Dept. of Biochemistry
Marischal College
University of Aberdeen
Aberdeen AB9 1AS
UNITED KINGDOM

Gerard Pierron
Inst. de Recherches
Scientifique
7, rue Guy Mocquet
94802 Villejuif Cedex
FRANCE

Harold P. Rusch
McArdle Laboratory
for Cancer Research
University of Wisconsin
Madison, WI 53706
USA

David C. Ruth*
Division of Nutritional
Sciences
Cornell University
323 Savage Hall
Ithaca, NY 14853
USA

Wilhelm Sachsenmaier
Institut für Experimentelle
Krebsforschung der
Universität Innsbruck
Fritz-Preglstrasse 3/VII
A-6020 Innsbruck
AUSTRIA

Helmut Sauer
Dept. of Biology
Texas A & M University
College Station, TX 77843
USA

Timothy Schedl
Molecular Biology Laboratory
University of Wisconsin
Madison, WI 53706
USA

Thomas Schreckenbach
Max-Planck-Institut
für Biochemie
D-8033 Martinsried bei München
FRG

T. M. Shinnick*
Scripps Clinic and Research
Foundation
La Jolla, CA 92037
USA

Gregory L. Shipley
Dept. of Biology
Texas A & M University
College Station, TX 77843
USA

Dorothee Staiger*
Max-Planck-Institut
für Biochemie
D-8033 Martinsried bei München
FRG

D. W. Stockem
Institut für Cytologie und
Mikromorphologie der
Universität Bonn
Ulrich-Haberland-Strasse 61a
5300 Bonn 1
FRG

Tsuneo Suzuki*
Dept. of Biology
Chiba University
Yayoi-cho
Chiba, 260
JAPAN

Glen Sweeney
Dept. of Biochemistry
University of Leicester
Leicester LE1 7RH
ENGLAND

A. Tessier*
Dept. de Biologie
Faculte des Sciences
Universite Laval
Quebec, G1K 7P4
CANADA

G. Turnock*
Dept. of Biochemistry
University of Leicester
Leicester LE1 7RH
ENGLAND

John Tyson
Dept. of Biology
Virginia Polytechnic
Institute
Blacksburg, VA 24061

Volker Vogt
Section of Biochemistry,
Molecular & Cell Biology
Wing Hall
Cornell University
Ithaca, NY 14853
USA

Friedrick Wanka
Laboratorium voor
Chemische Cytologie
University of Nijmegen
Toernooiveld
6525 Ed Nijmegen
THE NETHERLANDS

Jaap H. Waterborg
Dept. of Biochemistry
University of Nevada
Reno, NV 89557
USA

D. I. Watts*
Dept. of Biochemistry
University of Leicester
Leicester LE1 7RH
ENGLAND

Anne-K. Werenskiold
Max-Planck-Institut
für Biochemie
D-8033 Martinsried bei München
FRG

Francois-Xavier Wilhelm
Inst. de Biologie Moleculaire
et Cellulaire du CNRS
Laboratoire de Biophysique
15 rue Rene Descartes
67084 Strasbourg Cedex
FRANCE

M. L. Wilhelm*
Inst. de Biologie Moleculaire
et Cellulaire du CNRS
Laboratoire de Biophysique
15 rue Rene Descartes
67084 Strasbourg Cedex
FRANCE

K.-E. Wohlfarth-Bottermann
Inst. für Cytologie und
Mikromorphologie der
Universität Bonn
Ulrich-Haberland-Strasse 61a
5300 Bonn 1
FRG

Michel Wright
CNRS
Lab. Pharmacologie et
Toxicologie Fondamentales
205, Route de Narbonne
31400 Toulouse Cedex
FRANCE

*Contributors to this volume who did not attend the Workshop are indicated by an asterisk.

INDEX